AF799747

The Intersection of Trauma and Disaster Behavioral Health

Katie E. Cherry • Allison Gibson
Editors

The Intersection of Trauma and Disaster Behavioral Health

Springer

Editors
Katie E. Cherry
Department of Psychology
Louisiana State University
Baton Rouge, LA, USA

Allison Gibson
College of Social Work
University of Kentucky
Lexington, KY, USA

ISBN 978-3-030-51524-9 ISBN 978-3-030-51525-6 (eBook)
https://doi.org/10.1007/978-3-030-51525-6

© Springer Nature Switzerland AG 2021
This work is subject to copyright. All rights are reserved by the Publisher, whether the whole or part of the material is concerned, specifically the rights of translation, reprinting, reuse of illustrations, recitation, broadcasting, reproduction on microfilms or in any other physical way, and transmission or information storage and retrieval, electronic adaptation, computer software, or by similar or dissimilar methodology now known or hereafter developed.
The use of general descriptive names, registered names, trademarks, service marks, etc. in this publication does not imply, even in the absence of a specific statement, that such names are exempt from the relevant protective laws and regulations and therefore free for general use.
The publisher, the authors, and the editors are safe to assume that the advice and information in this book are believed to be true and accurate at the date of publication. Neither the publisher nor the authors or the editors give a warranty, expressed or implied, with respect to the material contained herein or for any errors or omissions that may have been made. The publisher remains neutral with regard to jurisdictional claims in published maps and institutional affiliations.

This Springer imprint is published by the registered company Springer Nature Switzerland AG
The registered company address is: Gewerbestrasse 11, 6330 Cham, Switzerland

Contents

Contributors

Lacey A. Bagley School of Family Life, Brigham Young University, Provo, UT, USA

Nikki D. Bellamy Substance Abuse and Mental Health Services Administration (SAMHSA), Center for Mental Health Services (CMHS), Rockville, MD, USA

Traci Birch LSU School of Architecture, Coastal Sustainability Studio, Louisiana State University, Baton Rouge, LA, USA

Lisa M. Brown Palo Alto University, Palo Alto, CA, USA

Sandra C. Brown Southern University and A & M College, Baton Rouge, LA, USA

Matthew R. Calamia Department of Psychology, Louisiana State University, Baton Rouge, LA, USA

Yi Ling Chan Department of Geography and Anthropology, Louisiana State University, Baton Rouge, LA, USA

Katie E. Cherry Department of Psychology, Louisiana State University, Baton Rouge, LA, USA

Sarah E. Choate School of Social Work, Louisiana State University, Baton Rouge, LA, USA

Amy L. Copeland Department of Psychology, Louisiana State University, Baton Rouge, LA, USA

Tamecia M. Curry School of Social Work, Louisiana State University, Baton Rouge, LA, USA

Molly Everett Davis Department of Social Work, George Mason University, Fairfax, VA, USA

Xi Du Department of Sociology, Louisiana State University, Baton Rouge, LA, USA

Emily M. Elliott Department of Psychology, Louisiana State University, Baton Rouge, LA, USA

Allison Gibson College of Social Work, University of Kentucky, Lexington, KY, USA

Isaac Henry Department of Planning, Roanoke County, VA, USA

Yaakov Hoffman Interdisciplinary Department of Social Sciences, Bar-Ilan University, Ramat Gan, Israel

Shuei Kozu Department of Social Work, Southern Connecticut State University, New Haven, CT, USA

Judy Kuriansky Teachers College, Columbia University, New York, NY, USA

Nancy Kusmaul School of Social Work, University of Maryland, Baltimore County, Baltimore, MD, USA

Emma L. Lucas Palo Alto University, Palo Alto, CA, USA

Aimee Moles Social Research and Evaluation Center, Louisiana State University, Baton Rouge, LA, USA

Kim Mosby Coastal Sustainability Studio, Louisiana State University, Baton Rouge, LA, USA

Mandana Mostofi Palo Alto University, Palo Alto, CA, USA

Marla Nelson Department of Planning and Urban Studies, University of New Orleans, New Orleans, LA, USA

Darlyne G. Nemeth The Neuropsychology Center of Louisiana, Baton Rouge, LA, USA

Yasuo Onishi Higashi Nippon Internatonal University, Iwaki, Fukushima, Japan

Judith R. Phillips Department of Psychology, California State University San Marcos, San Marcos, CA, USA

Amit Shrira Interdisciplinary Department of Social Sciences, Bar-Ilan University, Ramat Gan, Israel

Wanda Raby Spurlock Southern University and A & M College, Baton Rouge, LA, USA

Aaron F. Waters Department of Psychology, Louisiana State University, Baton Rouge, LA, USA

Scott E. Wilks School of Social Work, Louisiana State University, Baton Rouge, LA, USA

Alyssa Banford Witting, PhD School of Family Life, Brigham Young University, Provo, UT, USA

Dahyung Yang Korea National Institute of Forest Science, Seoul, South Korea

Haojie Zhu Department of Geography and Anthropology, Louisiana State University, Baton Rouge, LA, USA

About the Editors

Katie E. Cherry, PhD is a Developmental Psychologist and a Professor at Louisiana State University in Baton Rouge. Her research expertise is the psychology of aging with emphasis on cognition and successful aging. Her current research program focuses on the impact of natural disasters on healthy aging in later life. She has two edited books on disasters from a lifespan perspective and has over 170 peer-reviewed publications. Her work has been funded by the National Institute on Aging, the National Science Foundation, and the Louisiana Board of Regents. In 2002, she was awarded the Emogene Pliner Distinguished Professor of Aging Studies professorship for her contributions to the field of adult development and aging. She is a member of the Gerontological Society of America and is the current co-convener of a special interest group on Disasters and Older Adults.

Allison Gibson, PhD is an Assistant Professor in the University of Kentucky's College of Social Work in Lexington. Dr. Gibson's clinical and research focus is grounded in the biopsychological perspective for aging. As a social worker and researcher, Dr. Gibson's research focuses on the relationship between aging, health, crisis response, and social justice. Her research frequently examines individuals' social support, psychological well-being, and health outcomes in the context of disaster response and recovery. She is dedicated to improving the health and well-being of older adults. Dr. Gibson has advanced knowledge and translated empirical findings into interventions to support families confronting crises and challenging life events. She is a passionate advocate for social justice and gerontological social work education. She is a member of the Gerontological Society of America and the former co-convener of a special interest group on Disasters and Older Adults.

Part I
Overview and Theory

Chapter 1
The Intersection of Trauma and Disaster Behavioral Health: An Introduction

Katie E. Cherry and Allison Gibson

Introduction

An increasingly important issue for behavioral health professionals today is how to better understand and address the needs of individuals affected by disasters. Disasters bring catastrophic destruction, including loss of life, homes, and property, with immediate and long-lasting effects. The disaster science literature documents threats to health and wellbeing for directly affected individuals as well as those indirectly affected during the immediate impact period and also over time. Environmental disasters can be naturally occurring or man-made. Each year, people worldwide suffer various types of disasters—ranging from weather events, such as hurricanes, floods, earthquakes, and fires, to oil spills and leakage of nuclear materials. The topic of this volume is important, given that occurrence of disasters, large events that threaten lives and/or possessions of those in affected areas, is on the rise (Mooney & Dennis, 2018).

According to the US Federal Emergency Management Agency (FEMA), a natural disaster is defined as "An occurrence of severity and magnitude that normally results in deaths, injuries, and property damage and that cannot be managed through routine procedures and resources of government. It . . . requires immediate, coordinated, and effective response by multiple government and private sector organizations to meet human needs and speed recovery" (Oriol, 1999, p. 6). While the definition of disaster may be unambiguous, the disaster experience may be profoundly different among survivors who live through catastrophic environmental events. For instance, variables key to understanding disaster impacts include, but are

K. E. Cherry (✉)
Louisiana State University, Baton Rouge, LA, USA
e-mail: pskatie@lsu.edu

A. Gibson
University of Kentucky, Lexington, KY, USA

© Springer Nature Switzerland AG 2021

K. E. Cherry, A. Gibson (eds.), *The Intersection of Trauma and Disaster Behavioral Health*, https://doi.org/10.1007/978-3-030-51525-6_1

not limited to, where survivors live, their developmental stage in life (e.g., a child, adult, or elderly person), cultural ethos, and prevailing cultural practices. In addition, both risk factors and resilience resources may impact post-disaster outcomes over time.

The purpose of this volume is to bring together a diverse array of scholars whose professional expertise speaks to the topic of trauma and disaster behavioral health over the years after destructive environmental events. Readers will detect a lifespan developmental perspective throughout the chapters. For example, the disaster experience is likely to differ for survivors based on their age at the time of the event (Cherry, 2009). Consequently, developmental trends have implications for disaster planning and recovery that warrant careful consideration. To support this point, some chapters in this volume include behavioral data collected with samples of younger and older adults exposed to varying environmental events in the United States (i.e., the 2005 Atlantic Hurricanes Katrina and Rita; the 2007 wildfires in San Diego County, California; the 2016 flooding in Baton Rouge, Louisiana; and the 2017 Hurricane Irma in Florida and Harvey in Hurricane Houston, Texas) as well as to events abroad (i.e., missile fire in the 2014 Israel-Gaza conflict). In addition to developmental trends, most experts would agree that prior lifetime trauma can also influence the outcome for individuals facing disaster (Cherry, 2015). Readers will also detect a recurring emphasis throughout this volume on traumatic stress and/or previous disaster exposure in relation to current outcomes.

Trauma and Disaster Behavioral Health: Chapter Highlights

The volume is packaged in three main sections. Section I, entitled "Theoretical Perspectives," includes demographic trends across the lifespan and the general challenges of meeting recovery needs in the context of trauma and disaster response. Chapter authors **Wilks, Choate, Brown, Du, and Curry** provide a rich conceptual tour of trauma and resilience. As trauma-focused research becomes more established in the literature, it is necessary to provide such conceptual clarity to the varying constructs associated with trauma and resiliency. In another chapter, authors **Davis and Bellamy** cast a spotlight on trauma-informed principles and disaster behavioral health issues primarily among older adults. Their chapter explores how disaster behavioral health can integrate trauma-informed principles as a strategy to reduce the vulnerability of older adults. Chapter authors **Shrira and Hoffman** present unique data on the protective factors among older survivors of missile fire in Israel. Their data suggests that subjective age can buffer against the traumatic effects of disaster, a factor that is determined both by resources and by stressors. Authors **Banford Witting and Bagley** provide a unique glimpse into disasters from a perspective of family systems. They bring to light the importance of considering relational aspects of the disaster experience by reviewing the tenets and prior use of four prominent theoretical lenses used by disaster scholars in the past two decades. These lenses include the conservation of resources theory, ecological theory, attachment theory, and family stress theory. This section closes with **Kozu and Gibson's** elegant overview of grief in a disaster context with a developmental emphasis on young children. In addition, the authors discuss progress and limitations to current treatment approaches across the lifespan.

Section II of this volume focuses on direct practice, including the application of theory and/or methods to the resolution and prevention of psychosocial problems experienced by individuals, families, and groups in the context of disasters and trauma. Chapter authors **Mostofi and Brown** take a focused look at cultural competence during disaster recovery and emphasize that this is crucial in providing effective disaster behavioral health. Behavioral health providers can utilize and adapt these guidelines as needed to address the unique needs of their culturally diverse communities. In another chapter, **Cherry, Calamia, and Elliott** examine the challenges researchers face and when attempting to carry out "sudden science" in the wake of a natural disaster. Their chapter reflects on valuable lessons learned based on over a decade of research on post-disaster psychological reactions among directly and indirectly impacted individuals. Chapter author **Kusmaul** lays out for the reader the role of trauma-informed care in disasters and argues for more integration of trauma-informed care as a universal approach to treat all survivors. In the next chapter, **Phillips** reviews the critical role of social support as a protective factor in survivors' response, based on behavioral data collected after a series of wildfires in California. In this study, the researchers found that severity of exposure to the wildfires played a role in the frequency of social support received, with those who faced the closest exposure to the wildfire receiving more social support compared to others. Chapter authors **Lucas and Brown** focus on groups that are frequently vulnerable post-disaster, cover evidence-based treatment for mental health care after a disaster, and offer recommendations for future research. Authors **Spurlock and Brown** highlight the role of family nurse practitioners in providing post-disaster health care and discuss the education and training needs of nurses to ensure the nursing profession is disaster-ready. Section II closes with **Mosby's** insightful analysis of survivors after the devastating Hurricane Katrina in New Orleans, Louisiana, using a narrative approach based on direct interviews with displaced survivors to reveal the multidimensional impact of the disaster. This work promotes that the reduction of negative behavioral health consequences associated with disaster-related trauma is possible when communities embrace purposeful policy interventions.

Section III covers Policy and Community Practice, bringing a macro-level perspective to issues that affect communities, cities, and major institutions after a disaster. This section opens with a chapter by authors **Birch, Henry, and Nelson,** who present a thoughtful examination of disasters and human-made environment of communities and cities, highlighting the potentials and pitfalls of urban planning in post-disaster recovery. Their chapter considers the role that "place" plays in one's identity and the traumatic impact that can occur when one experiences loss of place post-disaster. In another chapter, authors **Moles, Birch, Chan, Yang, Zhu, and Cherry** bring a perspective of ecological systems theory to the creation of a community wellbeing index, based on examination of the disastrous flooding that devastated south Louisiana in 2016. Building upon the community-wide assessment of wellbeing presented in this chapter, the authors of the next chapter, **Cherry, Calamia, Birch, and Moles,** address the challenges of mental health assessment after this 2016 flood. They report behavioral data obtained from a companion online survey, which documents the persistence of mental health symptoms up to 3-years post-flood. The chapter by **Waters and Copeland** addresses substance abuse and behavioral health considerations in the wake of disaster. Throughout their chapter, these authors review how natural disasters affect behavioral health and consider a number of novel approaches to addressing these problems. Chapter authors **Nemeth, Kuriansky, and Onishi** cover the importance of emotional resilience post-disaster and highlight their work facilitating group wellness workshops. In the final chapter of this volume, **Kuriansky** brings the topic of disaster behavioral health onto a global stage. Her insights on international advocacy and policy work on issues that affect trauma and disaster response are based on a wealth of personal experience doing international advocacy work.

Conclusion

Many different types of disasters, both natural and man-made, have occurred in recent years. These events have claimed countless lives and forever altered trajectories of development for survivors and their families and communities. An alarming reality is that more disastrous environmental events are expected in the future (U.S. Department of Interior, 2020). The present volume conveys the state-of-the-art knowledge concerning trauma, resilience, and post-disaster behavioral health. Taken together, these chapters bring unique interdisciplinary perspectives that shed new light on the pressing and perennial challenges of meeting post-disaster recovery needs.

Given the increasing occurrence of disasters, among other tragic events that affect the lives of individuals and families, there is a great need for professionals, as well as the general public, to be prepared to face the challenges associated with catastrophic environmental events. The authors maintain that understanding the factors that promote survival and ensure safety will better prepare the global community for future disasters. The urgent need for systematic research about psychosocial issues, clinical interventions to mitigate adversity, and thoughtful disaster policies and planning emphasized in this volume can increase the likelihood of effective preparation and management of future catastrophic disasters on local, state, and national levels.

References

Cherry, K. E. (2009). *Lifespan perspectives on natural disasters: Coping with Katrina, Rita and other storms*. New York: Springer.

Cherry, K. E. (Ed.). (2015). *Traumatic stress and long-term recovery: Coping with disasters and other negative life events*. New York: Springer.

Mooney, C., & Dennis, B. (2018, January 8). *Extreme hurricanes and wildfires made 2017 the most costly U.S. disaster year on record*. The Washington Post. Retrieved from http://washingtonpost.com.

Oriol, W. E. (1999). *Psychosocial issues for older adults in disasters*. U.S. Department of Health and Human Services, Substance Abuse and Mental Health Services Administration, Center for Mental Health Services.

U.S. Department of Interior. (2020). *How can climate change affect natural disasters?* Retrieved from https://www.usgs.gov/faqs/how-can-climate-change-affect-natural-disasters-1.

Chapter 2
Overview on Trauma and Resilience

Scott E. Wilks, Sarah E. Choate, Sandra C. Brown, Xi Du, and Tamecia M. Curry

Introduction

Diagnostic criteria for reactions to trauma-related stress were introduced by the American Psychiatric Association (APA, 2013) in 1952 – as a personality disorder (Gerhart, Canetti, & Hobfoll, 2015). The modern perspective of trauma-related stress is a multifaceted set of reactions unique to the individual: Reactions may be mild or overwhelming; longevity of effects from these reactions may be acute or a longer, declining spiral of stress (Bonanno, 2004; Polusny & Follette, 1995). The domains of trauma are equally varied, including war, abuse, natural disasters, and other catastrophic events (Hobfoll, 2004). Accordingly, understanding trauma and its related effects is burdensome to any single chapter or article. Rather, in this chapter, we attempt to unravel the complexity of trauma and its related stress effects to explication of basic definitions, attributes, and magnitude or scope of the construct. Additionally, we do the same with resilience, for it is in the face of trauma that individuals demonstrate a remarkable, sometimes incredible, ability to overcome traumatic stress (Gerhart et al., 2015).

S. E. Wilks (✉) · S. E. Choate · T. M. Curry
School of Social Work, Louisiana State University, Baton Rouge, LA, USA
e-mail: swilks@lsu.edu

S. C. Brown
Southern University and A & M College, Baton Rouge, LA, USA

X. Du
Department of Sociology, Louisiana State University, Baton Rouge, LA, USA

© Springer Nature Switzerland AG 2021

K. E. Cherry, A. Gibson (eds.), *The Intersection of Trauma and Disaster Behavioral Health*, https://doi.org/10.1007/978-3-030-51525-6_2

Conceptual Explication of Trauma

Trauma Defined

Two forms of trauma, physical and psychological, may occur simultaneously or independently of one another. While physical trauma refers to severe injury to one's physical body, it has the potential to incite psychological trauma (Hobfoll, 2004). Multiple conceptualizations of each type exist within both biological and social sciences – particularly, psychological trauma (Gerhart et al., 2015). Trauma as it pertains to the mental health community refers to the process of exposure to a devastating event or experience, the results of such an occurrence as manifested by an individual's psychological response, or to the event itself.

Trauma as an event may be defined as an atypical and overwhelming event or experience, which supersedes common stressful occurrences, which in turn has the ability to cause distressing negative emotions of fear, shame, or powerlessness (Courtois, 1999). The Diagnostic and Statistical Manual of Mental Disorders, fifth Edition (DSM-5; APA, 2013), states that a traumatic event is one of "actual or threatened death, serious injury, or sexual violence" and may take the form of an individual directly experiencing the event, witnessing such an event, learning of the event happening to family or friends, or experiencing repeated or intense exposure to the event (APA, 2013, p. 271). Yet, traumatic events also have been addressed more generally – defined as any event that exceeds an individual's ability to cope with or respond to it (Horowitz, 1989).

Trauma also refers to the psychological effects and responses caused by such distressing events. Both the American Psychological Association (2019) as well as the Substance Abuse and Mental Health Services Administration (SAMHSA, 2014) adhere to this position. SAMHSA (2014) explained trauma as the actual result of an individual experiencing an event, single or multiple, or circumstances that are physically or mentally harmful and/or threatening. The event described must also have "adverse effects on the individual's functioning and physical, social, emotional, or spiritual well-being" (SAMHSA, 2014, p.7). Similarly, the American Psychological Association (2019) defined trauma as the emotional response to an abysmal event – such as a vehicle accident, sexual assault, or natural disaster. For the purpose of this discussion, *trauma* will be defined as a response, and the definition will refer to it as a psychological and/or emotional pathological response to a traumatic event or crisis that is significantly distressing or disturbing and exceeds one's ability to cope.

Common Attributes of Trauma

Trauma manifests following numerous types of adverse emotional and/or physical triggering events. Instances of such events include but are not limited to sexual abuse, natural disasters, domestic violence, child neglect or deprivation, physical injuries, school violence, or traumatic grief (Center for Substance Abuse Treatment, 2014). These events may take the form of a single isolated occurrence, known as acute trauma, or more than one traumatic event, known as complex trauma. Complex trauma may occur by experiencing the same type of trauma repeatedly, or by an individual experiencing multiple distinctive types of trauma (John et al., 2019).

Traumatic stress may present in a number of means depending on the individual and the severity of the trauma. After traumatic events occur, it is common for individuals to experience shock and denial; however, longer-term manifestations of trauma may include unstable emotions, avoidance of reminders of the event, or even physical symptoms such as nausea or shortness of breath (APA, 2019). As these symptoms may persist and inhibit normal daily functioning, they can become pathological. In order to appropriately diagnose trauma-related disorders, health practitioners rely on types and severity of symptoms.

The DSM-5 (APA, 2013) specifically recognizes trauma-related disorders as any psychological disorder in which a traumatic or stress-related event is listed explicitly as one of the diagnosing criteria and includes post-traumatic stress disorder (PTSD), acute stress disorder (ASD), reactive attachment disorder (RAD), and disinhibited social engagement disorder (DSED; see Kusmaul, this volume, Chap. 9). Each trauma-related diagnosis bears its own explicit diagnostic criteria. Even still, the sheer volume included in each disorder precludes a thorough review in this chapter. Accordingly, a brief overview of symptomology and prevalence of trauma-related disorders will be introduced.

PTSD is one of the most commonly known trauma-related diagnoses. To warrant the diagnosis of PTSD, one must meet the criterion of exposure to a specified traumatic event. Other diagnostic criteria include intrusion symptoms, such as recurrent memories or flashbacks, avoidance behaviors, persistent negative emotional state, distorted cognitions about the trauma, or hypervigilance. A PTSD diagnosis also specifies slightly varied diagnostic criteria based on the age of the individual and is separated into categories of ages six or younger and ages seven or older. Per the American Psychiatric Association (2013), Americans have an 8.7% chance of developing PTSD over one's lifetime. ASD carries similar diagnostic criteria to PTSD; however, the main difference is duration of symptom persistence – ASD only applies to the persistence of symptoms up to 1 month.

The RAD and DSED are trauma-related diagnoses for children, which stem from parental neglect or the inability to form stable attachments and do not require exposure to death, injury, or violence. RAD specifies that the disturbance must be present prior to age five and may include symptoms such as withdrawn behavior, minimal social and emotional responses, or episodes of irritability or sadness. More recent research has also suggested that children diagnosed with RAD are more prone to anger outbursts (Vasquez & Miller, 2018). To be diagnosed with RAD, children must have been exposed to neglect or deprivation, a constant change in caregivers, or other circumstances that inhibit a healthy attachment to caregivers. DSED exposure criteria are similar to that of RAD; however, the symptoms manifest as diminished reticence with unfamiliar persons.

Scope of Trauma

In the past, trauma was considered to be a rare phenomenon, but research statistics have shown that notion to be off the mark. Research reports from several countries suggest that the majority of adults will experience some type of traumatic event within their lifetimes (Magruder, McLaughlin, & Elmore Borbon, 2017). The most commonly reported type of traumatic event was witnessing someone being severely injured or killed, followed by disaster involvement (fire, flood, or other natural disaster) and involvement in a life-threatening accident or assault (Gerhart et al., 2015).

Given the heterogeneity of symptom profiles in the DSM-V (APA, 2013), Galatzer-Levy and Bryant (2013) calculated over 636,000 possible presentations of clinical post-trauma. In the United States, 61% of men and about 50% of women report exposure to one or more traumatic events; 90% of clients in behavioral health care settings report traumatic experience (SAMHSA, 2014). The chances of experiencing a traumatic event in general, as well as the chances of experiencing specific types of traumatic experiences, may be influenced by certain demographic variables, such as sex, socioeconomic status, or ethnicity (Magruder et al., 2017). The Center for Substance Abuse Treatment (2014) reported that overall, 83.7% of non-Latino White Americans reported experiencing some type of traumatic event, compared to 76.4% of African Americans, 68.2% of Latinos, and 66.4% of Asian Americans, Native Hawaiians, or Pacific Islanders. Their research also showed that the exposure to certain types of traumatic events is more common among specific ethnicities based on the National Epidemiologic Survey on Alcohol and Related Conditions. Specifically, non-Latino White Americans were the most likely to report an unexpected death, while African Americans were the most likely to report a traumatic event in the form of violent assault (Center for Substance Abuse Treatment, 2014).

Focusing more specifically on trauma prevalence in disaster contexts (e.g., weather-related catastrophes [floods, hurricanes, etc.], transportation accidents, school shootings, etc.), a national survey of American residents showed that 13% experienced a natural or human-generated disaster (Burkle Jr., 1996). A National Comorbidity Survey found that almost 20% of men and 15% of women reported exposure to one or more experiences of a natural disaster (Kessler, Sonnega, Bromet, Hughes, & Nelson, 1995). Quite often, these individuals face health-damaging stress, acute and long term, related to these disaster experiences. Galea and his colleagues (Galea, Nandi, & Vlahov, 2005) reviewed studies of PTSD after disasters. They found that adult survivors of disasters have a 30–60% prevalence rate of PTSD. Galea et al. also reported that rates of PTSD among children after disasters are largely inconsistent and dependent upon the disaster and other factors. For example, a 1984 Los Angeles school shooting led to 38% of children with PTSD symptoms, whereas a 1998 Illinois school shooting led to a much lower PTSD rate of 8% (Galea et al.). These researchers concluded the following:

> Arguably, the available information is sufficient to suggest plausible ranges of PTSD prevalence that can be expected in the first year after disasters, within exposure groups. The empirical evidence suggests that the prevalence of PTSD among direct victims of disasters is 30–40%, the prevalence among rescue workers is approximately 10–20%, and the prevalence in the general population is approximately 5–10%. (p. 84)

Paradigm Shift

The way trauma is viewed within American culture has changed significantly over the decades. Accounts of post-trauma symptoms have been noted as far back as ancient Greece. During the Civil War era, these symptoms were referred to as *soldier's heart* and *nostalgia*, while more recently, during the World War I era, the term *shell shock* was commonly used (Center for Substance Abuse Treatment, 2014). During these times, symptoms of PTSD were viewed as a psychological weakness or a flaw in character and at times symptoms were even attributed to homesickness.

Views on psychological trauma started to shift during the World War II era. Although soldiers were still being screened for moral weaknesses that would cause such symptoms, new basic interventions, such as allowing for rest periods before return to battle, were applied to alleviate psychological stress. From this rudimentary beginning, talk therapy as well as group therapy began to be utilized to treat PTSD symptoms. During the 1970s, the trauma intervention movement expanded. Finally, during the post-Vietnam 1980s, PTSD was recognized in the DSM's third edition as a legitimate psychological disorder among war veterans, and the fourth edition of the DSM further expanded what was constituted as trauma (James & Gilliland, 2013). These additions were the impetus of a new wave of trauma-related research among varying civilian populations, including disaster survivors.

Our discussion now moves to a strengths-based phenomenon that often is observed in the face of trauma and other adversity. The term, resilience (and similar root terms, e.g., resiliency, resilient), is mentioned frequently in literature, including the remainder of this chapter. This chapter offers a foundational understanding of resilience that may be imperative for resilience-focused research, as well as useful for professional practice with adversity-laden patients and clients.

Conceptualization of Resilience

"The power of ordinary magic." Such is the perspective from Ann Masten's (2001) groundbreaking work on human adaptation to adversity – a class of psychological with behavioral outcomes phenomena called *resilience*. This section delves into a straightforward understanding of resilience, including definitions and attributes of this concept, and the paradigm shift from a deficit-focused model to one of strengths-based on protective factors. We begin this section with definitions of resilience at the micro and macro levels.

Resilience Defined

A common understanding of resilience is the ability to overcome adversity, i.e., successful coping. This is not inaccurate, per se, yet this definition neglects a critical component of resilience: post-adversity hardiness. Chronologically, adversity occurs first, followed by successful coping and then a degree of post-adversity strength or hardiness from said coping. This indicates resilience. Resilience research expert Ann Masten (2001) astutely noted that in order to observe resilience, demonstrable adversity, or risk, must be present and observed. This risk construct is explicated in the next section, along with another attribute inextricably linked with resilience, protective factor.

Lazarus and Folkman's landmark model of stress appraisal, coping, and adaptation (Lazarus & Folkman, 1984) tells us that a person typically initiates a coping mechanism to manage an adverse event or stressor. If the coping mechanism achieves the desired result of successfully managing the stressor, then the person's perception of this stressor may change, in time, to a more manageable adverse event. Consequent to this successful coping and perception-of-stressor change, the person's sense of self-efficacy may be enhanced. In other words, the person may think the following: "I managed this stressor. I am proud of myself for successfully handling this stressful event. This feels great because it means I know I can handle it in the future!" This enhanced perspective of the adversity and the sense of strength to manage similar future adversities is, in essence, resilience.

So, what is the clear distinction between successful coping and resilience? Think of these constructs temporally linked with adversity (Wilks & Vonk, 2008). Coping is a durational strength, meaning that it is a time-limited process to manage a stressor *during* the stressor. In contrast, resilience is an outcome strength, a product *resultant* from successful coping to adversity. In Werner and Smith's (1992) groundbreaking, longitudinal study of resilience among Hawaiian (Kauai) children, amid a plethora of health and well-being risk factors, "one out of three of these children grew into competent, confident and caring adults" (Werner, 2005, p. 11). The researchers identified resilience as a determining factor to the health of these Hawaiian individuals. They identified resilience as an ongoing track record of successful adaptation for the individual repeatedly exposed to adversity, thereby lowering susceptibility for future stressful events (Wilks & Vonk, 2008).

Resilience Models and Attributes

Understanding theoretical models of resilience as a construct, in the models' basic form, requires three elements: risk, a protective factor, and a health outcome indicating resilience. We examine each of these elements, beginning with risk. The models show a negative relationship between risk and a resilience outcome, with a protective factor either mediating or moderating the deleterious effect of risk (see Figs. 2.1 and 2.2,

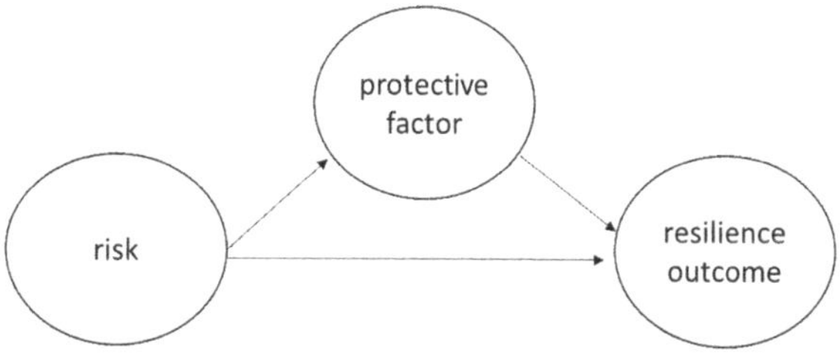

Fig. 2.1 Mediation model of resilience

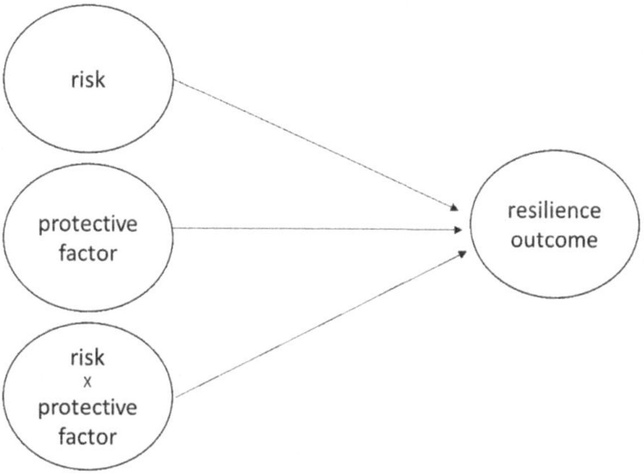

Fig. 2.2 Moderation model of resilience

respectively). Interested readers are directed to Baron and Kenny's (1986) foundational explanation of mediation versus moderation and to Wilks and Croom's (2008) illustrative examples of mediation versus moderation analyses using risk and resilience variables.

Risk Any condition that increases the likelihood of a problem or developing problem is called *risk* or risk factor in resilience literature (Greene, 2002). Risk has many synonyms: adverse event, adversity, burden, negative condition, problem, problematic condition, stress(or), to name a few. As the definition indicates, risk is not something a person normally wishes to have, with good reason. Rarely does one desire a condition that leads to problems, or to more problems one is already facing. Risk negatively influences health or well-being. Accordingly, risk negatively influences or lowers resilience. When risk is relieved, resilience has a change to emerge or flourish (Masten, 1994).

Protective Factor Modern resilience research pioneer Michael Rutter (1990) explained protective factors as internal assets and environmental conditions correlating to resilience. Protective factors encourage or promote resilience; they are not exclusively inherent or constant. Rutter identified a characteristic or condition as *protective* if it (a) reduces an individual's exposure to risk, (b) reduces negative subsequent reactions that follow risk, and (c) promotes self-esteem and self-efficacy through accomplishment. In both models, the presence of the protective factor buffers or ameliorates the otherwise direct, negative effect of risk on resilience when the protective factor is not present (see Figs. 2.1 and 2.2).

Resilience Outcome Rutter (1990) stated that resilience is "the positive pole of the ubiquitous phenomenon of individual difference in people's responses to stress and adversity" (p. 181). This positive pole of response to adversity has been observed and measured in a number of ways. Historically, researchers inferred resilience by increased levels of positive health and well-being outcomes (e.g., more happiness, higher marital satisfaction, greater self-esteem) or by lower levels of negative health/well-being outcomes (e.g., lower blood pressure level, decreased anxiety), in short, increased strengths and decreased or absence of limitations and risk conditions (West & Farrington, 1973). Such observation of resilience is fine, especially when a measure of resilience as a valid, autonomous construct is lacking. Modern resilience literature redressed this gap. Researchers and health practitioners no longer have to rely on health outcome scales or measures to infer resilience; rather, resilience can be explicitly measured. Popular measures of resilience include the *CD-RISC* (Connor & Davidson, 2003), Wagnild and Young's *Resilience Scale* (1993), the *Resilience Scale for Adults* (Friborg, Hjemdal, Rosenvinge, & Martinussen, 2003), and the *Brief Resilience Scale* (Smith et al., 2008).

Given the aforementioned brief explications of trauma and resilience, our discussion turns to the connection between these two phenomena. Adherent to this text's disaster-related trauma context, the following discussion emphasizes a similar theme among two populations most vulnerable to the impact of disasters (World Health Organization, n.d.): children and older adults.

Connection between Trauma and Resilience

Disaster-Related Trauma and Resilience among Children

Disaster disproportionately affects vulnerable groups in society. It has been well established that vulnerable or *at-risk* groups are likely to be less prepared for a natural disaster, are more susceptible during it, have higher mortality rates, and have poorer outcomes in the recovery period (Allen & Nelson, 2009; Stough, 2015; see Mosby, this volume, Chap. 13). Recent research has demonstrated the profound influence of disaster-related stressors on neurodevelopment in children. Kessel et al. (2018) examined whether exposure to Hurricane Sandy-related stressors altered children's brain response to emotional information and provided novel evidence that exposure to natural disaster-related stressors alters the trajectory of emotional reactivity to negative stimuli. Findings from this study are consistent with previous literature that reported the following: Early-life stress, including those influenced by natural disasters, enhances vulnerability to stressors in adulthood and alters neural reactivity to negative emotional information (Dich et al., 2015; Kujawa, MacNamara, Fitzgerald, Monk, & Phan, 2015). Enhanced reactivity to negative emotional information is associated with internalizing disorders in both children and adults and may be a marker of increased vulnerability for subsequent maladaptive outcomes later in life.

The impact of disasters on children depends on the meaning the child attaches to the disaster. Doostgharian (2009) described two major overall impacts of disasters among children: (1) emotional consequences of losing a family member and separation from friends and (2) the loss of protective functions. When children lose the protective functions of their parents, they can face major life challenges such as threats to fulfillment of basic needs, food, water, clothing, or shelter; permanent or temporary loss of homes and schools; and interruptions in community-based services. All of these protective functions are indeed protective factors of resilience. The resilience perspective in disaster literature tells us that it is paramount to strengthen supportive protective factors – secure and ongoing attachment relationships – within the ecology of the child, thereby restoring the child's sense of safety, life routine, agency, and self-efficacy (Stafford, Schonfeld, Keselman, Ventevogel, & Stewart, 2006).

Disaster-Related Trauma and Resilience among Older Adults

Resilience and vulnerability are essential in the study of disaster management. Numerous students suggest that both terms are opposites of the same continuum (Castleden, McKee, Murray, & Leonardi, 2011; Fu, Leoutsakos, & Underwood, 2014; Norris, Stevens, Pfefferbaum, Wyche, & Pfefferbaum, 2008). Yet, resilience is not necessarily the opposite of vulnerability. Overlap exists, as certain characteristics or attributes can make a person vulnerable and simultaneously affects that person's ability to adapt (Matyas & Pelling, 2015). Older adults, compared to younger counterparts, may be more resilient when faced with extraordinary life events. A lifetime of accumulated experience can serve as a psychological buffer that facilitates adaptive coping and promotes resilience (Cherry et al., 2010b).

Even so, vulnerable older populations are extremely susceptible to disasters, such as those with disabilities, dementia, or other cognitive impairment; and nursing home residents, many of whom maintain these diagnoses and impairments (see Allen & Nelson, 2009). They are at very high risk for a range of adverse outcomes, such as worsening disability, immobility, delirium, dehydration, malnutrition, pressure sores, and side effects or withdrawal from medications (Rothman & Brown, 2008). Kwan and Walsh (2017) studied older adults' disaster resilience. As is the case with many resilience studies, they observed the notable protective factor of social support as key to their bolstered resilience (see Phillips, this volume, Chap. 10). Brockie and Miller (2017) noted a similar theme. They mentioned that during floods, older adults in their sample recalled how critical decision-making (including decision to evacuate) stemmed from direct interactions with family and neighbors.

Additional studies with older adults in disaster-related settings show two integral components related to competent disaster preparedness (a protective factor of disaster-related resilience): instrumental, timely social support from family and neighbors; and the motivation to be valuable to their social support networks during these difficult times (Ashida, Robinson, Gay, & Ramirez, 2016; Loke et al., 2012).

Post-traumatic stress disorder (PTSD) and depression are among the most prevalent and commonly studied psychological sequelae of disasters (see Cherry, Calamia, Birch, & Moles, this volume, Chap. 16). Pietrak, Southwick, Tracy, Galea, and Norris (2012) examined the prevalence and correlates of disaster-related PTSD and depression in older persons affected by Hurricane Ike. Results revealed that being married/living with a partner was protective against depressive symptoms. However, previous studies have found that being married may actually increase disaster-related distress, as marital stress may intensify after a disaster (Norris et al., 2002). Prior disaster exposure was associated negatively with Ike-related PTSD symptoms. These findings provide some support for the *inoculation hypothesis*, which suggests that being familiar or experienced with a particular traumatic event may help one cope with a similar event in the future (Knight, Gatz, Heller, & Bengtson, 2000; Norris & Murrell, 1988). Findings from this study revealed that prior disaster exposure seemed to increase vulnerability to depressive symptoms in the aftermath of a newly experienced disaster. More research is needed to determine how to utilize older adults with disaster experience as a resource – a resilience protective factor with experiential knowledge and social support – to others in disaster recovery efforts while alleviating their own negative reactions to the disaster event (see Cherry, Calamia, & Elliott, this volume, chapter 8).

Trauma, Resilience, and Behavioral Health

Complex trauma involves multiple and/or chronic interpersonal traumatic experiences typically occurring within or around the home or other caregiving system (Cook et al., 2005). Exposure to complex trauma can begin in early childhood, interfering with many aspects of the child's development, including problems with regulation of emotions and behaviors, problems with attachment and interpersonal functioning, issues with attention/cognition and dissociation, and difficulties with biological/physiological functioning and sense of self (Cook et al., 2005).

The key to understanding resilience in youth with complex trauma is identifying and building protective factors (Kisiel, Summersett-Ringgold, Weil, & McClelland, 2017). Hypothesizing that protective factors buffer the negative impact of trauma on mental health and daily functioning, Kisiel et al. (2017) assessed 7483 youth entering an intensive, child welfare stabilization program to conduct a longitudinal examination of individual, child-level strengths in relation to complex trauma exposure, traumatic stress symptoms, risk behaviors, and other mental health needs. This study measured potential protective factors in terms of child, caregiver, and environmental resources that promote healthy development and well-being. Examples of areas of strengths assessed include the following: family relationships, interrelationship, educational setting, vocational resources, coping skills, talents and interests, and community ties and support. Findings of this study demonstrated the effect of child strengths in decreasing negative outcomes. Approximately 90% of risk attitudes and behaviors (e.g., suicide risk, danger to others, runaway, and delinquency) at initial assessment were buffered by protective factors at closing assessment, and traumatic stress symptoms also were moderately stable over time. Protective factors of coping skills, family strengths, interpersonal strengths, and educational background contributed the most to the child health and daily abilities outcomes (Kisiel et al., 2017). This pattern highlights the critical role of protective factors relating to traumatic stress and resilience among youth in child welfare.

Evidence suggests that many people with traumatic loss or grief, including disaster-related trauma, can continue to have positive emotional experiences and show only minor and transient disruptions in their ability to function (Bonanno, 2004; Bonanno, Galea, Bucciarelli, & Vlahov, 2007; see Kozu & Gibson, this volume, Chap. 6). Bereavement – the period of grief and mourning caused by a person's reaction of loss – often displays the negative behavioral health symptomology resultant from a traumatic event. Yet, an individual observed as bereaved is identified as *resilient* when level of negative psychological symptoms does not significantly exceed the level of similar symptoms observed in individuals of similar traumatic contexts outside of the bereavement period (Bonanno, Moskowitz, Papa, & Folkman, 2005). Resilient bereaved individuals are noted as better adjusted and possessing more positive interpersonal traits (e.g., friendly, thoughtful, supportive) compared to non-resilient counterparts. Resilience does not ensure that individuals and families are immune to loss-related distress; rather, their ability to maintain functioning, during bereavement or otherwise, holds relatively firm in the face of the traumatic event (Mancini & Bonanno, 2006).

Can individuals suffer from trauma-related, negative health outcomes while also maintaining and demonstrating their resilience? Yes! For example, studies show that those who provide care to others during and after traumatic events also experience intense psychological distress, the effects of which may endure for months or even years (e.g., Parkes et al., 1969; Parkes and Weiss, 1983; Stroebe et al., 1993). Yet, caregivers also demonstrate an incredible amount of resilience. In Folkman, 1997, a renowned stress and adaptation psychologist, Susan Folkman, investigated positive psychological states and coping with severe stress among a large sample of HIV/AIDS caregivers. Her findings highlighted an interesting, somewhat paradoxical psychological phenomenon with caregivers. They reported co-occurrence of positive and negative psychological states in the midst of enduring profoundly stressful circumstances. Their resilience amid ill effects of trauma was illuminating. Caregivers dealing with their respective partners' deaths expectedly reported dealing with more negative than positive emotions in their narratives; yet, a substantial percentage of these emotions (40%) were reported as positive (Stein & Levine, 1987).

Another example of resilience in tandem with trauma-related, negative health symptomology arose in a landmark resilience study of adult women exposed to one or more potentially traumatic events (PTEs; Rusch, Shvil, Szanton, Neria, & Gill,

2015). Rusch and her colleagues reported findings that confirmed deeply held premises of resilience theory: Exposure to trauma (i.e., risk) weakens psychological defenses and jeopardizes resilience. Further, the presence of protective factors, such as social support and mastery to leave a vulnerable situation, buffers deleterious effects of the traumatic stressor. Yet, it should be noted that, even with diminished levels of resilience consequential to trauma and a lack of protective factors, these women nonetheless aggregately reported *some* level of resilience. This speaks to modern resilience theory pioneer Ann Masten's (2015) assertion that there is no question of whether one has resilience; rather, the question is one's level of resilience and how to elevate it with protective factors.

Tsirigotis and Luczak (2017) investigated ego resilience pertaining to female survivors of domestic violence. Ego resilience is strongly connected with elements of mental and behavioral health (Alessandri, Vecchione, Caprara, & Letzring, 2011). Tsirigotis and Luczak found that the survivors' resilience varied based upon the relationship established with the perpetrator. Findings showed a statistically significant difference in resilience scores of women based on the survivor's relationship to the perpetrator. A parent as perpetrator led to lower survivor resilience. Intimate partner violence was linked with the highest levels of resilience. Intimate partner violence was shown to weaken or damage resilience to a lesser extent than paternal violence or marital violence due to the survivor's awareness that the relationship can be terminated easier than possibly from a father or husband. This study demonstrated that protective factors of sense of self-control, awareness of personal choice, and availability of alternative options can influence deeply one's personal resilience and behavioral health.

While there is a wide range of reactions to trauma, it is fairly common to see resilience even after the most pernicious stressor events or disasters (Mancini & Bonanno, 2006). Bonanno et al. (2007) explored this possibility in a study of resilience among those who personally experienced September 11, 2001 (i.e., 9.11), terrorist attack in New York City. Their results demonstrated that almost two-thirds of the samples in the contiguous New York area exhibited resilience at approximately 6 months after terrorist attack, and the resilient group experienced a genuinely healthy course of adjustment. Protective factors to a positive health outcome post-disaster were further identified by Bonanno et al. (2010), including no income loss as a result of the disaster, pre-disaster good health, and high levels of social support.

Concluding Thoughts

The word *resilience* is Latin in origin, from *salire*, meaning to help, later changing to *resilire*, meaning to leap back or recoil. Word origins are meaningful, as it gives us an understanding of the evolution of a construct. Indeed, resilience is, at its core, a phenomenon of help to the individual's health and well-being amid adversity. And, indeed, in times of stress, the individual often leaps back or recoils. The recoil is common and expected, for adversity is just that – adverse. As previously mentioned, Masten (2001) reminded us to view this risk or adversity as a necessary, albeit unfortunate, part of life within a resilience-focused perspective. Think of risk as a *necessary evil* from which resilience can emerge and elevate. Individuals who endure trauma, including those recovering from disasters, show remarkable resilience by coping, surviving, and moving forward with renewed hardiness. Yet, the trait of resilience is not remarkable. It is not some Darwinist characteristic for the strongest of the strong. Resilience is ordinary magic, existing within each and every person. For researchers and health professionals working with trauma-laden clients or patients, the question should not be, "Does this person have resilience to successfully manage this trauma?" Rather, as the pioneers of modern resilience research (Masten, Rutter, and Werner and Smith) relayed to us, a more accurate question should be, "What level of resilience does this person already have, and how may I assist this person to strengthen it?"

Disasters reveal strengths and limitations among varying social levels, from individuals to societies (Cherry, Allen, & Galea, 2010a). In terms of health interventions, the Mental Health and Psychosocial Support in Emergency Settings' pyramid (IASC, 2007) informs health practitioners that, amid disaster-related crises, four layers of integrated and holistic services should be addressed, in the following order of priority: (1) social considerations in basic services and security, (2) strengthening community and family supports, (3) psychosocial supports (e.g., support groups), and (4) clinical services (e.g., counseling or psychotherapy). This model of intervention promotes agency of disaster survivors (Stafford et al., 2006). It is imperative to promote agency with this population. By enhancing the ability to control one's functioning in the course of events, in this context, disaster-related events, survivors are enabled to become self-sufficient and manage their own vulnerabilities (Grove, 2014; Stafford et al., 2006). In the face of disaster, management of vulnerability and ultimately self-sufficiency reflects that remarkable, ordinary magic of resilience.

References

Alessandri, G., Vecchione, M., Caprara, G., & Letzring, T. D. (2011). The Ego resiliency scale revised: A crosscultural study in Italy, Spain, and the United States. *European Journal of Psychological Assessment, 28*, 139–146. https://doi.org/10.1027/1015-5759/a000102

Allen, P. D., & Nelson, H. W. (2009). Disaster services with frail older persons: From preparation to recovery. In K. E. Cherry (Ed.), *Lifespan perspectives on natural disasters* (pp. 153–169). New York: Springer.

American Psychiatric Association (APA). (2013). *Diagnostic and statistical manual of mental disorders* (5th ed.). Arlington, VA: American Psychiatric Publishing.

American Psychological Association. (2019). *Trauma*. Retrieved from https://www.apa.org/topics/trauma

Ashida, S., Robinson, E. L., Gay, J., & Ramirez, M. (2016). Motivating rural older residents to prepare for disasters: Moving beyond personal benefits. *Ageing Society, 36*(10), 2117–2140.

Baron, R. M., & Kenny, D. A. (1986). The moderator-mediator variable distinction in social psychological research: Conceptual, strategic, and statistical considerations. *Journal of Personality and Social Psychology, 51*, 1173–1182.

Bonanno, G. A. (2004). Loss, trauma, and human resilience: Have we underestimated the human capacity to thrive after extremely aversive events? *American Psychologist, 59*(1), 20–28. https://doi-org.libezp.lib.lsu.edu/10.1037/0003-066X.59.1.20

Bonanno, G. A., Galea, S., Bucciarelli, A., & Vlahov, D. (2007). What predicts psychological resilience after disaster? The role of demographics, resources, and life stress. *Journal of Consulting & Clinical Psychology, 75*(5), 671–682. Retrieved from. https://doi-org.libezp.lib.lsu.edu/10.1037/0022-006X.75.5.671

Bonanno, G. A., Moskowitz, J. T., Papa, A., & Folkman, S. (2005). Resilience to loss in bereaved spouses, bereaved parents, and bereaved gay men. *Journal of Personality and Social Psychology, 88*, 827–843.

Bonanno, G. A., Brewin, C. R.; Kaniasty, K., & La Greca, A. M. (2010). Weighing the costs of disaster: Consequences, risks, and resilience in individuals, families, and communities. *Psychological Science in the Public Interest, 11*(1), 1–49. https://doi.org/10.1177/1529100610387086

Brockie, L., & Miller, E. (2017). Older adults' disasters lifecycle experience of the 2011 and 2013 Queensland floods. *International Journal of Disaster Risk Reduction, 25*, 10.

Burkle Jr., F. M. (1996). Acute-phase mental health consequences of disasters: Implications for triage and emergency medical services. *Annals of Emergency Medicine, 28*, 119–128.

Castleden, M., McKee, M., Murray, V., & Leonardi, G. (2011). Resilience thinking in health protection. *Journal of Public Health, 33*(3), 369–377.

Center for Substance Abuse Treatment. (2014). *Trauma-informed care in behavioral health services*. Rockville, MD: Substance Abuse and Mental Health Services Administration.

Cherry, K. E., Allen, P. A., & Galea, S. (2010a). Older adults and natural disasters: Lessons learned from hurricanes Katrina and Rita. In P. Dass-Brailsford (Ed.), *Crisis and disaster counseling: Lessons learned from hurricane Katrina and other disasters* (pp. 115–130). Thousand Oaks, CA: Sage.

Cherry, K. E., Su, L. J., Welsh, D., Galea, S., Jazwinski, S. M., Silva, J. L., et al. (2010b). Cognitive and psychosocial consequences of hurricanes Katrina and Rita among middle-aged, older, and oldest-old adults in the Louisiana healthy aging study. *Journal of Applied Social Psychology, 40*(10), 2463–2487.

Connor, K. M., & Davidson, J. R. (2003). Development of a new resilience scale: The Connor-Davidson resilience scale. *Depression and Anxiety, 18*, 76–82. https://doi.org/10.1002/da.10113

Cook, A., Spinazzola, J., Ford, J., Lanktree, C., Blaustein, M., Cloitre, M., et al. (2005). Complex trauma in children and adolescents. *Psychiatric Annals, 35*, 390–398.

Courtois, C. (1999). *Recollections of sexual Abuse: Treatment, principles, and guidelines*. New York, NY: Norton.

Dich, N., Hansen, A. M., Avlund, J., Lund, R., Morten, E. L., Bruunsgaard, H., et al. (2015). Early life adversity potentiates the effects of later life stress on cumulative physiological dysregulation. *Anxiety, Stress, & Coping, 28*, 372–390.

Doostgharian, T. (2009). Children affected by earthquakes and their immediate emotional needs. *International Journal of Social Work, 52*, 96–106.

Folkman, S. (1997). Positive psychological states and coping with severe stress. *Social Science & Medicine, 45*(8), 1207–1221. https://doi.org/10.1016/s0277-9536(97)00040-3

Friborg, O., Hjemdal, O., Rosenvinge, J. H., & Martinussen, M. (2003). A new rating scale for adult resilience: What are the central protective resources behind healthy adjustment? *International Journal of Methods in Psychiatric Research, 12*(2), 65–76.

Fu, C., Leoutsakos, J. M., & Underwood, C. (2014). An examination of resilience cross-culturally in child and adolescent survivors of the 2008 China earthquake using the Connor-Davidson resilience scale (CD-RISC). *Journal of Affective Disorders, 155*, 149–153.

Galatzer-Levy, I. R., & Bryant, R. A. (2013). 636,120 ways to have posttraumatic stress disorder. *Perspectives on Psychological Science, 8*(6), 651–662.

Galea, S., Nandi, A., & Vlahov, D. (2005). The epidemiology of post-traumatic stress disorder after disasters. *Epidemiologic Reviews, 27*, 78–91. https://doi.org/10.1093/epirev/mxi003

Gerhart, J. I., Canetti, D., & Hobfoll, S. E. (2015). Traumatic stress in overview: Definition, context, scope, and long-term outcomes. In K. E. Cherry (Ed.), *Traumatic stress and long-term recovery: Coping with disasters and other negative life events* (pp. 3–24). New York: Springer.

Greene, R. (2002). Human behavior theory: A resilience orientation. In R. R. Greene (Ed.), *Resiliency: An integrated approach to practice, policy, and research* (pp. 1–27). Washington, DC: NASW Press.

Grove, K. (2014). Agency, affect, and the immunological politics of disaster resilience. *Society and Space*. https://doi.org/10.1068/d4813

Hobfoll, S. E. (2004). *Stress, culture, and community: The psychology and philosophy of stress*. New York: Springer.

Horowitz, M. J. (Ed.). American Psychiatric Association Task Force on Treatments of Psychiatric Disorders. (1989). *Treatments of psychiatric disorders: A task force report of the American Psychiatric Association*. Washington, DC: American Psychiatric Association.

Inter-Agency Standing Committee. (2007). *IASC guidelines on mental health and psychosocial support in emergency settings*. Geneva: IASC.

James, R., & Gilliland, B. (2013). *Crisis intervention strategies*. Belmont, CA: Brooks Cole.

John, S. G., Brandt, T. W., Secrist, M. E., Mesman, G. R., Sigel, B. A., & Kramer, T. L. (2019). Empirically-guided assessment of complex trauma for children in foster care: A focus on appropriate diagnosis of attachment concerns. *Psychological Services, 16*(1), 120–133. https://doi.org/10.1037/ser0000263

Kessel, E. M., Nelson, B. D., Kujawa, A., Hajcak, G., Kotov, R., Bromet, E. J., … Klein, D. N. (2018). Hurricane Sandy exposure alters the development of neural reactivity to negative stimuli in children. *Child Development, 89*(2), 339–348. Retrieved from. https://doi-org.libezp.lib.lsu.edu/10.1111/cdev.12691

Kessler, R. C., Sonnega, A., Bromet, E., Hughes, M., & Nelson, C. B. (1995). Posttraumatic stress disorder in the National Comorbidity Survey. *Archives of General Psychiatry, 52*, 1048–1060.

Kisiel, C., Summersett-Ringgold, F., Weil, L., & McClelland, G. (2017). Understanding strengths in relation to complex trauma and mental health symptoms within child welfare. *Journal of Child and Family Studies, 26*(2), 437–451. Retrieved from. https://doi-org.libezp.lib.lsu.edu/10.1007/s10826-016-0569-4

Knight, B. G., Gatz, M., Heller, K., & Bengtson, V. L. (2000). Age and emotional response to the Northridge earthquake: A longitudinal analysis. *Psychology and Aging, 15*, 627–634.

Kujawa, A., MacNamara, A., Fitzgerald, K., Monk, C., & Phan, K. L. (2015). Enhanced neural reactivity to threatening faces in anxious youth: Evidence from event-related potentials. *Journal of Abnormal Child Psychology, 80*, 381–389.

Kwan, C., & Walsh, C. A. (2017). Seniors' disaster resilience: A scoping review of the literature. *International Journal of Disaster Risk Reduction, 25*, 259–273.

Lazarus, R. S., & Folkman, S. (1984). *Stress, appraisal, and coping*. New York: Springer.

Loke, A.Y., Lai, C.K., Fund, O.W. (2012). At-home disaster preparedness of elderly people in Hong Kong. *Geriatrics & Gerontology International, 12*(3), 524–531.

Magruder, K. M., McLaughlin, K. A., & Elmore Borbon, D. L. (2017). Trauma is a public health issue. *European Journal of Psychotraumatology, 8*(1), 1375338. https://doi.org/10.1080/20008198.2017.1375338

Mancini, A. D., & Bonanno, G. A. (2006). Resilience in the face of potential trauma: Clinical practices and illustrations. *Journal of Clinical Psychology, 62*(8), 971-985. Retrieved from https://doi-org.libezp.lib.lsu.edu/10.1002/jclp.20283

Masten, A. S. (1994). Resilience in individual development: Successful adaptation despite risk and adversity. In M. Wang & E. Gordon (Eds.), *Educational resilience in inner-city America: Challenges and prospects* (pp. 3–25). Hillsdale, NJ: Lawrence Erlbaum.

Masten, A. S. (2001). Ordinary magic. Resilience processes in development. *American Psychologist, 56*(3), 227–238. https://doi.org/10.1037//0003-066X.56.3.227

Masten, A. S. (2015). *Ordinary magic. Resilience in development*. New York: The Guilford Press.
Matyas, D., & Pelling, M. (2015). Positioning resilience in the post-2015 disaster risk management policy landscape: Integrating resistance, persistence and transformation. *Disaster Journal, 39*, 1–18.
Norris, F. H., Stevens, S. P., Pfefferbaum, B., Wyche, K. F., & Pfefferbaum, R. L. (2008). Community resilience as a metaphor, theory, set of capacities, and strategy for disaster readiness. *American Journal of Community Psychology, 41*, 127–150.
Norris, F. H., Friedman, M. J., Watson, P. J., Byrne, C. M., Diaz, E., & Kaniasty, K. (2002). 60,000 disaster speak: Part I. an empirical review of the empirical literature, 1981-2001. *Psychiatry, 65*, 207–239.
Norris, F. H., & Murrell, S. A. (1988). Prior experience as a moderator of disaster impact on anxiety symptoms in older adults. *American Journal of Community Psychology, 16*, 665–683.
Parkes, C. M., & Weiss, R. S. (1983). *Recovery from bereavement*. New York: Basic Books.
Parkes, C. M., Benjamin, B., & Fitzgerald, R. G. (1969). Broken heart: A statistical study of increased mortality among widowers. *British Medical Journal, 1*, 740–743.
Pietrak, R. H., Southwick, S. M., Tracy, M., Galea, S., & Norris, F. H. (2012). Posttraumatic stress disorder, depression, and perceived needs for psychological care in older persons affected by hurricane Ike. *Journal of Affective Disorders, 138*, 96–103.
Polusny, M. A., & Follette, V. M. (1995). Long-term correlates of child sexual abuse: Theory and review of the empirical literature. *Applied and Preventive Psychology, 4*(3), 143–166.
Rothman, M., & Brown, L. M. (2008). The vulnerable geriatric casualty: Medical needs of frail older adults during disasters. *Generations*, 16–20.
Rusch, H. L., Shvil, E., Szanton, S. L., Neria, Y., & Gill, J. M. (2015). Determinants of psychological resistance and recovery among women exposed to assaultive trauma. *Brain and Behavior,5*(4). https://doi.org/10.1002/brb3.322
Rutter, M. (1990). Psychosocial resilience and protective mechanisms. In J. E. Rolf, A. S. Masten, D. Cicchetti, K. H. Nuechterlein, & S. Weintraub (Eds.), *Risk and protective factors in the development of psychopathology*. New York, NY: Cambridge University Press.
SAMHSA. (2014). *Trauma-informed care in behavioral health services*. Retrieved from https://store.samhsa.gov/product/TIP-57-Trauma-Informed-Care-in-Behavioral-Health-Services/SMA14-4816
Smith, B. W., Dalen, J., Wiggins, K., Tooley, E., Christopher, P., & Bernard, J. (2008). The brief resilience scale: Assessing the ability to bounce back. *International Journal of Behavioral Medicine, 15*, 194–200. https://doi.org/10.1080/10705500802222972
Stafford, B., Schonfeld, D., Keselman, L., Ventevogel, P., & Stewart, C. L. (2006). *The emotional impact of disaster on children and families*. Retrieved from https://www.aap.org/en-us/Documents/disasters_dpac_PEDsModule9.pdf
Stein, N. L., & Levine, L. (1987). Thinking about feelings: The development and organization of emotional knowledge. In R. Snow & M. Farr (Eds.), *Aptitude, learning and instruction*. Hillsdale, NJ: Erlbaum.
Stroebe, M. S., Stroebe, W., & Hansson, R. O. (eds.) (1993). *Handbook of bereavement*. New York: Cambridge University Press.
Substance Abuse and Mental Health Services Administration, Trauma and Justice Strategic Initiative. (2014). *SAMHSA's concept of trauma and guidance for a trauma-informed approach*. Rockville, MD: Substance Abuse and Mental Health Services Administration.
Tsirigotis, K., & Łuczak, J. (2017). Resilience in women who experience domestic violence. *Psychiatric Quarterly, 89*(1), 201–211. https://doi.org/10.1007/s11126-017-9529-4
Vasquez, M., & Miller, N. (2018). Aggression in children with reactive attachment disorder: A sign of deficits in emotional regulatory processes? *Journal of Aggression. Maltreatment & Trauma, 27*(4), 347–366. https://doi.org/10.1080/10926771.2017.1322655
World Health Organization, Wagnild, G. M., & Young, H. M. (1993). *Development and psychometric evaluation of the resilience scale*. Journal of Nursing Measurement, *1*(2), 165–178.

Werner, E. E. (2005). Resilience and recovery: Findings from the Kauai longitudinal study. *Research, Policy, and Practice in Children's Mental Health, 19*(1), 11–14.
Werner, E. E., & Smith, R. S. (1992). *Overcoming the odds: High risk children from birth to adulthood*. Cornell University Press.
West, D. J., & Farrington, D. P. (1973). *Who becomes delinquent? Second report of the Cambridge study in delinquent development*. London: Heinemann Publishers.
Wilks, S. E., & Croom, B. (2008). Perceived stress and resilience in Alzheimer's disease caregiving: Testing moderation and mediation models of social support. *Aging & Mental Health, 12*, 357–364. https://doi.org/10.1080/13607860801933323
Wilks, S. E., & Vonk, M. E. (2008). Private prayer among Alzheimer's caregivers: Mediating burden and resiliency. *Journal of Gerontological Social Work, 50*, 113–131. https://doi.org/10.1300/J083v50n3_09
World Health Organization. (n.d.). *Environmental health in emergencies: Vulnerable groups*. Retrieved from www.who.int/environmental_health_emergencies/vulnerable_groups/en/
Stough, L. M. (2015). World report on disability, intellectual disabilities, and disaster preparedness. Costa Rica as a case example. *Journal of Policy and Practice in Intellectual Disability, 12*(2), 138–146.

Chapter 3
Integrating Trauma-Informed Principles into Disaster Behavioral Health Targeting Older Adults

Molly Everett Davis and Nikki D. Bellamy

Introduction

There is a critical need to understand the challenges faced in the provision of disaster services to older adults. News stories and media images, which surface from many different disasters, often show older adults who are suffering or dying while waiting for assistance. As an example, consider the St. Rita's nursing home tragedy in the hours after the 2005 Hurricane Katrina made landfall. As the flood waters began to recede, a group of first responders searching for survivors came upon a nursing home with residents sitting outside of the facility in sweltering heat. These residents had survived the severe flooding following levee breaches that had resulted in the drowning deaths of over 30 residents. These residents found themselves in this predicament because a decision had been made that evacuation was a greater threat to the residents than remaining in the facility (Cherry, 2020). This example underscores the need for continuing efforts on the part of first responders and emergency management officials to develop plans to evacuate nursing home facilities effectively, to understand managing risk with older adults, and to save lives is an ongoing challenge (Gibson & Hayunga, 2006).

The views expressed herein represent the opinions and analyses of the individual authors and may not necessarily reflect the opinions, official policy, or position of the Substance Abuse and Mental Health Services Administration (SAMHSA). The author completed this article based on previous work in her private capacity while working at the respective institution.

M. E. Davis (✉)
Department of Social Work, George Mason University, Fairfax, VA, USA
e-mail: mdavi7@gmu.edu

N. D. Bellamy
Substance Abuse and Mental Health Services Administration (SAMHSA), Center for Mental Health Services (CMHS), Rockville, MD, USA

© Springer Nature Switzerland AG 2021

K. E. Cherry, A. Gibson (eds.), *The Intersection of Trauma and Disaster Behavioral Health*, https://doi.org/10.1007/978-3-030-51525-6_3

In fact, there is a great deal of evidence that suggests that disasters adversely influence older adults to a greater extent than other groups. Older adults are recognized as generally least prepared for disasters (Shih et al., 2018). This fact has led to increased numbers of disaster-related deaths and a higher level of risks in multiple types of disasters and in every phase of disaster evacuation, rescue, and recovery for older adults (American Psychological Association, 2018).

This chapter will highlight the factors that support the need for a trauma-informed approach to disaster behavioral health as it specifically relates to older adults. To date, we have not made significant progress in implementing trauma-informed approaches to reduce the psychological, physiological, and social impact of the disaster (Campbell, 2011). This chapter documents the need for differential consideration of the increased vulnerability experienced by older adults during a disaster, highlighting the intersection between trauma and disaster behavioral health. The complex relationship between assessing pre-disaster vulnerability and promoting post-disaster resilience outcomes provides new understanding for promoting disaster recovery. A model for trauma-informed risk stratification based upon developing a vulnerability profile is one strategy for improving potential disaster-related outcomes for older adults. This approach has special value as it relates to reducing post-disaster risk for post-traumatic stress disorder (PTSD). We will explore recommended strategies for implementing a different approach toward understanding the vulnerability of older adults during a disaster. Finally, the authors will propose inclusion of this approach in the training of first responders and disaster behavioral health personnel to promote a trauma-informed strategy in the planning and emergency management of disasters as it relates to older adults.

Disaster Behavioral Health

The prevalence and unpredictability of disasters and the numbers of older adults who are potentially impacted by these events suggest that a population health approach might be needed (see Johnson & Galea, 2009). Population health is defined as an approach to health that aims to improve the health of the entire population and to reduce health inequities among population groups (Cohen et al., 2014). Disaster data suggests that the risk to older adults during a disaster is disproportionate and special efforts to reduce this disparity is needed. This vulnerability to disasters is based upon characteristics of older adults that may increase the risk for harm during a disaster and impact post-disaster outcomes.

Disaster behavioral health should reflect a public health and population health response to an emergency event. Along with reducing disproportionate risk to members of the population, disaster behavioral health utilizes tools such as psychological first aid and other strategies to manage the emotional, cognitive, developmental, and social impact of the disaster response (Speier, Osofsky, & Osofsky, 2009). It should reflect key elements of preparation and response commensurate with an understanding of the target population at the greatest vulnerability and risk. A number of studies has clearly indicated that older adults are a population at risk and vulnerable (Parker et al., 2015).

Disaster behavioral health intervention provides psycho-education information on hazards and threats designed to reduce risk and vulnerability. The goal of these activities is to promote resilience and recovery of disaster survivors. In order to accomplish these goals, there should be a recognition of the need for individualized plans to screen for those who may be at a higher risk for long-term adverse reactions. In addition, this screening can support the referral to appropriate medical psychological or other services based upon survivor's profile (see Spurlock & Brown, this volume, Chap. 12).

In order to increase the efficacy in serving older adults after a disaster, an approach like risk stratification may be very helpful. Risk stratified care management (RSCM) is the process of assigning a health/behavioral risk status to older survivors and using the person's status to direct and improve care. This approach is predicated upon being able to identify the person's risk status. Secondly, the development of a personalized care plan is based upon the use of a screening or risk assessment tools. The creation of risk assessment tools to determine an individual's vulnerability profile provides a means of personalizing the understanding of who the older person is and the unique experiences that will most contribute to disaster risk.

Understanding Vulnerability

There is a growing attention being directed toward what is being called the "vulnerability perspective" in disasters. Some view this as a dominant view in the field of disasters. This perspective focuses less on the kind of disaster in assessing the probable outcomes and more concerning the level of vulnerability of groups being impacted by the disaster. The greatest impact of a disaster is when it targets a vulnerable population. The limited and compromised ability to manage even a minor disaster is impacted more by their vulnerability status rather than the actual strength or magnitude of a disaster. This principle suggests a differential impact of disasters based upon vulnerabilities that may exist prior to the disaster, during the disaster, and post-disaster. For example, a review of the disasters that cause the highest level of mortalities revealed surprising results. Over the period of the last 20 years, the most common natural disaster has been flooding accounting for 43% of all recorded events (Wallemacq & House, 2018). The UN Office of Disaster Risk Reduction reports that 91% of all disasters were caused by floods, storms, droughts, heatwaves, and other extreme weather events (2018). In addition, winter weather (snow, ice, blizzards, etc.) also contributes to higher mortality in older adults during disasters. The conclusion however in assessing the human costs of disasters according to this report suggests, "vulnerability to risk and degrees of suffering are determined by levels of economic development, rather than simple exposure to natural hazards per se" (Wallemacq & House, 2018, p. 6,7). When poverty is combined with old age and disability status, there are increasing levels of vulnerability experienced by those impacted by a disaster (Table 3.1).

Poverty, age, and disability produce more adverse effects on disaster outcomes regardless of the disaster (Rodríguez, Wachtendorf, & Russell, 2004). These results

Table 3.1 Deaths from natural disasters by type, 1970–2004

	Number	Percent
All deaths from natural disasters	19,959	100.0
Heat or drought	3906	19.6
Severe weather (severe storm/thunderstorm, fog, hail, wind)	3762	18.8
Winter weather	3612	18.1
Flooding	2788	14.0
Tornado	2314	11.6
Lightning	2261	11.3
Coastal (storm surge, rip current)	456	2.3
Hurricane or tropical storm	304	1.5
Geophysical (earthquake, tsunami, volcano)	302	1.5
Mass movement (avalanche, landslide)	170	0.9
Wildfire	84	0.4

Source: Adapted from Borden and Cutter (2008), Fig. 1 and Table 3. https://doi.org/10.1186/1476-072X-7-64, licensed under the terms of the Creative Commons Attribution License (http://creativecommons.org/licenses/by/2.0)

suggest the central role in the vulnerability of the group as a key factor in disaster mortality outcomes rather than the kind of disaster.

Vulnerability is a term that is defined as "the characteristics of a person or group and their situation that influences their capacity to anticipate, cope with, resist and recover from the impact of natural hazards, in the context of disasters" (Wisner, Blaikie, Cannon, & Davis, 2004, p. 11). Vulnerability is dynamic and varies across different population groups. A review of data from multiple disasters indicate a consistent vulnerability of older adults during a disaster, warranting special efforts to mediate the risk (Al-Rousan, Rubenstein, & Wallace, 2014). Because vulnerability can vary, the assessment of each disaster, its characteristics, and the demographics of populations at most risk is critical.

The use of the term **vulnerable** to describe the status of older adults during a disaster is appropriate because the broad risk for adverse disaster outcomes may be fairly disproportionate compared to other groups. There should be, however, a recognition that older adults are diverse, and the age range associated with the term older adults may span for a period of 20–40 years (APA, 2017). Vulnerability should not be evenly applied across all older adults. We might think about applying the concept of vulnerability factors. Vulnerability factors is defined as variables that when combined can increase the probability of producing the likelihood of an outcome and provide a means of addressing the diversity among older adults (Davis, 2017). This allows for consideration of specific factors that may define and impact vulnerability status requiring reasonable accommodations prior to, during, and post-disaster. This known vulnerability provides grounds for planning and implementation and strategies in order to address known risk. Yet, often disaster behavioral health plans and emergency management plans often do not sufficiently address vulnerability factors that combine to increase the vulnerability of older adults and reduce risk factors. Without modification of these strategies to assess and address vulnerabilities of older adults, they will continue to be more adversely impacted by disasters than other groups. The International Federation of Red Cross and Red Crescent Societies (IFRC) suggest two questions that are relevant in determining vulnerability (2019).

1. To what threat or hazard is the group vulnerable?
2. What makes the group vulnerable to the threat or hazard?

Many research studies have clearly indicated such vulnerability for older adults in disasters exists, thus justifying the use of strategies to counteract vulnerabilities. We clearly know that disasters are one specific threat/hazard for which older adults are vulnerable. The evidence is clear. In response to the second question, what makes the group vulnerable, there are a variety of different risk factors that contribute to their vulnerability, such as physical status, cognitive impairment, mobility challenges, social isolation, poverty, prior exposure, etc. This chapter adds traumatic life events to the list of known threats and hazards, suggesting a need to utilize a trauma-informed approach to disaster behavioral health planning and emergency management strategies (Centers for Disease Control, 2012a, 2012b).

The realization that a vulnerability status exists suggests efforts should be directed toward counteracting the vulnerability by specifically addressing vulnerability factors as they are relevant to those impacted by the disaster. Firstly, considering specific strategies to reduce the impact of the hazard (disaster) where possible through mitigation, prediction, warning, and preparedness is necessary. Secondly, building capacities to withstand and cope with hazards should be ongoing. Finally, efforts should be made to tackle the root cause of the vulnerability, e.g., poverty, discrimination, poor access to resources, education, and training, to mention a few (IFRC, 2019). This chapter provides strategies to clearly identify the threats/hazards that contribute to older adults being vulnerable in disasters. It also, at a micro level, identifies the threats and hazards through an assessment process. The goal of this strategy is to implement action plans to mitigate threats and hazards, as they are relevant to the particular disaster profile. This tool allows for the development of a vulnerability profile for a disaster event.

Older adults are particularly difficult to manage vulnerabilities that might be prominent during a disaster. The variation that exists among older adult populations requires consideration of many factors (Wisner et al., 2004). This variation is not necessarily chronologically age based. There are 75-year-olds who live in the community with active self -sufficient lives. In fact, it is estimated that 95% of older adults live in the community. The remaining 5% that reside in long-term care facilities are the most vulnerable and require protections to survive a disaster. The chronological age may be less relevant than functional status. Those who are nonambulatory and confined to a wheelchair or required to have life-sustaining medical equipment pose the greatest risk and vulnerability during a disaster (see Allen & Nelson, 2009, for discussion).

It is therefore critical to have the ability to assess vulnerability factors within a population on an ongoing basis, by first responders and other disaster professionals who will be engaged in the disaster management and recovery process. Although this appears to present a significant challenge, in reality, the need to manage vulnerability is an important element of disaster management and recovery. How do we assess and manage vulnerability concerning older adults, prior to, during, and after a disaster? No matter the context of the disaster, older adults present an additional level of vulnerability that should be a component of the disaster management plan (Wisner, 2004a; 2004b; Campbell, 2011).

Risk and Vulnerability of Older Adult Disaster Survivors

As we set our focus specifically on older adults, again, we realize that not all have the same level of vulnerability to disasters. Older adults vary in their roles and statuses and are diverse in their life experiences. For example, older adults who live in Florida on the coastline may have experienced many hurricanes and have developed specific strategies to protect themselves during these storms. Their knowledge and prior preparation may reduce their vulnerability during a hurricane as compared with individuals with no prior knowledge, experience, or preparation. For the same reason, an older adult living at the poverty level or with chronic illnesses may experience increased vulnerability to disasters because of these conditions.

Older adults according to a number of studies might experience increased vulnerability in some areas during and immediately after a disaster (Bei et al., 2013; Malik et al., 2017; Weisler & Barbee, 2006; but see Cherry, Calamia, Birch, & Moles, this volume, Chap. 16). For instance, these are periods of time in which older adults may be less likely to be able to access services, medical care, and the support they need to stabilize their health and recovery status. Because older adults are more likely to be socially isolated than some groups due to multiple chronic conditions, limitations in daily activities, changing physical and cognitive conditions, vision and hearing changes, they are more likely to be disconnected from support systems and resources (Levac, Toal-Sullivan, & O'Sullivan, 2011). Assessment of the older adult becomes critical to determining the particular vulnerability factors that might be in operation. This chapter suggests the development of a vulnerability profile based upon a consideration of multiple vulnerability factors that might vary with the type of disaster as well as the individual, social, and environmental context.

While chronological age is not a significant risk factor alone in producing vulnerability, often other factors associated with advanced age begin to increase risk and, thus, vulnerability. Factors such as having a greater likelihood of chronic health conditions, cognitive changes that may impact decision-making, social isolation, disability, and changes in hearing, vision, and mobility all serve to increase risks during a disaster.

The definition of vulnerability suggests that there can be differential risk to specific hazards. The term risk is based upon a determination of the likelihood of harmful consequences (injury, death, economic disruptions, and lifelong impact) of an event or exposure to a hazard. Putting the concept of vulnerability and risk together, we can conclude that there are circumstances and conditions that pose varying levels of risk to certain people, groups, or situations. The greater the risk factors, the greater the vulnerability of people, or groups in certain situations (Gelkopf, Berger, Bleich, & Silver, 2012). The number of older adults who die or are impacted by disasters suggests that special consideration of their risk and vulnerability should be a part of emergency management and disaster planning activities. It is widely recognized that older adults are vulnerable to disasters (Bei et al., 2013; Malik et al., 2017). Pre-disaster conditions such as having multiple chronic illnesses, limitations in ambulation and daily activities, difficulties in vision and hearing, prior traumatic lifetime events, and cognitive impairments that impacts decision-making are conditions related to increased vulnerability.

The absence of support systems or problems with social isolation also represent risk factors. Many older adults rely upon the use of support services such as meals on wheels, which often are paused during and after a disaster (Benson & Aldrich, 2007). To the extent that their day-to-day activities are discontinued, there is a greater likelihood of vulnerability due to disruptions. This disruption factor may be significant as a vulnerability factor for some older adults. Significant changes in routines that impact the ability to meet basic needs, such as those associated with having a safe place to sleep, access to routine medicines and health care, family support, and access to sufficient financial resources, can have a disruptive impact. The compounding effects of these variables often result in the older adult not being able to evacuate if necessary, without support or assistance. The inability to get to safety will always be a risk associated with increased vulnerability during disasters.

An additional risk factor faced by older adults is related to the fact that many older adults do not engage in any preparation for a disaster. This also includes facilities where there are not sufficient plans for disasters (Al-Rousan et al., 2014). It is clear that the increased numbers of disasters being experienced and lack of adequate preparation, training, and planning by older adults themselves increases vulnerability. The failure of facilities that provide housing for older adults to engage in emergency planning for disasters is also a factor that increases vulnerability. The disproportionate impact of disasters on older adults is reflected in a variety of different statistics such as the number of older adults age 60 and older who died during the Hurricane Katrina. Older adults represented only 11.7% of the population, although they represented 74% of those who died (Adams, Kaufman, van Hattum, & Moody, 2011).

This chart represents common vulnerability factors that influence older adults during disasters. Some of the factors increase vulnerability when they exist prior to the disaster; others are more related to risk mid- or during the disaster, and others have a post-disaster impact (Fig. 3.1).

Trauma History Contributes to Vulnerability

It is increasingly recognized that lifetime exposure to traumatic events is not uncommon. Studies have concluded that the experience of lifetime traumatic events occurs in the general population with up to 90% of the population. A large survey of over 24 countries representing almost 69,000 respondents, across six continents, reported

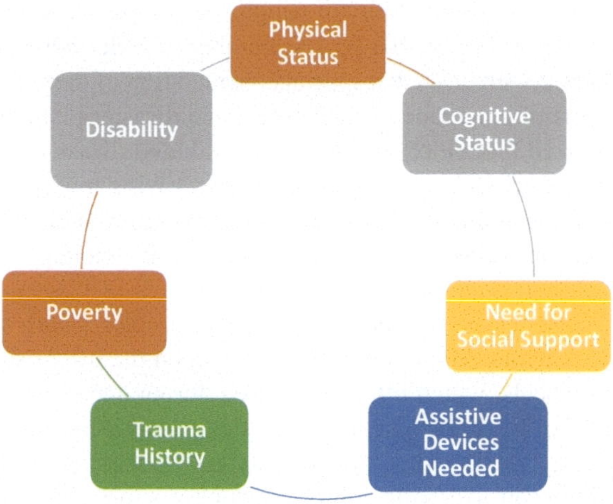

Fig. 3.1 Factors that impact vulnerability of older adults during disasters (Davis, 2017)

over 70% of the respondents had a traumatic event exposure (Benjet et al., 2016). Other general population studies in developed countries has indicated 28–90% of the respondents having experienced traumatic events (Roberts, Gilman, Breslau, Breslau, & Koenen, 2010; Ogle, Rubin, & Siegler, 2014).

Most common lifetime experiences of trauma are unexpected death of a loved one, motor vehicle accidents, and interpersonal violence (Benjet et al., 2016). Men and women can differ in the kinds of events they experience, but both have traumatic event exposure (Hatch & Dohrenwend, 2007). Studies have also found that many multiple exposures to traumatic events are common. It is recognized that there are a number of variables that are relevant in this discussion about traumatic event exposure. What are the psychological consequences of exposure to different kinds of traumatic events? Are there traumatic events that have a more significant impact on older adults? These are important questions that need to be explored. Studies suggest that it is the cumulative effect of exposure to traumatic events throughout a person's lifetime that has an impact on older adults' health and mental health status and increases the greater likelihood of developing post-traumatic stress reactions (PTSD) (Ogle et al., 2014). In addition, there is a clear linkage between the accumulation of lifetime trauma and psychological distress and psychiatric disorders (Turner & Lloyd, 1995).

Brown, Rothman, and Norris (2007) suggest that pre-disaster health status, social networks, and health sustaining networks are good predictors of post-disaster needs for resources. A key element of this chapter suggests that using a trauma-informed approach would provide for a better means of assessment pre-disaster, during disaster, and post-disaster. Although individual knowledge of an older adult's pre-disaster status may not be readily known, the trauma-informed approach is designed to provide some universal guidelines due to the prevalence of traumatic life events among the general population. A trauma-informed approach is based upon an awareness of the pervasiveness of the experience of traumatic events and a recognition of the potential impact of such events on individuals, prompting greater sensitivity to avoiding re-traumatizing disaster survivors.

It should be clear that disasters by their nature have the potential to be viewed as a traumatic event. Disaster exposures can have a long-lasting consequence on the physical and mental status over the life course (Keinan, Shrira, & Shmotkin, 2011). The experience of a traumatic event like disasters paired with the consequences of the aging process may create a condition of double jeopardy.

Why a Trauma-Informed Approach to Disaster Behavioral Health

There is an increasing momentum to view the experience of traumatic events in the general population from a public health perspective. This perspective is particularly valuable as it is compatible with the trauma-informed approach (see Kusmaul, this volume, Chap. 9). As we have increased our ability to understand how pervasive the experience of trauma actually is, and its connection to behavioral, physical health, and well-being, many organizations have moved toward becoming trauma-informed. By using this approach, they are attempting to be more sensitive and responsive to people they serve who may have experienced trauma. The rationale for supporting a trauma-informed approach in disaster behavioral health is based upon an understanding that disasters have the potential to be experienced as traumatic for many. It influences diverse groups, populations, services, and settings, increasing vulnerability and risk. The Substance Abuse Mental Health Services Administration (SAMHSA) has created a definition of trauma-informed care, although there are several perspectives on trauma-informed care.

There are two important definitions that help to clarify perspectives on being trauma informed. "Trauma Informed Care involves the adoption of principles and practices that promote a culture of safety, empowerment and healing based upon the prevalence and impact of trauma" (SAMHSA). A trauma-informed approach is based upon the recognition that consideration of the impact of trauma is a necessary component of effective behavioral health service delivery. It is based upon an understanding of the impact of trauma as an almost universal experience that may have long-term implications influencing individuals of all ages, families, organizations, and communities (Davis & Bellamy, 2017). The concept of a trauma-informed approach is based upon the presence of the six principles identified by SAMHSA.

There are a number of common behavioral reactions to disasters. These reactions can be associated with other health-related issues. Research links the experience of trauma with other health care needs after an event whose effects can sometimes last a few years. Trauma can be linked to cardiovascular, musculoskeletal, and neurological illness and post-traumatic stress disorder (PTSD), anxiety, depression, and substance abuse disorders. It is not uncommon that disasters displace individuals and families causing them to be separated from their support systems and communities. First responders are often negatively impacted by exposure to widespread destruction, hazardous materials, and need for a support system. A recent news report indicated that 200 firefighters who dug through the World Trade Center ashes and dust have now died as a result of illnesses directly tied to the work they did. Several forms of cancer and respiratory illness have been characterized as the World Trade Center illness (CNN, 2019).

Disaster behavioral health and the trauma-informed approach represent similar perspectives.

- Both recognize the potential impact of disasters as an event is often experienced as traumatic (Benjet et al., 2016; Breslau, Peterson, & Schultz, 2008).
- Pre-disaster life events are recognized to impact behavior during the disaster and post-disaster impacting physical and mental health outcomes (Ogle et al., 2014).
- Post-disaster PTSD (post-traumatic stress disorders) risk is related to lifetime trauma exposure to multiple events.
- A history of pre-disaster symptoms can predict the presence of post-disaster symptoms (Shmotkin & Litwin, 2009).
- Cumulative trauma or multiple forms of trauma over a lifetime has been linked to a variety of different psychological outcomes (Breslau et al., 2008; Krause, 2004).

As we examine how a trauma-informed approach provides insight into disaster behavioral health service delivery, we can identify several areas of importance in regard to older adults. Most prominently, exposure to trauma across the lifetime is linked to the aging process and the status of older adults.

- The greater the exposure to trauma, the greater the degree of impairment and the likelihood of PTSD (dose response relationship). Age-related changes in health and mobility may increase risk during a disaster and may increase the impact of cumulative trauma exposure on the individual (Krause, Shaw, & Cairney, 2004).
- Older adults may show resilience to a disaster event at the same time vulnerability to cumulative exposure and post-traumatic outcomes (Palgi, Shrira, & Shmotkin, 2015).
- Low social support is a strong predictor of PTSD among trauma-exposed adults (Brewin, Andrews, & Valentine, 2000; Ozer, Best, Lipsey, & Weiss, 2003).

Screening for Trauma

A key element of a trauma-informed approach is the early identification of exposure to lifetime trauma. A number of studies establishing the dose-response relationship or presence of cumulative trauma suggest that the level of trauma exposure may increase vulnerability to adverse reactions (Jones et al., 2009). Research results demonstrate clear relationships between many traumatic events and the effect of accumulated lifetime trauma experiences being associated with both psychological distress and psychiatric disorder (Turner, 1995: Jones et al., 2009).

The presence of a history of trauma may be related to other health and behavioral health disorders. Screening for trauma allows for a consideration of these factors and how they impact the individual. When trauma-informed screening is implemented, it involves:

- Developing assessment protocols that screen and identify level of exposure to trauma experiences.
- Determination of timing of screening pre- or post.
- Using trauma screening as a risk assessment tool.
- Creation of an effective intervention and referral system when appropriate.

Developing a Vulnerability Profile

A vulnerability profile is useful for understanding and training disaster and emergency management personnel. It is an identification of relevant risk factors that can affect the vulnerability level of a person or groups pre-disaster, during disaster, and post-disaster. Because this chapter seeks to highlight and provide a rationale for special consideration of older adults during a disaster, several risk factors will be identified that are commonly associated with older adults. The vulnerability profile suggests the higher the number and strengths of risk factors commonly operating, the greater the vulnerability and the increased likelihood that these factors will influence the outcome (Donner & Rodríguez, 2008). Risk assessment allows for a consideration of the relevant factors that are operating with different groups, with different demographic characteristics under different disaster conditions. Being able to assess the level of vulnerability being faced by first responders and others on the scene of the disaster is extremely helpful. Risk assessment is dynamic and requires a clear understanding of relevant individual and group characteristics that can be described as *vulnerability factors*. Vulnerability factors are demographic or circumstantial risk factors that increase the likelihood for injury and death after a disaster (Johnson & Galea, 2009), as discussed more fully in the next section.

Vulnerability and Resilience

There is a complex relationship between vulnerability and resilience among older adults experiencing disasters. (Palgi et al., 2015). It is not uncommon that older adults sometimes face the prospect of a disaster without concern, based upon their prior lived experiences with disasters. A history of successful survival bolsters a sense of confidence, beyond the circumstances, that survival is very likely to occur. This may be seen in many examples of disasters, but for this discussion we will use the Hurricane Katrina disaster. It is clear from the statistics that the mortality rate for older adults was significant because of the storm. Seventy-four to seventy-five percent of the dead were people over age 60 (see Johnson & Galea, 2009).

Several factors affected older adults who did not survive because of the storm. The inability for these disaster victims to evacuate successfully was a major factor in their death. Many were trapped in their homes with rescue not occurring quickly enough. Unable to evacuate prior to the storm's landfall left them stranded within their own homes when the levees broke. The poor planning for evacuation led to only some older adults being able to make it to the Super Bowl facility or the Convention Center, neither a suitable site for sheltering. The pre-disaster status of poverty, social isolation, chronic illnesses, and learned helplessness, based upon dependence on the system, contributed to high levels of vulnerability to the conditions being faced during Katrina (see Mosby, this volume, Chap. 13). Katrina truly represented a system that imploded. Many older adults died because the system could not generate a way to provide for access to medical care, location of families and relatives, provision of safe places for sheltering, and a coherent plan to take care of its elderly residents with dementia and the gravely ill.

The experience of Katrina represented a highly traumatic event in the lives of many people ranging from children to older adults (Jones et al., 2009). Being displaced far from home, losing touch with family, poor access to medical care, observing dead bodies floating in the water, many of whom were older adults, hopelessness, and despair produced a formula for multiple trauma experiences. It truly represented a population health traumatic event, which most people will never forget.

A Trauma-Informed Vulnerability Profile

This chapter has sought to highlight several important points about the need to incorporate principles of a trauma-informed approach into disaster behavioral health. This is particularly true with specialized groups such as older adults. It has been documented that disasters can have a more significant impact on older adults because of a number of factors, explored in this chapter. Mobility impairment and reduction in strength and capability, cognitive impairment, chronic disabilities, social isolation, and a trauma history of lifetime adversity are some that have been explored. It has been established that there are significant risk factors that increase the vulnerability in a disaster impacting older adults, although older adults also can demonstrate resilience in the ability to overcome challenges faced in their past (see Palgi et al., 2015). There are several moderating variables that impact whether trauma exposure has a positive impact and promotes resilience or has a negative deleterious impact. Prior exposure related to past experiences of trauma has been identified as one of these variables. The level of exposure and the strength of the exposure to traumatic experiences will impact the outcomes (Palgi et al., 2015). Secondly, an accumulation of lifetime adverse circumstances is associated with increased risk for PTSD (Pan American Health Organization, 2012). When there is repeated exposure to traumatic experiences and the length of time of the trauma or disaster is extensive (Shrira, Ben-Ezra, Spalter, Kave, & Shmotkin, 2011), adverse consequences can occur. This data supports the need for a trauma history and screening and lifetime assessment of adversity.

SAMHSA has developed six principles of trauma-informed care (see Kusmaul, this volume, Chap. 9). These six principles can provide guidance in integrating trauma principles into the discussion of older adults and the vulnerabilities that may impact the delivery of disaster behavioral health services. These six principles are being used to develop a vulnerability profile that can inform thinking and planning by first responders, emergency management planners, disaster behavioral health personnel, and others who have a central role in disasters. The trauma-informed approach is designed to change the service delivery to older adults, to reflect what is known about risk factors and vulnerability that impacts them. This strategy provides a lens through which older adults can be viewed by considering the potential impact of having experienced a traumatic disaster experience.

Developing a vulnerability profile also allows for a strategy to analyze risks/threats and hazards that combine to create a condition of vulnerability. Vulnerability profiles can be created to address individual, groups, families, community, or nations. Specifically, as this relates to disaster behavioral health, older adults on an individual level are the focus in a person-centered profile. Figure 3.2 presents a visual of the factors that must be considered in assessing vulnerability.

- Identify population: What is the vulnerable population? Consider children, adolescents, and older adults.
- Person/individual factors: Examine the presence of social skills, support factors.
- Social and economic factors: Education level, poverty, ethnicity.

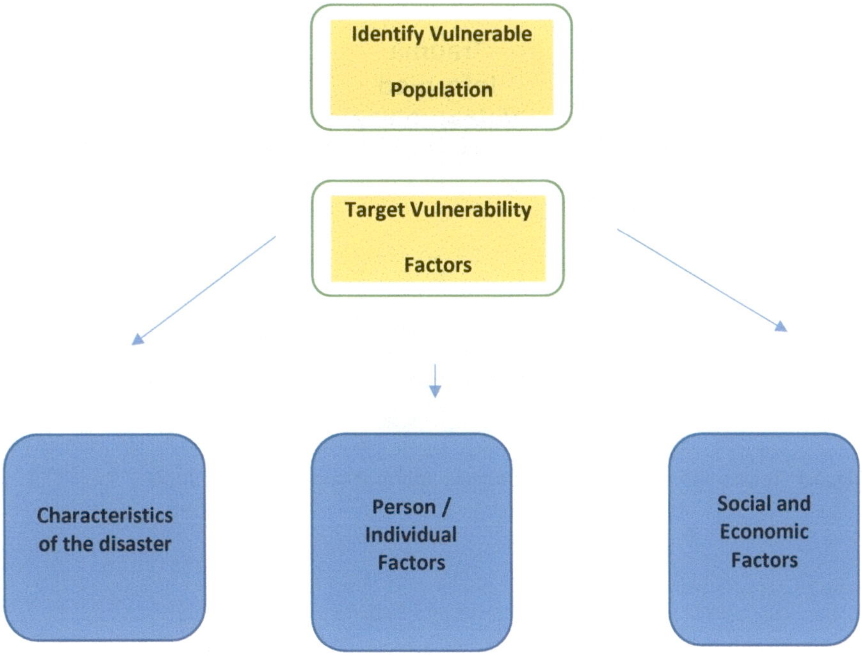

Fig. 3.2 Vulnerability factors

Core Elements of the Person-Centered Vulnerability Profile

The Substance Abuse Mental Health Services Administration "prepares states, territories, tribes, and local entities to deliver an effective mental health and substance abuse response to disasters" (SAMHSA, 2011). The next chart pairs the six principles of trauma-informed care with important disaster response data that would allow for considering ways to integrate trauma-informed concepts into the disaster response (Fig. 3.3) (Table 3.2).

The person-centered vulnerability profile is designed to provide a means of assessing the individual's risk and vulnerability. This profile is useful pre-, mid-, and post-disaster in planning and providing disaster behavioral health services. It begins by identifying the vulnerable population and targeting vulnerability factors.

Fig. 3.3 Trauma-informed principles of trauma-informed care (SAMHSA, 2014)

Table 3.2 Trauma-informed principles linked to disaster planning. Developed by Molly Everett Davis

Trauma-informed principle	Implications for disaster	Key issues	Action plans sample
Safety	Consider requirements to ensure the safety of older adults who may have physical challenges and need more support and evaluation of risk factors based upon the conditions of the disaster. Special attention directed toward evacuations.	Evacuation capability Physical status Mobility Support system Medication Trauma history Cognitive status Disaster risk factors	Community-based evacuation plans Identify resources and community supports Consider trauma history and sensitivities
Trust and transparency	Contact with older adults whether pre- disaster, during disaster, or post-disaster should be respectful and open to support engaging them in a process of collaboration and joint planning to manage disaster experience	Trauma history Trauma screening Openness and honesty Develop a plan early Be prepared	Incorporate trauma history in pre-, mid-, and post-disaster planning Include older adult in disaster behavioral health services and planning
Peer support	Reduce social isolation by using peer support strategies to increase community awareness, joint development of emergency plans, evacuation plans	Sharing planning resources Telephone trees Social activities to share resources Practice drills Mutual support	Utilize peer support in disaster planning Maintain peer support where possible during the disaster
Collaboration	Shared decision-making and collaborative planning that includes key stakeholders and representatives from vulnerable group	Partnerships Shared disaster planning Building community connections Resource development	Collaboration with family and community resources in pre-, mid-, and post-disaster planning and resource development
Voice and choice	Recognize strengths and experiences; support advocacy, empowerment, and promotion of resiliency in disaster survivors; incorporate choice in development of emergency plan; identify champions to disseminate messages about planning	Promoting self-advocacy Inclusion Shared voice in planning Choice and preferences allowed Involvement of family Individualized approach Skill development Use of strengths	Utilize inclusive strategies in safety planning, evacuation strategies, and post-disaster resource utilization

(continued)

Table 3.2 (continued)

Trauma-informed principle	Implications for disaster	Key issues	Action plans sample
Culture history and gender	Challenging stereotypes and bias	Consideration of cultural context Promote communication Overcoming barriers of bias Culture-sensitive planning Integration of tradition and rituals	Develop pre-, mid-, and post-disaster planning that is culturally sensitive Employ cultural humility in meeting the disaster-related needs

Person-Centered Vulnerability Profile (Table 3.3)

Table 3.3 Vulnerability checklist items. Developed by Molly Everett Davis

Vulnerability factors	Implications for disaster	Action plan	Strength rating of the variable 1–5 points 1 weak……5 strong
Advanced age with increased chronic illnesses and cognitive changes	Slowed reaction times, inability to understand warnings, social isolation, and lack of social support		
Sensory deficits hearing, vision, smelling, touch	Impact ability to heed disaster warning, vision impairment, hearing impairment, and difficulty following evacuation instructions		
Frailty – Weakness and inability to move independently without support	Reduction of stamina, may need assistance to take care of basic needs		
Impaired mobility – Requires support to aid mobility	Relies on mobility aids, such as wheel chairs and walkers; can be unable to evacuate buildings or climb stairs		
Climate factors (heat, cold, rain, snow, ice, volcanic, fire)	Risk of dehydration, heat strokes, hypothermia, exposure		
Poverty – Unable to access basic needs to include ability to evacuate (insufficient access to resources)	Unable to access core needs for food, medicine refills, insurance for recovery, need assistance in obtaining resources. Impacts dependence on governmental systems to provide assistance		

Conclusion

Disasters are often experienced as a traumatic life event. While some older adults are resilient in the face of disaster (see Wilks et al., this volume, Chap. 2), other older adults have a special vulnerability to disasters and their long-lasting impact. Recognition of the potential traumatic impact of disasters on older adults is critical. A trauma-informed approach is based upon the recognition that first responders and other disaster response professionals should implement strategies to avoid re-traumatization and the long-term impact of the experience of disasters. This chapter suggest that assessment of vulnerability factors to create a vulnerability profile can be useful in preparing for the disaster prior to, during, and after the disaster.

Older adults may have increased vulnerability due to a number of factors; however, the best practice consideration will assess vulnerability by assessing the uniqueness of the older adult population being impacted by the disaster and the characteristics of the disaster and its potential outcome. Consequently, disaster response professionals should have training on gerontology and also trauma to ensure that older adults survive disasters and to promote resilience (see Spurlock & Brown, this volume, Chap. 12). Risk stratification, a process to systematically determine and assess risk based upon an understanding of characteristics of those potentially impacted by the disaster, is critical to overall disaster planning. This chapter highlights the important need to engage in this kind of process, along with an understanding of changes associated with the aging process, to ultimately reduce the potential of negative outcomes and increase the potential for resilience among older adult disaster survivors. This approach allows for providing disaster behavioral care in ways in which re-traumatization and long-lasting negative outcomes to include post-traumatic stress disorders are reduced.

A new approach to disaster behavioral health is needed, which reflects an integration of an increased understanding of older adults and the aging process, along with understanding assessment of vulnerability, the principles of trauma-informed care, and the promotion of resilience. The combined impact of each of these components will strengthen the ability of disaster professionals to serve older adults prior to, during, and after a disaster in ways that support increased resilience and reduce risk for post-traumatic stress disorders.

References

Adams, V., Kaufman, S. R., van Hattum, T., & Moody, S. (2011). Aging disaster: Mortality, vulnerability, and long-term recovery among Katrina survivors. *Medical Anthropology, 30*(3), 247–270. https://doi.org/10.1080/01459740.2011.560777

Allen, P. D., & Nelson, H. W. (2009). Disaster services with frail older persons: From preparation to recovery. In *Lifespan Perspectives on Natural Disasters* (pp. 153–169). New York, NY: Springer.

Al-Rousan, T. M., Rubenstein, L. M., & Wallace, R. B. (2014). Preparedness for natural disasters among older US adults: A Nationwide survey. *American Journal of Public Health, 104*(3), 506–511. https://doi.org/10.2105/ajph.2013.301559

American Psychological Association. (2017). *Multicultural Aging Efforts in APA*. https://www.apa.org/pi/aging/resources/multicultural-efforts.

American Psychological Association. (2018). *Older adults and disasters: How to be prepared and assist others*. Retrieved from http://www.apa.org/pi/aging/resources/older-adults-disasters.pdf

Bei, B., Bryant, C., Gilson, K.-M., Koh, J., Gibson, P., Komiti, A., et al. (2013). A prospective study of the impact of floods on the mental and physical health of older adults. *Aging & Mental Health, 17*(8), 992–1002. https://doi.org/10.1080/13607863.2013.799119

Benjet, C., Bromet, E., Karam, E. G., Kessler, R. C., Mclaughlin, K. A., Ruscio, A. M., et al. (2016). The epidemiology of traumatic event exposure worldwide: Results from the world mental health survey consortium. *Psychological Medicine, 46*(2), 327–343. https://doi.org/10.1017/s0033291715001981

Benson, W. F., & Aldrich, N. (2007). *CDC's disaster planning goal: Protect vulnerable older adults*. Retrieved from http://healthbenefitsabcs.com/home/wp-content/uploads/2015/06/disaster_planning_goal.pdf

Borden, K. A., & Cutter, S. L. (2008). Spatial patterns of natural hazards mortality in the United States. *International Journal of Health Geographics, 7*(1), 64. https://doi.org/10.1186/1476-072x-7-64

Breslau, N., Peterson, E. L., & Schultz, L. R. (2008). A second look at prior trauma and the post traumatic stress disorder effects of subsequent trauma: A prospective epidemiological study. *Arch General Psychiatry, 65*(4), 431–437. https://doi.org/10.1001/archpsyc.65.4.431

Brewin, C. R., Andrews, B., & Valentine, J. D. (2000). Meta-analysis of risk factors for posttraumatic stress disorder in trauma exposed adults. *Journal Consulting Clinical Psychological, 68*(5), 748–766.

Brown, L. M., Rothman, M., & Norris, F. (2007). Issues in mental health care for older adults after disasters. *Generations, 31*(4), 21–26.

Campbell, J. (2011). Breaking the silence: Trauma informed behavioral healthcare. *National Council Magazine, 2*, 12.

Centers for Disease Control and Prevention. (2012a). *Identifying Vulnerable Older Adults and Legal Options for Increasing Their Protection During All-Hazards Emergencies: A Cross-Sector Guide for States and Communities*. Retrieved from https://www.cdc.gov/aging/emergency/pdf/guide.pdf

Centers for Disease Control and Prevention. (2012b). *Identifying Vulnerable Older Adults and Legal Options for Increasing Their Protection During All-Hazards Emergencies: A Cross-Sector Guide for States and Communities*. Retrieved from https://www.cdc.gov/aging/emergency/pdf/guide.pdf

Cherry, K. E. (2020). *The other side of suffering: Finding a path to peace after tragedy*. New York, NY: Oxford University Press.

CNN. (2019). *The 200th firefighter just died from a world trade center related illness*. Retrieved from https://www.cnn.com/2019/07/18/us/200-firefighters-died-world-trade-center- illnesses-trnd/index.html

Cohen, D., Huynh, T., Sebold, A., Harvey, J., Neudorf, C., & Brown, A. (2014). The population health approach: A qualitative study of conceptual and operational definitions for leaders in Canadian healthcare. *SAGE Open Medicine, 2*, 205031211452261. https://doi.org/10.1177/2050312114522618

Davis, M. (2017). *Promoting SBIRT in an Interprofessional setting with vulnerable populations, the BIG initiative, national SBIRT*. ATTC and NORC at the University of Chicago.

Davis, M., & Bellamy, N. (2017, July). *A Trauma Informed Approach to Disaster Behavioral Health. Paper presented at the meeting of the international world congress on geriatric and gerontology*. San Francisco, CA.

Donner, W., & Rodríguez, H. (2008. Population composition, migration and inequality: The influence of demographic changes on disaster risk and vulnerability. *Social Forces*, *87*(2), 1089–1114. Retrieved from http://muse.jhu.edu/journals/social_forces/v087/87.2.donner.html

Gelkopf, M., Berger, R., Bleich, A., & Silver, R. C. (2012). Protective factors and predictors of vulnerability to chronic stress: A comparative study of 4 communities after 7 years of continuous rocket fire. *Social Science Medicine, 74*(5), 757–766. https://doi.org/10.1016/j.socscimed.2011.10.022

Gibson, M. J., & Hayunga, M. (2006). *We can do better: Lessons learned for protecting older persons in disasters*. Washington D.C: AARP Public Policy Institute.

Hatch, S. L., & Dohrenwend, B. P. (2007). Distribution of traumatic and other stressful life events by race/ethnicity, gender, SES and age: A review of the research. *American Journal of Community Psychology, 40*(3–4), 313–332. https://doi.org/10.1007/s10464-007-9134-z

International Federation of Red Cross. (2019). *What Is Vulnerability?* Retrieved from https://www.ifrc.org/en/what-we-do/disaster-management/about-disasters/what-is-a-disaster/what-is-vulnerability/

Johnson, J., & Galea, S. (2009). Disasters and population health. In K. E. Cherry (Ed.), *Lifespan perspectives on natural disasters: Coping with Katrina, Rita and other storms* (pp. 281–326). New York, NY: Springer Science+Business Media.

Jones, R. T., Burns, K. D., Immel, C. S., Moore, R. M., Schwartz-Goel, K., & Culpepper, B. (2009). The impact of hurricane Katrina on children and adolescents: Conceptual and methodological implications for assessment and intervention. In K. E. Cherry (Ed.), *Lifespan perspectives on natural disasters: Coping with Katrina, Rita, and other storms* (pp. 65–94). New York, NY: Springer.

Keinan, G., Shrira, A., & Shmotkin, D. (2011). The association between cumulative adversity and mental health: Considering dose and primary focus of adversity. *Quality of Life Research, 21*(7), 1149–1158. https://doi.org/10.1007/s11136-011-0035-0

Krause, N. (2004). Lifetime trauma, emotional support and life satisfaction among older adults. *The Gerontologist, 44*(5), 615–623. https://doi.org/10.1093/geront/44.5.615

Krause, N., Shaw, B. A., & Cairney, J. (2004). A descriptive epidemiology of lifetime trauma and the physical health status of older adults. *Psychology and Aging, 19*(4), 637–648. https://doi.org/10.1037/0882-7974.19.4.637

Levac, J., Toal-Sullivan, D., & O'Sullivan, T. L. (2011). Household emergency preparedness: A literature review. *Journal of Community Health, 37*(3), 725–733. https://doi.org/10.1007/s10900-011-9488-x

Malik, S., Lee, D. C., Doran, K. M., Grudzen, C. R., Worthing, J., Portelli, I., et al. (2017). Vulnerability of older adults in disasters: Emergency department utilization by geriatric patients after hurricane Sandy. *Disaster Medicine and Public Health Preparedness, 12*(2), 184–193. https://doi.org/10.1017/dmp.2017.44

Ogle, C. M., Rubin, D. C., & Siegler, I. C. (2014). Cumulative exposure to traumatic events in older adults. *Aging & Mental Health, 18*(3), 316–325. https://doi.org/10.1080/13607863.2013.832730

Ozer, E. J., Best, S. R., Lipsey, T. L., & Weiss, D. S. (2003). Predictors of posttraumatic stress disorder and symptoms in adults: A meta-analysis. *Psychological Bulletin, 129*: (pp.52-73).

Palgi, Y., Shrira, A., & Shmotkin, D. (2015). Aging with trauma across the lifetime and experiencing trauma in old age: Vulnerability and resilience intertwined. In K. E. Cherry (Ed.), *Traumatic stress and long-term recovery: Coping with disasters and other negative life events* (pp. 293–308). New York, NY: Springer Publishing Company.

Parker, G., Lie, D., Siskind, D. J., Martin-Khan, M., Raphael, B., Crompton, D., et al. (2015). Mental health implications for older adults after natural disasters – A systematic review and meta-analysis - CORRIGENDUM. *International Psychogeriatrics, 28*(1), 11–20. https://doi.org/10.1017/s1041610215001465

Roberts, A. L., Gilman, S. E., Breslau, J., Breslau, N., & Koenen, K. C. (2010). Race/ethnic differences in exposure to traumatic events, development of post-traumatic stress disorder, and treatment-seeking for post-traumatic stress disorder in the United States. *Psychological Medicine, 41*(1), 71–83. https://doi.org/10.1017/s0033291710000401

Rodríguez, H., Wachtendorf, T., & Russell, C. (2004). Disaster research in the social sciences: Lessons learned, challenges, and future trajectories. *International Journal of Mass Emergencies and Disasters, 22*(2), 117–135.

Shih, R., Acosta, J., Chen, E., Carbone, E., Xenakis, L., Adamson, D., et al. (2018). Improving disaster resilience among older adults: Insights from public health departments and aging-in-place efforts. *Rand Health Quarterly, 8*(1), 3. https://doi.org/10.7249/rr2313

Shmotkin, D., & Litwin, H. (2009). Cumulative adversity and depressive symptoms among older adults in Israel: The differential roles of self oriented verses other oriented events of potential trauma. *Social Psychiatry Epidemiology, 44*(11), 989–997.

Shrira, A. Palgi,Y. Ben-Ezra, M., Spalter, T., Kave, G. Shmotkin, D. (2011). For better and for worse: The relationship between future expectations and functioning in the second half of life. Journals of Gerontology Series B: Psychological Sciences and Social Science. 66B(2): 195–203, March 2011. doi: https://doi.org/10.1093/geronb/gbq103,

Speier, A. H., Osofsky, J. D., & Osofsky, H. J. (2009). Building a disaster mental health response: Louisiana and hurricane Katrina. In K. E. Cherry (Ed.), *Lifespan perspectives on natural disasters: Coping with Katrina, Rita and other storms* (pp. 241–260). New York, NY: Springer.

Substance Abuse and Mental Health Services Administration (2011). *Disaster training and technical assistance* (HHS Publication No. SMA 13–4801). Retrieved from https://store.samhsa.gov/system/files/sma11-4627.pdf

Substance Abuse and Mental Health Services Administration. *Trauma-Informed Care in Behavioral Health Services. Treatment Improvement Protocol (TIP) Series 57*. HHS publication no. (SMA) 13–4801. Rockville, MD: Substance Abuse and Mental Health Services Administration, 2014.

Turner, R. J., & Lloyd, D. A. (1995). Lifetime traumas and mental health: The significance of cumulative adversity. *Journal of Health and Social Behavior, 36*(4), 360. https://doi.org/10.2307/2137325

Wallemacq, P., & House, R. (2018). *Economic Losses, Poverty & Disasters 1998–2017*. Retrieved from https://www.unisdr.org/files/61119_credeconomiclosses.pdf

Weisler, R. H., & Barbee, J. G. (2006). Mental health and recovery in the Gulf coast after hurricanes Katrina and Rita. *JAMA, 296*(5), 585–588.

Wisner, B. (2004a). *Ambiguity and change: Humanitarian NGOs prepare for the future. A report prepared for: World vision, CARE, Save US, Mercy Corps, Oxfam USA, Oxfam GB & Catholic Relief*. Retrieved from Tufts University.

Wisner, B. (2004b). *Ambiguity and change: Humanitarian NGOs prepare for the future*. Retrieved from Tufts University, The Feinstein International Famine Center website: https://fic.tufts.edu/assets/Ambiguity-and-Change-Humanitarian-NGOs-Prepare-for-the-Future.pdf

Wisner, B., Blaikie, P., Cannon, T., & Davis, I. (2004). *At risk: Natural hazards, people's vulnerability, and disasters* (2nd ed.). London: Routledge.

Chapter 4
PTSD-Depression Comorbidity and Health among Older Adults Exposed to Missile Attacks: The Buffering Effect of Young Subjective Age

Amit Shrira and Yaakov Hoffman

Introduction

Older adults generally feel younger than their current chronological age, and the tendency to feel younger than one's age increases across the lifespan (Chopik, Bremner, Johnson, & Giasson, 2018). In fact, one's subjective age rather than one's chronological age is a better predictor for health outcomes in the second half of life (Kotter-Grühn, Kornadt, & Stephan, 2015). A younger subjective age is associated with better physical functioning and longevity (Westerhof et al., 2014). Feeling younger is also negatively related to biological markers of inflammation (Stephan, Sutin, & Terracciano, 2015) and accelerated senescence (Lahav, Avidor, Stein, Zhou, & Solomon, 2020). A younger subjective age is further associated with lower mental distress, higher well-being (Keyes & Westerhof, 2012), as well as with less anxiety and depression (Shrira, Bodner, & Palgi, 2014).

Feeling younger than one's age also serves as a protective buffer against effects of trauma and trauma-related symptoms, such as post-traumatic stress disorder (PTSD) symptoms, in older adults (Stephan, Sutin, & Terracciano, 2018). Thus, those who feel younger have lower PTSD symptom levels (Avidor, Benyamini, & Solomon, 2016; Hoffman, Shrira, Cohen-Fridel, Grossman, & Bodner, 2016; Palgi, 2016; Solomon, Helvitz, & Zerach, 2009). This was also true for trauma resulting from rocket and missile fire, where a younger subjective age was associated with lower PTSD levels (Hoffman, Shrira, & Grossman, 2015).

However, previous research examining subjective age's buffering against detrimental PTSD effects did not take into account the PTSD-depression comorbidity, despite it being among the most frequent PTSD comorbid conditions, with approximately half of those suffering from PTSD are also suffering from depression

A. Shrira (✉) · Y. Hoffman
Interdisciplinary Department of Social Sciences, Bar-Ilan University, Ramat Gan, Israel
e-mail: amit.shrira@biu.ac.il

© Springer Nature Switzerland AG 2021

K. E. Cherry, A. Gibson (eds.), *The Intersection of Trauma and Disaster Behavioral Health*, https://doi.org/10.1007/978-3-030-51525-6_4

(Horesh et al., 2017). This is important, as noted by Flory and Yehuda (2015), since the PTSD-depression comorbidity imposes significantly greater difficulties in physical, cognitive, social, and behavioral domains than does PTSD alone. Accordingly, suffering simultaneously in the present from both disorders constitutes more stress than suffering from PTSD alone. In addition, the difference between comorbid and PTSD alone is not an artifact of overlapping symptoms, as the conditions present distinctive biological patterns, such as lower amygdala activation (Kemp et al., 2007) and lower epigenetic methylation (Yehuda et al., 2015) in PTSD alone versus comorbidity. We propose that a young subjective age, being critically relevant to older adults, may constitute a pivotal buffering effect even for comorbid PTSD-depression condition.

The Psychological Meaning of Subjective Age in the Context of Trauma

There are different approaches to the underlying meaning of subjective age; it has been suggested to be a self-attribute (Diehl et al., 2014) or a biopsychosocial marker of aging (Stephan et al., 2018). In any event, these different theoretical approaches are addressing how subjective age is generated, yet from the present review, it seems that having a younger subjective age is a potent resource. Favorable aging perceptions have been deemed a resource (Wurm, Tomasik, & Tesch-Römer, 2008). Stephan, Caudroit, and Chalabaev (2011) have endorsed this idea with regard to subjective age, whereby having a younger subjective age reflects an individual with higher psychological resources, for example, higher sense of well-being, vitality, life satisfaction, and meaning in life (see also Westerhof & Wurm, 2015). Others have argued that subjective age is not a resource per se, but rather a more complex entity that is determined both by resources and levels of stress.

The latter approach is based on several theories building on one another. First, the subjective weathering hypothesis (Foster, Hagan, & Brooks-Gunn, 2008) claims that experiencing trauma at a younger age may later lead to an older subjective age if one's emotional/cognitive maturity is not on par with current demands. Drawing on this theory, Palgi (2016) suggested that dealing with the demands of trauma exposure in addition to coping with the simultaneous aging difficulties, might increase one's subjective age. Building on Palgi (2016), Hoffman, Shrira, et al. (2016) suggested that feeling younger than one's age reflects a perception that one has greater resources than the demands/stressors experienced by past or present life events. As further outlined in the following section, there is a significant difference between these two latter approaches, namely, whether subjective age is a resource per se or whether it is a product of resources and stressors.

The Current Study

One manner to distinguish between these possibilities (subjective age being a resource alone or being a product of both resource and stress) is by the following question: Will subjective age moderate the association between high level of comorbid stress and physical health? In other words, will a young subjective age protect even those experiencing comorbidity (PTSD-depression) symptoms from greater physical impairment? Figure 4.1 provides some hypothetical data to illustrate the different possible outcomes that might be obtained if subjective age is best conceptualized as a product of resources or both resources and stress. For example, if subjective age is a resource, then there are two potential outcomes. The first option is that subjective age is not strong enough a resource to buffer against such a debilitating condition of comorbidity (see Fig. 4.1, comorbidity slope a vs. one disorder and no disorder slopes). In other words, the level of distress in comorbidity is beyond a threshold for a young subjective age to buffer. The second option is that subjective age is a sufficiently potent resource, so that it may provide some level of buffering against the detrimental effects of a comorbid condition similarly the way it buffers against a single disorder. In other words, we should observe parallel slopes for physical health by subjective age for single and comorbid conditions (see Fig. 4.1, comorbidity slope b vs. one disorder and no disorder slopes). This option along with the first option should hold true if subjective age is only a resource. Yet, if subjective age is the product of both resources and stress, then we should observe an interaction, whereby the effect of subjective age on physical health for the comor-

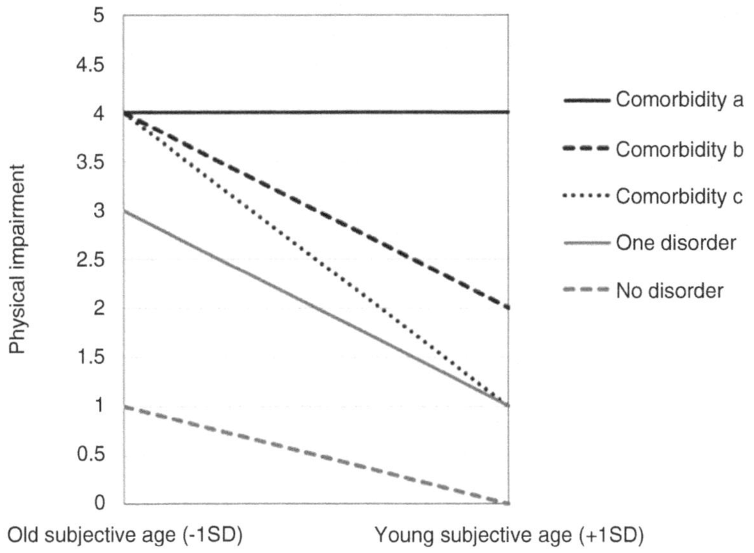

Fig. 4.1 Possible moderation effects of subjective age on the relationship between comorbidity and physical impairment

bid condition should be stronger, meaning that the benefit derived from a young subjective age should be relatively greater for the comorbid participants versus other conditions (see Fig. 4.1, comorbidity slope c vs. one disorder and no disorder slopes).

In summary, we wish to examine if a young subjective age is robust enough to buffer against the potential detrimental effects of suffering from a comorbid condition on physical health. The result pattern should reveal the following: first, if subjective age is indeed able to protect against such impairments or not, and, second, even if a young subjective age is sufficient to buffer against such effects, will this buffering be additive, in which case subjective age may be a mere resource (comorbidity slope b in Fig. 4.1), or will this effect be over-additive, in which case subjective age would constitute a product of resources and stress (comorbidity slope c in Fig. 4.1).

To address the current question of how a young subjective age may buffer against the PTSD-depression comorbidity, we used data from the Israeli component of the Survey of Health, Ageing and Retirement in Europe (SHARE-Israel). The SHARE items are administered across many European countries, including Israel (Litwin, 2009), to persons dwelling in the community who are older than 50. Their spouses were also surveyed, and spouses could be younger (Börsch-Supan et al., 2013); thus, we only included participants above 50. The SHARE study includes two components. The first component is common to all participating countries; these items are administered via a personal interview (a computer-assisted interview, CAPI). The second component is individual to each and every country and is called a "drop-off" questionnaire, a supplemental collection of items, which is completed by paper-and-pencil questionnaire and sent back. This drop-off unit is typically of specific interest to a given country. In the current study, we refer to items within the second component (see below) that queried about trauma following missile attacks on Israeli civilians.

The current study focused on the fourth data wave (2015). We selected this data wave as the items queried about trauma following the 2014 Protective Edge Operation (the 2014 Israel-Gaza conflict). Protective Edge Operation lasted 60 days, during July–August 2014. More than 4382 rockets were launched at Israel from the Gaza Strip; according to official sources (e.g., Israel Defense Forces online information, 2014), during this time, more than 70% of the Israeli population was living in the rocket range. Despite the large number of rockets fired on a relatively small location, the number of civilian casualties was relatively low (six Israeli civilians were killed and 837 wounded). This was largely due to the Iron Dome defense system, which intercepted the deadly missiles (e.g., Hoffman, Cohen-Fridel, Bodner, Grossman, & Shrira, 2016).

The current study utilized the drop-off questions and additional items from the main SHARE questionnaire. Therefore, the current SHARE-Israel data included ratings of PTSD, depression, subjective age, and physical health completed by older adults exposed to missile attack, and should thus reveal if the greater suffering endured in the comorbid group may be mitigated by a young subjective age. Based on these data, we divided participants into four groups: those suffering from neither PTSD nor depression, only-PTSD, only-depression, and PTSD-depression comorbidity. In the next sections, we present additional details about SHARE-Israel and the relevant measures it included.

Method

Participants and Procedure

The number of participants whom completed the personal interview in the fourth Israeli SHARE wave of 2015 was 2035, while 1810 (89%) participants completed the second component (drop-off questionnaire). Thus, the current sample comprised respondents who completed the drop-off unit and whose age was 50 or above, totaling 1793 participants. The mean age was 69.65 (*SD* = 9.49, range = 50–105), 57.6% were women, 86.6% were veteran Israeli Jews, 71.4% were married, and education level was 1.92 (*SD* = 0.75). For further details, see Table 4.2. Results of those participants who did and did not, complete the drop-off unit were compared (attrition analysis).[1] SHARE-Israel received ethical approval by the Institutional Review Board of the Hebrew University of Jerusalem.

Measures

Independent variable Exposure to warfare events during the 2014 Israel-Gaza conflict was assessed through a list of 12 occurrences, which respondents might have undergone. They were asked to indicate, regarding each event, whether they have experienced it between June and August 2014. The events listed in this variable included if one felt physical danger to his/herself or to a close other; if one was injured; if one's property or workplace was damaged, or in danger of being damaged; if one was exposed to other persons who were injured. Questions also included items addressing if one's daily routine was interrupted for more than a week or if one was forced to leave home. Frequency of each exposure item is depicted in Table 4.1. Although not all of these occurrences fulfilled the DSM criteria for

[1] Such comparisons were conducted by generating a dummy variable (code: 0 = no completion; 1 = completing drop-off unit). We applied bivariate analyses to check for differences across sociodemographic and health variables, *t*-tests to assess differences between groups in continuous variables, and χ^2 analyses to check for differences across variables of a categorical nature. Effect sizes for these analyses were also computed (φ, Cramer's V, and Cohen's *d*). For φ and Cramer's V, 0.1, 0.3, and 0.5 were, respectively, considered small, medium, and large effects; for Cohen's *d*, these values were, respectively, 0.2, 0.5, and 0.8. The results show that respondents from the drop-off were younger ($p < 0.05$) (Cohen's *d* = 0.21), more likely to be veteran Israeli Jews (Cramer's V = 0.17) and to be married (φ = −0.06). Drop-off completers were also less depressed (Cohen's *d* = 0.24), had less medical conditions (Cohen's *d* = 0.21), had less disability (Cohen's *d* = 0.70), and had higher grip strength (Cohen's *d* = 0.34). They did not differ from non-completers in terms of gender and education level.

Table 4.1 Frequencies of exposure types

Exposure type	*n* (%)
You were personally injured	10 (0.6)
A person close to you was killed	27 (1.5)
A person close to you was injured	29 (1.6)
You were not personally injured but were in danger of physical injury	259 (14.5)
A person close to you was in danger of physical injury	317 (17.8)
There was damage to your personal property (e.g., home, car)	23 (1.3)
There was danger of damage to your personal property (e.g., home, car)	266 (15.0)
There was damage to your place of work or business	26 (1.5)
There was danger of damage to your place of work or business	100 (5.6)
You were exposed to people who were injured	182 (10.2)
Your daily routine was seriously disrupted for a week or more	429 (24.1)
It was necessary to leave your home for a week or more	31 (1.7)

Note. The total number of respondents ranged from 1775 to 1782 in the various items

traumatic exposure (APA, 2013), the vast majority of Israelis were exposed to missile attacks during the 2014 Israel-Gaza conflict (Shrira et al., 2018).
Participants whose responses indicated trauma exposure were instructed to respond to the PTSD questionnaire, which was measured by the four-item SPAN (Meltzer-Brody, Churchill, & Davidson, 1999). Previous studies (e.g., Brewin, 2005) have established that this brief PTSD measure is comparable in its sensitivity to the longer PTSD measures. Participants received instructions to respond to the PTSD symptoms of the SPAN during the last week based on their experiences during the aforementioned 2014 Israel-Gaza (Operation Protective Edge) conflict. The SPAN items were responded to using a scale of 0–4, ranging from "not at all" to "extremely." The scale had good internal consistency ($\alpha = 0.81$). A cutoff score of 5 was used to determine PTSD (Meltzer-Brody et al., 1999).

Depressive symptoms were assessed via the Euro-D scale for late-life depression (Prince et al., 1999). The scale is composed of 12 "yes" or "no" questions about symptoms experienced in the past month, such as feeling depressed or a loss of interest. The scale has a range of 0–12. This measure had good internal reliability in the current sample (Kuder-Richardson $\rho = 0.75$). A cutoff score of 4 was used to determine depression (Prince et al., 1999).

Combining the dichotomous scores of PTSD and depression yielded four groups: The first group comprising 67.7% ($n = 1159$) of the sample suffered from neither disorder, that is, if respondents were below the cutoff point for both SPAN and Euro-D. The PTSD-only group comprised of 5.2% ($n = 89$) of the sample and the depression-only group comprised of 21.0% ($n = 360$) of the sample; these groups were, respectively, reserved for being above the SPAN or Euro-D cutoff points. Finally, the fourth PTSD-depression comorbid group comprised 6.1% ($n = 104$) of the sample; these participants scored above both the SPAN and Euro-D cutoff points.

Moderator Subjective age was assessed by a single item asking participants how old they felt most of the time (Stephan et al., 2015). The subjective age score was computed by subtracting respondents' felt age from their chronological age and dividing this outcome by one's chronological age. Higher subjective age values reflect a younger age identity, such that the deviation from the chronological age is expressed in percentiles. For example, a 50-year-old respondent who feels 45 has a subjective age score of 0.1, indicating that she/he feels 10% younger than her/his chronological age (cf. Stephan et al., 2015). In line with previous studies (e.g., Stephan et al., 2015), responses 3 *SD*s above or below the mean were considered outliers, leading to the exclusion of 42 participants.

Physical Outcomes Medical conditions were assessed by a sum of listed illnesses that participants reported to have been diagnosed by a physician (see Shrira, Palgi, Ben-Ezra, & Shmotkin, 2011). The illnesses consisted of heart attack or any other heart problem, including congestive heart failure, high blood pressure or hypertension, high blood cholesterol, or a stroke or cerebrovascular disease; diabetes or high blood sugar; chronic lung disease such as chronic bronchitis or emphysema; or cancer or malignant tumor, including leukemia or lymphoma, but excluding minor skin cancers; stomach or duodenal ulcer; peptic ulcer, Parkinson disease; cataracts; hip fracture or other fractures; Alzheimer's disease, dementia, organic brain syndrome, senility, or any other serious memory impairment; rheumatoid arthritis, osteoarthritis, or other rheumatism; and chronic kidney disease (the possible range was 0–16).

Physical disability was assessed by (1) difficulties in basic activities in daily living (ADL, adapted from Katz, Downs, Cash, & Grotz, 1970), and (2) difficulties in instrumental activities in daily living (IADL, adapted from Lawton & Brody, 1969). ADL difficulties included six functions: dressing, crossing a small room, bathing, getting in or out of bed, eating, and toileting. The IADL difficulties included nine functions: using a map, preparing meals, daily shopping, using the telephone, taking medications, doing housework, handling personal finances, leaving the house independently and accessing transportation services, and doing personal laundry. Difficulties in all of the disability functions were rated with a dichotomized answer (not having difficulties/having difficulties). For the combined scales, internal reliability measured by Kuder-Richardson's ρ was .93. Disability score was the sum of activities with reported limitation (the possible range was 0–15; cf. Shrira et al., 2011).

Grip strength was assessed by giving participants a dynamometer (Smedley, S Dynamometer, TTM, Tokyo, 100 kg) and asking to squeeze it as hard as they could while standing or sitting with their elbow at a 90-degree angle pressed against their body. The grip strength was recorded twice for each hand, and the final value used for the analysis was the maximal grip strength obtained from either one of the four trials. Values were recorded as missing if the measurement differed by more than 20 kg for one hand or if only one measurement in one hand was obtained. Valid values were defined as greater than zero and smaller than 100 kg. Values outside this range were also defined as missing (Andersen-Ranberg, Petersen, Frederiksen, Mackenbach, & Christensen, 2009).

Covariates Several sociodemographic covariates were also considered in the main analyses due to their known associations with physical health in old age. Age was used as a continuous variable. Gender was coded as 1 (men) and 2 (women). Baseline education was recorded as one of the seven levels according to the International Standard Classification of Educational Degrees (ISCED-97; United Nations Educational, Scientific and Cultural Organisation (UNESCO), 1997). The seven levels were subsequently transformed into three levels, reflecting elementary, high school, and secondary education level. Marital status was coded as 1 (divorced, never married, widowed) and 2 (married or registered partnership). Origin was coded as 1 (veteran Jewish Israelis), 2 (Arab Israelis), and 3 (immigrants from former USSR).

Results

Descriptive Statistics

Table 4.2 presents the correlations between the study variables. PTSD and depressive symptoms (continuous scores) showed moderate correlation with each other. PTSD symptoms showed weak relationships with the physical outcomes, whereas depressive symptoms showed moderate to strong relationships with physical outcomes. On average, respondents felt 9% younger than their age. A young subjective age was related to lower depressive symptoms and was weakly related to lower physical impairment.

We next assessed between group differences in background characteristics using one-way ANOVAs and χ^2 tests. These analyses revealed that respondents without either psychiatric disorders were significantly chronologically younger compared to the other groups ($\eta^2 = 0.04$, $p < 0.0001$). Respondents suffering from depression-only had lower education level ($\eta^2 = 0.03$, $p < 0.0001$) and an overrepresentation of unmarried (Cramer's V = 0.20, $p < 0.0001$) and Israeli Arabs (Cramer's V = 0.27, $p < 0.0001$) compared to the other groups. Respondents with comorbidity were more likely to be female (Cramer's V = 0.12, $p < 0.0001$) and had an overrepresentation of immigrants from the former USSR (Cramer's V = 0.27, $p < 0.0001$).

The four groups also differed in subjective age ($\eta^2 = 0.01$, $p < 0.0001$). Post hoc Bonferroni tests (*ps* < 0.05) showed that respondents with depression-only felt significantly older ($M = 0.07$, $SD = 0.14$) compared to respondents with PTSD-only ($M = 0.14$, $SD = 0.10$) and those suffering from neither disorder ($M = 0.10$, $SD = 0.13$). Those with PTSD-depression comorbidity felt significantly older ($M = 0.08$, $SD = 0.11$) than those with PTSD-only.

Table 4.2 Descriptive statistics and correlations for the study variables

	M	*SD*	1	2	3	4	5	6	7	8	9	10	11
1. PTSD symptoms	1.13	2.59	–										
2. Depressive symptoms	2.51	2.42	0.25[a]	–									
3. Subjective age	0.09	0.13	0.04	−0.11[a]	–								
4. Medical conditions	1.99	1.87	0.20[a]	0.41[a]	−0.09[a]	–							
5. Disability	1.32	2.97	0.13[a]	0.46[a]	−0.21[a]	0.41[a]	–						
6. Grip strength	27.84	10.53	−0.12[a]	−0.26[a]	0.12[a]	−0.19[a]	−0.32[a]	–					
7. Age	69.65	9.49	0.08[b]	0.27[a]	0.02	0.41[a]	0.46[a]	−0.26[a]	–				
8. Gender	57.6%	–	0.06[c]	0.13[a]	−0.007	0.01	0.03	−0.66[a]	−0.05[c]	–			
9. Origin (Jews)	86.6%	–	0.13[a]	−0.10[a]	0.07[b]	0.005	−0.04	−0.04	0.02	0.02	–		
10. Education	1.92	0.75	0.005	−0.21[a]	0.03	−0.09[a]	−0.16[a]	−0.003	−0.07[b]	0.06[b]	0.31[a]	–	
11. Marital status (married)	71.4%	–	−0.03	−0.20[a]	−0.007	−0.16[a]	−0.22[a]	0.24[a]	−0.31[a]	−0.21[a]	−0.07[b]	0.06[b]	–
12. Exposure	0.92	1.60	0.60[a]	0.18[a]	0.12[a]	0.16[a]	0.05[c]	−0.009	0.04	−0.02	0.18[a]	0.01	0.01

[a]$p < 0.001$; [b]$p < 0.01$; [c]$p < 0.05$

Main Analyses: Group Differences

ANCOVAs (analysis of covariances) were performed to examine group differences in physical outcomes while controlling for covariates (age, gender, education, marital status, and origin). The groups differed in number of medical conditions ($F[31574] = 41.15, p < 0.0001, \eta^2 = 0.07$), disability ($F[31574] = 39.16, p < 0.0001, \eta^2 = 0.07$) and grip strength ($F[31350] = 14.55$, $p < 0.0001$, $\eta^2 = 0.03$). Post hoc Bonferroni tests ($ps < 0.05$) showed that those with depression-only and those with comorbidity were significantly higher in number of medical conditions than PTSD-only and those with no disorder. Those with comorbidity reported more disability relative to all other groups, and likewise those with depression-only reported more disability than PTSD-only and those with neither disorder. Those suffering from no disorder had significantly more grip strength than the three other groups.

Main Analyses: Group X Subjective Age Interaction

In order to assess whether subjective age moderated the effect of group on physical outcomes, we performed a regression analysis for each of the three outcome measures using PROCESS (Hayes, 2013). The multicategorical variable was coded into three dummy variables using Helmert coding: D_1 referred to comparing those without disorder versus the other three groups, D_2 referred to comparing those with PTSD-only to the two depression groups (depression-only and comorbidity), and D_3 referred to comparing depression-only to comorbidity.

Table 4.3 presents the results from these regression analyses. One salient emerging result from all three regressions is that although the level of physical outcomes was worse for the comorbidity condition for participants who had an older subjec-

Table 4.3 Regression analyses predicting physical outcomes by study group, subjective age, and their interaction

	Medical conditions B (SE)	Disability B (SE)	Grip strength B (SE)
D_1	0.93 (0.14)[a]	1.47 (0.23)[a]	−3.24 (0.69)[a]
D_2	1.12 (0.32)	2.36 (0.49)[a]	−2.46 (1.45)
D_3	0.83 (0.22)[a]	1.52 (0.34)[a]	−0.85 (1.02)
Subjective age	−1.99 (0.58)[a]	−6.16 (0.90)[a]	13.75 (2.63)[a]
D_1 X subjective age	−1.44 (0.86)	−3.95 (1.33)[b]	7.41 (3.92)[c]
D_2 X subjective age	−1.35 (1.89)	−6.88 (2.93)[c]	21.25 (8.53)[c]
D_3 X subjective age	−4.17 (1.50)[b]	−7.53 (2.32)[b]	14.28 (6.80)[c]
R^2	0.34	0.30	0.58
N	1451	1451	1246

Note. D_1 refers to comparing those without disorder versus the other three groups, D_2 refers to comparing those with PTSD-only to the two depression groups (depression-only and comorbidity), D_3 refers to comparing depression-only to comorbidity. Analyses controlled for age, gender, education, origin, and marital status

[a]$p < 0.001$; [b]$p < 0.01$; [c]$p \leq 0.05$

tive age, when subjective age was younger for this condition, these differences diminished (and even disappeared in some indices).

Figure 4.2a presents the interaction between group and subjective age for medical conditions. As can be seen, respondents with comorbidity had the highest level of

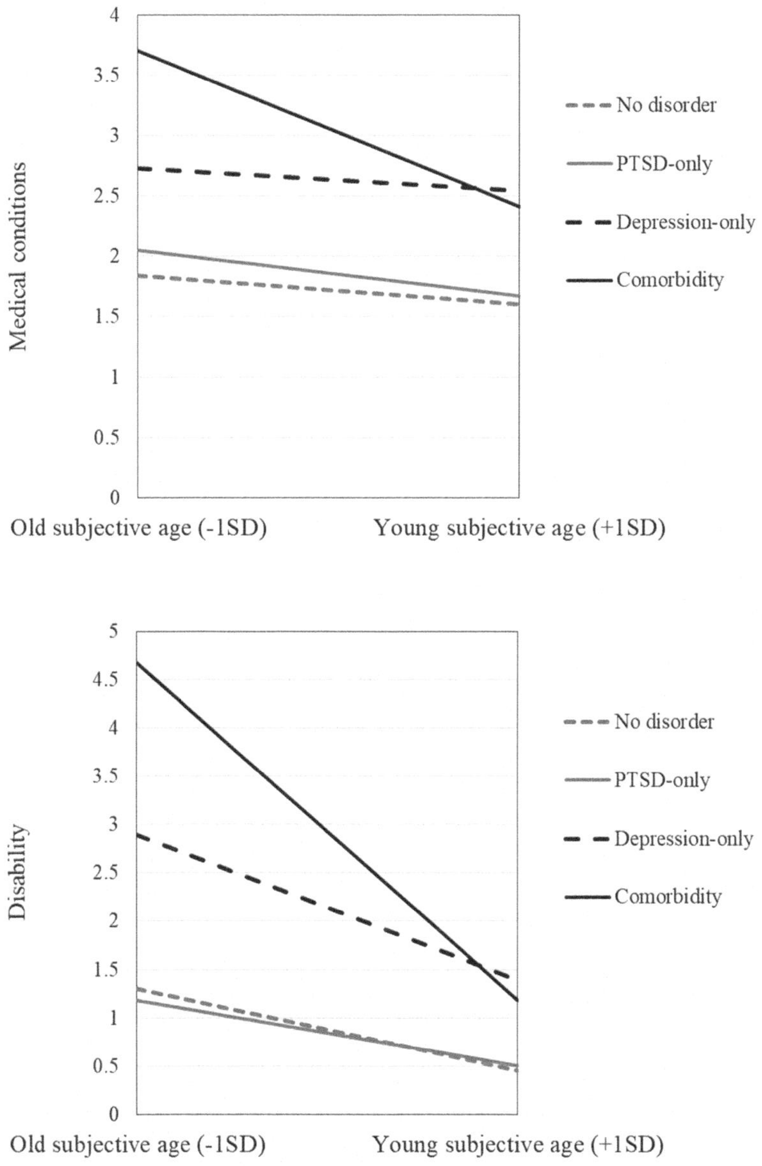

Fig. 4.2 (**a**) The interaction between study groups and subjective age on medical conditions. (**b**) The interaction between study groups and subjective age on disability. (**c**) The interaction between study groups and subjective age on grip strength

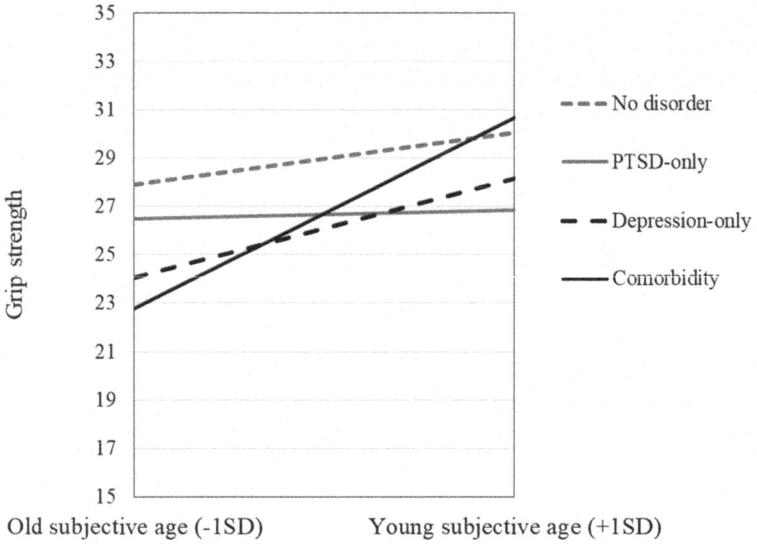

Fig. 4.2 (continued)

medical conditions when they felt older than their age; however, when moving to those who felt younger, their medical condition level dropped to a level equal to that of those with depression-only. Figure 4.2b further shows a similar pattern; again, respondents with comorbidity feeling older had the highest level of disability, yet when moving to those who felt younger, this difference significantly decreased. Figure 4.2c shows that the relationship between subjective age and grip strength was the strongest among respondents with comorbidity. Therefore, when looking at those feeling older than their age, respondents with comorbidity had weaker grip strength relative to most groups, yet among those who felt younger than their age, respondents with comorbidity had the strongest grip strength relative to most other groups.

Supplementary Analyses

In additional regression analyses, we tested the conventional method, whereby PTSD is not further subdivided into those with and without depression, in order to address in further resolution the potential differences between the conventional and the present method. Applying these analyses with the conventional dichotomous PTSD measure (0 = no-PTSD, 1 = PTSD regardless of their depression levels) revealed a significant interaction, whereby PTSD and subjective age conjointly predicted medical conditions (B = -6.57, se = 1.74, $p = 0.0002$). The effect of PTSD on medical conditions was 1.75 (se = 0.32, $p < 0.0001$) among those feeling older than their age (−1 *SD*), and the effect decreased to 0.01 (se = 0.30, $p = 0.96$) among those feeling younger than their age (+1 *SD*). There was a significant interaction between PTSD and subjective age predicting disability (B = –3.45, se = 1.12, $p = 0.002$). The effect of PTSD on disability was 1.04 (se = 0.21, $p < 0.0001$) among those feeling older than their age, and the effect decreased to 0.12 (se = 0.19, $p = 0.51$) among those feeling younger than their age. There was no significant interaction between PTSD and subjective age predicting grip strength (B = 7.38, se = 4.96, $p = 0.13$). As evident, the effects were stronger when PTSD included those with comorbid condition (PTSD-only along with PTSD-depression comorbidity), yet in participants with pure PTSD (without depression) effects were much weaker, attesting to the importance of the current separation method (into those with PTSD-only vs. those with PTSD-depression comorbidity), as approximately half of those diagnosed with PTSD also had depression.[2]

Discussion

The main result is that even when PTSD was accompanied by depression, constituting a high level of stress, a young subjective age was a potent buffer. In other words, having a young subjective age was associated with a mitigated–to-null-effect of the comorbid stressor on physical health indices. The former mitigation effect was shown for two self-report indices of physical health, namely, one's number of medical conditions and one's reported level of disability. Results showing the null stress effect for the comorbid condition were observed for individuals with a younger subjective age in the grip strength performance. It is important to note that grip strength does not rely on self-report and thus is not subjected to the same response biases as the two former physical health indices.

The result pattern seems compatible with the comorbidity c slope (Fig. 4.1). In other words, in line with Hoffman, Shrira, et al. (2016), a young subjective age is not only a resource capable of buffering effects of strong stressors but also reflects stress levels. This was demonstrated by the current interactions showing that in addition to resources, subjective age's impact was dependent on stress level. Such results essentially support the notion that subjective age may be determined both by one's level of stress endured and one's level of resources available for coping with such stress (see Hoffman, Shrira, et al., 2016; Palgi, 2016; see also the weathering hypothesis, Foster et al., 2008).

According to the earlier result pattern (comorbidity c slope, Fig. 4.1; see also Fig. 4.2a–c), the very same subjective age, for example, someone who feels 5%

---

[2]As the PTSD symptom level was significantly higher in the PTSD-depression comorbidity group compared to the PTSD-only group, we reanalyzed our data while controlling for PTSD level. The interactions showed the same pattern to those reported in Table 4.3. We also controlled for depression in the previous wave of SHARE-Israel to account for the possibility of chronic depression driving our results. The interactions showed the same pattern to those reported in Table 4.3 with a single exception of a nonsignificant interaction D_2 X subjective age predicting grip strength (further details can be supplied upon request).

older than their age, may have distinct meanings for people with different stress levels. For persons with lower stress levels, an older age identity may not be so detrimental, yet for persons with higher levels of stress, the very same subjective age is more detrimental. According to the aforementioned model, this is straightforward, for example, if Bob (PTSD-only or depression-only) and Marie (PTSD and depression) are at the same age, yet feel 5% older, this must mean that they have different levels of coping resources, otherwise they would not have the same subjective age (Bob's resource > Marie's resources). The current findings indicate that these functional differences abate as the subjective age gets younger. Following, discerning one's subjective age is not trivial, that is, not merely to ask "how old one feels," as although such a query will elicit a response, the meaning of this response may be complex, as it is the product of both resources and stressors. Not only does feeling younger have different meanings at different ages (Shrira, Segel-Karpas, Bodner, & Palgi, 2020) but also the current results indicate that even persons with the very same chronological age and the very same subjective age may feel and behave in a differential manner, depending on their resources and stress levels.

The results have an important theoretical implication. Namely, in several studies (e.g., Segel-Karpas, Palgi, & Shrira, 2017; Shrira, Palgi, et al., 2018), subjective age as a main effect tended to be smaller than its effect as a moderator. Yet, if indeed subjective age is comprised of both resources and stress, it is fairly straightforward why main effects may be weak, as two people with the same subjective age may show different patterns of function, counteracting potential effects. It is also understood why as a moderator, subjective age can flush out effects, especially if the other independent variable in the model is related to stress or resources.

It might be claimed that the current results were not driven by comorbidity per se, but rather because of the elevated PTSD symptoms found in the comorbid group. Toward this end, we ran the statistical models again while controlling for PTSD symptom level. The results were the same, indicating that although indeed PTSD-symptom levels were higher in the comorbid group, this was not driving the current results.

Accordingly, these results also have a very strong implication for PTSD research, in general, and especially in older adults. Although it is known that PTSD-depression comorbidity is approximately 45–55% of those suffering from PTSD (Flory & Yehuda, 2015), PTSD research has not typically differentiated between persons suffering from PTSD and persons suffering from both disorders. This is quite problematic on many levels, as indicated by the current results, whereby the concomitants of PTSD, such as physical health, seem to be greatly impacted by whether a person is suffering from PTSD-only or from PTSD-depression comorbidity. By not differentiating, we are actually mixing both groups into one and are reporting an average between lower (PTSD) and higher (comorbid) impairments. Lack of such differentiation is essentially challenging, and forces us to reexamine if what we know about PTSD applies to pure PTSD, comorbid PTSD, or both disorders. In addition, this lack of differentiation seems even more critical for older adults who typically have a constant younger subjective age. In such persons, these conditions (PTSD-only and PTSD-depression comorbidity), respectively, reflect different levels of distress/stress and thus give different meanings to the very same subjective age.

The current results also have practical implications. First, with regard to screening, querying about subjective age without gathering information about both resources and stress levels may be very limiting. Second, with regard to interventions, conceptualizing subjective age in a unidimensional fashion, such as it reflects resources, would suggest that an older adult with an old subjective age would benefit from interventions to induce a younger subjective age. As noted by Kotter-Grühn et al. (2015): "the goal of future research should be to prevent the development of older subjective ages rather than trying to make people feel younger (most of whom already feel younger without interventions)" (p. 91). Yet, according to the current results, there seem to be additional groups, such as those who even have a young subjective age but are suffering from a comorbid condition, who may require interventions to induce an even younger subjective age or elevate their resources. This is due to especially robust effects of a young subjective age in this group.

The current study had several limitations. First, this was a cross-sectional study, which had a relatively smaller number of participants in the PTSD-only and comorbid groups. Second, the current PTSD measure although reliable and valid (SPAN) is only one PTSD measure, and brief at that. Although it did produce valid results in the current study (e.g., approximately half the participants were comorbid), it would be important to replicate these results with additional PTSD measures. Third, the impairment observed in the PTSD-only group was relatively minor. Although this is probably also due to the separation of PTSD-only from depression, it may also reflect the minor level of stress. Namely, while the majority of the country was exposed to missile fire during the 2014 Israel-Gaza conflict (Shrira, Palgi, et al., 2018), relatively few persons were killed or injured, as well as people feeling protected by the Iron Dome system (Hoffman, Cohen-Fridel, et al., 2016). Finally, the current subjective age measure was based on a single item. Recent research suggests that it is important to query about subjective age across domains (e.g., Hoffman, Shrira, et al., 2016). A potentially important direction for future research would be to assess subjective ages across cognitive and/or affective dimensions (e.g., "how old do you feel cognitively," "how old do you feel emotionally," etc.)

In summary, the current results show that having a young subjective age can be a potent buffer even when facing strong levels of stress. Future research could also address possible links between a young subjective age and other available coping mechanism that older adults may engage in when faced with stress. For example, older adults who feel young may turn to selection, optimization, and compensation (Baltes, 1997; Freund & Riediger, 2003). Such mechanisms may be manifest, for example, in the current context by *selecting* particularly social interactions that may provide instrumental or emotional support during or following missile attacks, *optimizing* performance under attack (e.g., by repeatedly practicing seeking shelter), and *compensating* for one's slow reactions during attacks by remaining close to sheltered places. Most importantly, these data reveal the underlying structure of subjective age being comprised of both resources and stress. Emerging from this account is the demonstrated notion that the same subjective age may mean different things for different persons (depending on resources and stressors). As subjective age is an important construct in the second half of life (Kotter-Grühn et al., 2015), especially in the aftermath of trauma (Hoffman, Shrira, et al., 2016; Shrira, Palgi, et al., 2018), these results are important for understanding how older adults cope with trauma. That is, the same level of subjective age may have different meanings and ramifications for different stress levels. While this notion could benefit from future research, the current data suggest that one can potentially implement different intervention programs with older populations to both reduce stressors and to bolster a young age identity. This would in turn endow older adults with more opportunities to overcome the challenges presented by their stressors, such as disasters.

Acknowledgements This paper uses data from SHARE Wave 6 (https://doi.org/10.6103/SHARE.w6.700), see Börsch-Supan et al. 2013 for methodological details. The SHARE data collection has been funded by the European Commission through FP5 (QLK6-CT-2001-00360), FP6 (SHARE-I3: RII-CT-2006-062193,COMPARE:CIT5-CT-2005-028857,SHARELIFE:CIT4-CT-2006-028812), FP7 (SHARE-PREP: GA N°211909, SHARE-LEAP: GA N°227822, SHARE M4: GA N°261982) and Horizon 2020 (SHARE-DEV3: GA N°676536, SERISS: GA N°654221) and by DG Employment, Social Affairs & Inclusion. Additional funding from the German Ministry of Education and Research, the Max Planck Society for the Advancement of Science, the U.S. National Institute on Aging (U01_AG09740-13S2, P01_AG005842, P01_AG08291, P30_AG12815, R21_AG025169, Y1-AG-4553-01, IAG_BSR06-11, OGHA_04-064, HHSN271201300071C) and from various national funding sources is gratefully acknowledged (see www.share-project.org).

References

American Psychiatric Association. (2013). *Diagnostic and statistical manual of mental disorders (DSM-5R)*. American Psychiatric Pub: Washington, DC.

Andersen-Ranberg, K., Petersen, I., Frederiksen, H., Mackenbach, J. P., & Christensen, K. (2009). Cross-national differences in grip strength among 50+ year-old Europeans: Results from the SHARE study. *European Journal of Ageing, 6*, 227–236. https://doi.org/10.1007/s10433-009-0128-6

Avidor, S., Benyamini, Y., & Solomon, Z. (2016). Subjective age and health in later life: The role of posttraumatic symptoms. *The Journals of Gerontology: Series B, 71*, 415–424. https://doi.org/10.1093/geronb/gbu150

Baltes, P. B. (1997). On the incomplete architecture of human ontogeny: Selection, optimization, and compensation as foundation of developmental theory. *American Psychologist, 52*, 366–380. https://doi.org/10.1037/0003-066X.52.4.366

Börsch-Supan, A., Brandt, M., Hunkler, C., Kneip, T., Korbmacher, J., Malter, F., et al. (2013). Data resource profile: The survey of health, ageing and retirement in Europe (SHARE). *International Journal of Epidemiology, 42*, 992–1001. https://doi.org/10.1093/ije/dyt088

Brewin, C. R. (2005). Systematic review of screening instruments for adults at risk of PTSD. *Journal of Traumatic Stress, 18*, 53–62. https://doi.org/10.1002/jts.20007

Chopik, W. J., Bremner, R. H., Johnson, D. J., & Giasson, H. L. (2018). Age differences in age perceptions and developmental transitions. *Frontiers in Psychology, 9*, 67. https://doi.org/10.3389/fpsyg.2018.00067

Diehl, M., Wahl, H. W., Barrett, A. E., Brothers, A. F., Miche, M., Montepare, J. M., et al. (2014). Awareness of aging: Theoretical considerations on an emerging concept. *Developmental Review, 34*, 93–113. https://doi.org/10.1016/j.dr.2014.01.001

Flory, J. D., & Yehuda, R. (2015). Comorbidity between post-traumatic stress disorder and major depressive disorder: Alternative explanations and treatment considerations. *Dialogues in Clinical Neuroscience, 17*, 141–150.

Foster, H., Hagan, J., & Brooks-Gunn, J. (2008). Growing up fast: Stress exposure and subjective 'weathering' in emerging adulthood. *Journal of Health and Social Behavior, 49*, 162–177. https://doi.org/10.1177/002214650804900204

Freund, A. M., & Riediger, M. (2003). Successful aging. In R. M. Lerner, M. A. Easterbrooks, & J. Mistry (Eds.), *Handbook of psychology: Developmental psychology* (Vol. 6, pp. 601–628). New York: Wiley.

Hayes, A. F. (2013). *An introduction to mediation, moderation, and conditional process analysis: A regression-based approach*. New York, NY: Guilford Press.

Hoffman, Y., Cohen-Fridel, S., Bodner, E., Grossman, E., & Shrira, A. (2016). Confidence in the" Iron dome" missile defense system combined with a sense of resilience reduced the effect of exposure on posttraumatic stress disorder symptoms after missile attacks. *The Journal of Clinical Psychiatry, 77*, 407. https://doi.org/10.4088/JCP.15l10024

Hoffman, Y., Shrira, A., & Grossman, E. S. (2015). Subjective age moderates the immediate effects of trauma exposure among young adults exposed to rocket attacks. *Psychiatry Research, 229*, 623–624. https://doi.org/10.1016/j.psychres.2015.07.080

Hoffman, Y. S., Shrira, A., Cohen-Fridel, S., Grossman, E. S., & Bodner, E. (2016). Posttraumatic stress disorder symptoms as a function of the interactive effect of subjective age and subjective nearness to death. *Personality & Individual Differences, 102*, 245–251. https://doi.org/10.1016/j.paid.2016.07.017

Horesh, D., Lowe, S. R., Galea, S., Aiello, A. E., Uddin, M., & Koenen, K. C. (2017). An in-depth look into PTSD-depression comorbidity: A longitudinal study of chronically-exposed Detroit residents. *Journal of Affective Disorders, 208*, 653–661. https://doi.org/10.1016/j.jad.2016.08.053

Israel Defense Forces online information. (2014). Taken from http://www.idfblog.com/operationgaza/2014

Katz, S., Downs, T. D., Cash, H. R., & Grotz, R. C. (1970). Progress in development of the index of ADL. *The Gerontologist, 10*, 20–30. https://doi.org/10.1093/geront/10.1_Part_1.20

Kemp, A. H., Felmingham, K., Das, P., Hughes, G., Peduto, A. S., Bryant, R. A., et al. (2007). Influence of comorbid depression on fear in posttraumatic stress disorder: An fMRI study. *Psychiatry Research, 155*, 265–269. https://doi.org/10.1016/j.pscychresns.2007.01.010

Keyes, C. L., & Westerhof, G. J. (2012). Chronological and subjective age differences in flourishing mental health and major depressive episode. *Aging & Mental Health, 16*, 67–74. https://doi.org/10.1080/13607863.2011.596811

Kotter-Grühn, D., Kornadt, A. E., & Stephan, Y. (2015). Looking beyond chronological age: Current knowledge and future directions in the study of subjective age. *Gerontology, 62*, 86–93. https://doi.org/10.1159/000438671

Lahav, Y., Avidor, S., Stein, J. Y., Zhou, X., & Solomon, Z. (2020). Telomere length and depression among ex-prisoners of war: The role of subjective age. *The Journals of Gerontology: Series B, 75*, 21–29. https://doi.org/10.1093/geronb/gby006

Lawton, M. P., & Brody, E. M. (1969). Assessment of older people: Self-maintaining and instrumental activities of daily living. *The Gerontologist, 9*, 179–186. https://doi.org/10.1093/geront/9.3_Part_1.179

Litwin, H. (2009). Understanding aging in a middle eastern context: The SHARE-Israel survey of persons aged 50 and older. *Journal of Cross-Cultural Gerontology, 24*, 49–62. https://doi.org/10.1111/j.1747-0285.2012.01428.x.Identification

Meltzer-Brody, S., Churchill, E., & Davidson, J. R. T. (1999). Derivation of the SPAN, a brief diagnostic screening test for post-traumatic stress disorder. *Psychiatry Research, 88*, 63–70. https://doi.org/10.1016/S0165-1781(99)00070-0

Palgi, Y. (2016). Subjective age and distance-to-death moderate the association between posttraumatic stress symptoms and posttraumatic growth among older adults. *Aging & Mental Health, 20*, 948–954. https://doi.org/10.1080/13607863.2015.1047320

Prince, M. J., Reischies, F., Beekman, A. T., Fuhrer, R., Jonker, C., Kivela, S. L., et al. (1999). Development of the EURO-D scale-a European Union initiative to compare symptoms of depression in 14 European centres. *The British Journal of Psychiatry, 174*, 330–338. https://doi.org/10.1192/bjp.174.4.330

Segel-Karpas, D., Palgi, Y., & Shrira, A. (2017). The reciprocal relationship between depression and physical morbidity: The role of subjective age. *Health Psychology, 36*, 848–851. https://doi.org/10.1037/hea0000542

Shrira, A., Bodner, E., & Palgi, Y. (2014). The interactive effect of subjective age and subjective distance-to-death on psychological distress of older adults. *Aging & Mental Health, 18*, 1066–1070. https://doi.org/10.1080/13607863.2014.915925

Shrira, A., Palgi, Y., Ben-Ezra, M., & Shmotkin, D. (2011). How subjective Well-being and meaning in life interact in the hostile world? *The Journal of Positive Psychology, 6*, 273–285. https://doi.org/10.1080/17439760.2011.577090

Shrira, A., Palgi, Y., Hoffman, Y., Avidor, S., Bodner, E., Ben-Ezra, M., et al. (2018). Subjective age as a moderator in the reciprocal effects between posttraumatic stress disorder symptoms and self-rated physical functioning. *Frontiers in Psychology, 9*. https://doi.org/10.3389/fpsyg.2018.01746

Shrira, A., Segel-Karpas, D., Bodner, E., & Palgi, Y. (2020). Subjective age and emotion covariation: Findings from two daily experience studies. *The Journals of Gerontology: Series B, 75*, 991–1000. https://doi.org/10.1093/geronb/gby125

Solomon, Z., Helvitz, H., & Zerach, G. (2009). Subjective age, PTSD and physical health among war veterans. *Aging & Mental Health, 13*, 405–413. https://doi.org/10.1080/13607860802459856

Stephan, Y., Caudroit, J., & Chalabaev, A. (2011). Subjective health and memory self-efficacy as mediators in the relation between subjective age and life satisfaction among older adults. *Aging & Mental Health, 15*, 428–436. https://doi.org/10.1080/13607863.2010.536138

Stephan, Y., Sutin, A., & Terracciano, A. (2015). How old do you feel? The role of age discrimination and biological aging in subjective age. *PLoS One, 10*, e0119293. https://doi.org/10.1371/journal.pone.0119293

Stephan, Y., Sutin, A. R., & Terracciano, A. (2018). Determinants and implications of subjective age across adulthood and old age. In C. Ryff & R. F. Krueger (Eds.), *The Oxford handbook of integrative health science* (pp. 87–96). New York, NY: Oxford University Press.

United Nations Educational, Scientific and Cultural Organisation (UNESCO). (1997). *International standard classification of education 1997*. Geneva: UNESCO.

Westerhof, G. J., Miche, M., Brothers, A. F., Barrett, A. E., Diehl, M., Montepare, J. M., et al. (2014). The influence of subjective aging on health and longevity: A meta-analysis of longitudinal data. *Psychology and Aging, 29*, 793–802. https://doi.org/10.1037/a0038016

Westerhof, G. J., & Wurm, S. (2015). Longitudinal research on subjective aging, health, and longevity: Current evidence and new directions for research. *Annual Review of Gerontology and Geriatrics, 35*, 145–165. https://doi.org/10.1891/0198-8794.35.145

Wurm, S., Tomasik, M. J., & Tesch-Römer, C. (2008). Serious health events and their impact on changes in subjective health and life satisfaction: The role of age and a positive view on ageing. *European Journal of Ageing, 5*, 117–127. https://doi.org/10.1007/s10433-008-0077-5

Yehuda, R., Flory, J. D., Bierer, L. M., Henn-Haase, C., Lehrner, A., Desarnaud, F., et al. (2015). Lower methylation of glucocorticoid receptor gene promoter 1F in peripheral blood of veterans with posttraumatic stress disorder. *Biological Psychiatry, 77*, 356–364. https://doi.org/10.1016/j.biopsych.2014.02.006

Chapter 5
Theoretical Overview of Disaster Stressors and Responses: Relational and Clinical Implications

Alyssa Banford Witting and Lacey A. Bagley

Introduction

Natural disasters are varied and are generally defined as an event that has rapid and profound effects on the natural environment, causing significant loss financially or through death (Alexander, 2001). Their frequency is expected to increase in the coming years as it has in the previous few decades (Botzen & Van Den Bergh, 2009; Krausmann, Cozzani, Salzano, & Renni, 2011). Disasters often incur death and exact environmental tolls. Additionally, the financial impact of disasters for individuals, families, and communities can be high and difficult to recover from. Despite some uniformity in destruction *types*, losses from disaster events are not equal for all people affected by them. Wealthier nations, including those with better developed institutions, generally suffer fewer losses in terms of death, environmental degradation, and economic damage (Kahn, 2005; Padli, Habibullah, & Baharom, 2010; Raschky, 2008). Thus, socially, is it well understood that mass traumas including war and disasters are not equivalent in their effect on everyone struck by them (Hyndman, 2008; Rahman, 2013). The lethality and long-term consequences of natural disasters are worse for women, minorities, impoverished individuals, and other marginalized and vulnerable groups (Morrow, 1999; Reid, 2013; Walters & Gaillard, 2014).

Research on the mental health of disaster-affected populations garners hundreds of studies over the past few decades (e.g., Newman et al., 2014; Tang, Liu, Liu, Xue, & Zhang, 2014). Two methodological features of this body of literature have characterized research findings, namely, (1) a clear focus on mental health symptoms in *individuals* after natural disaster (Bonanno, Westphal, & Mancini, 2011; Clayton et al., 2015; Doherty & Clayton, 2011) and (2) a preeminent focus on the diagnoses and symptoms of post-traumatic stress disorder (PTSD) (Benight & McFarlane, 2007).

A. B. Witting (✉) · L. A. Bagley
School of Family Life, Brigham Young University, Provo, UT, USA
e-mail: alyssa_banfordwitting@byu.edu

© Springer Nature Switzerland AG 2021

K. E. Cherry, A. Gibson (eds.), *The Intersection of Trauma and Disaster Behavioral Health*, https://doi.org/10.1007/978-3-030-51525-6_5

In recent years, review of research has also begun to conclude that many individuals who are affected by disasters are able to adapt, adjust, and go on with their lives without lasting mental health detriment, or at least not a PTSD diagnosis (Bonanno et al., 2011). Yet, we certainly cannot suggest that traumatic stress is not a salient factor in disaster-affected populations. Indeed, contextual factors post-disaster have begun to elucidate our understanding of the mechanisms, which may serve to exacerbate acute trauma symptoms or contribute to resilience over time (Bonanno et al., 2011; Javidi & Yadollahie, 2012; Kristensen, Weisæth, & Heir, 2010; Nygaard & Heir, 2012); this is an area steeped for continued inquiry.

As such, the interdisciplinary field of disaster scholars may benefit from novel definitions of how traumatic stress may manifest in relational processes. In other words, statistical analyses have evolved in the past 15 years to allow for increasingly sophisticated techniques to allow family-oriented scholars to include data from more than one partner (Kenny, Kashy, & Cook, 2006). Thus, gathering and utilizing data from more than one family member is methodological advancement that disaster scholarship can benefit from now that data analytic infrastructure is becoming available to handle it. Including data from multiple family members and nesting individuals within family and community units will allow for disaster scholars to better understand the effects of disaster trauma in a contextualized way that is not siloed. Disaster studies utilizing couples, families, or communities as the level of analysis are still few in number (Bonanno & Mancini, 2012) and may hold promise in helping us deepen our understanding of the process of traumatic stress post-disaster. It may be that relational and systemic research with an emphasis on the nuanced influence of disasters beyond how they contribute to variation in diagnosable symptoms is a timely and helpful direction for scholarship in this area to grow. Although not the norm, family systems have not been ignored in disaster research either. In fact, families have been the focus (to one degree or another) of hundreds of research studies within the disaster scholarship canon in the past two decades (Banford Witting, Under review).

The purpose of this chapter is to examine methodological patterns in this burgeoning body of research and describe the implications for our understanding of traumatic stress through approaching research with families in disaster from various theoretical lenses. The emphasis on the importance of theory in relational disaster scholarship is a response to an additional trend in the disaster research. Namely, findings in disaster scholarship are often driven by the nature of disasters. Meaning, the rapid onset of many natural disasters has led to a body of literature that is quite event-reactive and not always theoretically driven (see Cherry, Calamia, & Elliott, Chap. 8, this volume). However, an increased body of theoretically driven inquiry is necessary to advance the field. Benight and McFarlane (2007) aptly point out that in our endeavors to study disaster outcomes, it is easy for planful research to be swallowed up in the urgency and haste of post-disaster response. It is necessary to build scholarly agendas driven by literature gaps that supersede any given situation, consider what we have yet to learn about human response to trauma and loss, and allow those larger ideas to guide inquiry. Doing so will ensure that research findings are not reactive and idiosyncratic but intentional and able to drive the field forward (Benight & McFarlane, 2007).

Our primary contention is to suggest that such general questions about adaptation and resilience are fruitfully conceptualized through the intentional use of theory in programmatic research. Kirmayer, Kienzler, Afana, and Pedersen (2010) have suggested that disaster research within psychiatry has often been approached from one of the three lenses (exemplifying broader questions about threat and loss): (1) a focus on symptoms and diagnostics, (2) a resource-based lens through which the operative definition of stress is loss (Hobfoll, 1989), and (3) a socio-ecological perspective situating a person in ecological layers, all affected by disaster and recursive in their effect on each other. Benight and McFarlane (2007) have also offered a review of the range of theoretical frameworks used in disaster scholarship, with some overlap to Kirmayer et al.'s (2010) analysis, namely, (1) diathesis-stress frameworks analyzing the preexisting vulnerabilities of disaster-affected populations, (2) the conservation of resources theory (Hobfoll, 1989), and (3) agentic social cognitive theories which, like socio-ecological models, help to elucidate the confluence of environmental, individual, and behavioral factors in recovery.

In disaster research, specifically carried with families or family-related topics in the past 20 years, we have noted a few similar theoretical patterns. For instance, several studies within the body of family disaster research have been approached from a resource lens (Bokszczanin, 2008; Brock et al., 2014; Jones, Ribbe, Cunningham, Weddle, & Langley, 2002; McDermott & Cobham, 2012; Rivera, 2012; Robertson, Morse, & Baird-Thomas, 2009; Sattler, 2006; Suar, Sekhar Das, Alat, & Kumar, 2017) and an ecological framework (Somasundaram, 2007; Beaton et al., 2008; Kamo et al., 2011; Kilmer & Gil-Rivas, 2014; Preston, Chadderton, Kitagawa, & Edmonds, 2016; Danielson et al., 2017). However, other theories have also been used to organize disaster research with families, which are more systemic in nature and overtly conceptualize the processes within family relationships, namely, attachment theory (Cohan & Cole, 2002; Gallagher et al., 2017; Juth, Silver, Seyle, Widyatmoko, & Tan, 2015; Osofsky, Kronenberg, Bocknek, & Cross Hansel, 2015) and the family stress theory (Felix, You, Vernberg, & Canino, 2013; Lowe, Rhodes, & Scoglio, 2012; Margolin, Ramos, & Guran, 2010; Peek & Marlatt, 2011; Scaramella, Sohr-Preston, Callahan, & Mirabile, 2008).

In the section that follows, a review of the tenets of each theoretical approach as they relate to disaster scholarship with families is presented. This is necessarily a brief review, with empirical findings structured around each theory. The scholarly and clinical implications of approaching disaster research with families from each theory are also highlighted. Finally, the need for continued theory-driven research with family and community systems affected by disaster will be discussed.

Four Prominent Theories to Study Disaster

Conservation of Resources Theory

Conservation of resources (COR): An overview of tenets The development of COR theory was a process of critique and adaptation of physics-based stress theories refined to include the nuanced context of human stress. Foundational to its other tenets, COR theory proposes that people seek pleasure and success. Second, at its core, stress is described as a reaction to loss in the environment. Accordingly, stress includes threat of a net loss of resources, or the realized net loss of resources, and/or a lack of resource gain following the investment of resources. Third, resources are defined as having both instrumental and symbolic value encompassing a few typologies, namely, objects, conditions, personal characteristics, and energies. Taken together "the model's basic tenet is that people strive to retain, protect, and build resources and that what is threatening to them is the potential or actual loss of these valued resources" (Hobfoll, 1989, p. 516).

Loss is central to threat and, therefore, central to stress. In the face of loss of resources, individuals will strive to minimize the net loss of resources by calling on other resources to offset deprivations. Additionally, those with fewer resources may be more prone to additional loss, this notion of one loss contextually leading to another is called a loss spiral. Typical first reactions to loss may be direct replacement, such as getting remarried after divorce. If direct replacement is not possible, symbolic replacement or indirect replacement may be employed. Examples of this may include someone losing self-esteem and then manipulating their environment in such a way that they will receive praise.

Loss can be dampened, if the value of a lost resource can be negotiated down. An example of this could be the loss of a good grade; if the recipient of a low grade lowers the value placed on formal education, then the loss will not seem so bad. Additionally, one may attempt to shift the focus of attention to reinterpret the stress as a positive challenge to be overcome. It is clear to see from these descriptions that norms and social values certainly guide a person's appraisal of events, loss, and resources. However, COR is known for being less focused on individual appraisal than objective assignment of value to resources that may be lost, gained, or protected. Social construction about the value of resources is contextually and culturally driven and, in some cases, may mean that "rather than combating the stressor or enduring the stress, people could merely alter their interpretation of events and their consequences" (Hobfoll, 1989, p. 519). Lastly, when individuals and communities are not under stress, they are motivated to continue to gain resources; however, the attainment of resources outside of a stressful situation tends to be less compelling than a loss on of the same magnitude. Hobfoll suggested in his original thesis (1989) that as a field, we need to identify which resources would be most helpful for which kind of losses, maximizing our return on investment. Hobfoll et al. (2007) documented myriad resources that may be lost or threatened in the context of natural disaster and need direct response to ameliorate, including calmness, connectedness, safety, self and community efficacy, and hope. However, we know less about the resources that may be threatened and how to intervene within the family system. Notwithstanding, several researchers have employed COR as a theoretical structure within which to examine a variety of outcomes following a natural disaster, including symptoms of traumatic stress as a function of loss (Gerhart, Canetti, & Hobfoll, 2015).

Using COR to conceptualize disaster research with families COR theory has been utilized in several ways, most of them being related to two discrete functions, namely, (1) to define resources within and outside the family specific to a disaster context and (2) to explain associations between defined resource loss and distress as well as resource gain or protection and well-being in the family. Several examples of the use of COR to define constructs and explain findings are offered, for example, with various samples.

For example, Jones et al. (2002) examined the psychosocial functioning of children after a fire and ultimately utilized COR to examine what loss means in natural disaster. Using COR theory, these researchers expanded the definition of loss in the disaster context beyond perhaps the most common loss considered in the disaster context. That is, they conceptualized the loss of home to more broadly include the stress induced from loss of time, damage of health, loss of other property, and other internal and external resources. This use of COR theory aids us in broadening attention to the contextual factors that bring salience to the definition of stress post-disaster.

McDermott and Cobham (2012) wanted to understand whether parents' perception of their parenting associated with child post-traumatic stress symptoms 3months post-disaster. The researchers essentially utilized COR to design research hypotheses that supported their findings that the loss of resources (disaster-related losses and perturbations) would drive changes in the way parents think about and approach their parenting, which would, in turn, lead to a loss in well-being of their children in the form of higher levels of post-traumatic stress symptoms following a disaster.

Another example of utilizing COR theory in relationally related disaster research includes Bokszczanin's (2008) study, wherein it was reported that lack of parental support, family conflict, and overprotectiveness are risk factors for post-traumatic stress disorder (PTSD) symptoms in adolescents after a natural disaster. The application of the theory is most clearly seen in the way Bokszczanin (2008) suggests that lack of parental support, family conflict, and overprotectiveness symbolize and embody a loss of family resources. This loss was then interpreted to influence or at least associate with the adolescent's stress experience and to produce lasting consequences.

COR theory has also been used to define protective factors in the disaster literature, such as income and social status. Suar et al. (2017) found that those with higher income and social status were more likely to receive material and informational support post-disaster, which was linked to lower levels of PTSD symptoms. COR has also been used to position support of various kinds as a key resource post-disaster. For instance, Brock et al. (2014) found that flood-related stress interacted with support received to weaken the association between stress and maternal depression through 30 months post-partum. Furthermore, Walsh, Gibson, and Brown (2016) synthesized a body of extant disaster literature using COR to suggest that older adults are uniquely susceptible to anxieties and stressors related to financial resource loss post-disaster.

Along similar lines, COR theory was used by Robertson et al. (2009) to contextualize the finding that family-resource loss was the most impactful predictor of post-disaster anxiety and depression in female juvenile offenders. Furthermore, they used COR to make meaning of the finding that greater resource loss (as defined in the study) was associated with greater symptoms and that resource gain was associated with fewer symptoms. COR theory has been used in concert with other theories to define constructs and make sense of research findings as well. For instance, Sattler (2006) utilized COR and the family resiliency model in a post-earthquake study. Sattler concluded that (1) resource loss was the greatest predictor of psychological distress in his sample and that (2) family coping and social support were also significant predictors.

Implications for research and clinical practice: COR While this is not an exhaustive review of the use of COR in the disaster literature, it is offered as a cross-section of examples of how the theory has been used in disaster research studies, wherein a family sample, family-related variable(s), or family-related question has been addressed in the past 20 years. Researchers tended to use COR to define the constructs in their studies as resources or resource loss. They also used COR theory to make sense of deleterious associations between those losses and poor outcomes. Inherent in these uses is also the assumption that researchers conducting these studies also used COR to conceptualize their research questions; in other words, a focus on loss has a prominent presence in the way that scholars in disaster research relating to the family are approaching inquiry.

From a research perspective, there are patterns in the way COR is being used and potential other ways that it could be to advance the field. For instance, researchers could be thought of as asking "what constitutes a resource or a loss for families facing disaster?" and "how does resource loss relate to functioning?" While replicating studies using these basic questions guided by COR will undoubtedly lead us to helpful new conclusions, the theory could be used in even more nuanced ways to help researchers ask questions about the broader processes of loss in context. For example, COR theory has the depth to drive such programmatic rather than situation specific questions such as "How much loss is too much to bear after disaster?" and "Which resources are most important to families after disaster in a given cultural context?" These deepened directions for basic research also lend opportunities for considering clinical process research and intervention with families who have faced tremendous loss in disaster. COR may help researchers and clinicians to see various circumstances, possessions, and processes as "resources." Utilizing the language of resources may be helpful in aiding relational systems to cope after trauma (Banford Witting & Busby, 2018). Furthermore, if clinicians carry the assumption that resources and loss are operative in distress, they might be enabled to conceptualize family problems in creative ways. As Bokszczanin (2008) suggested that lack of parental support, family conflict, and overprotectiveness substantiate family loss after a disaster; a clinician might see three separate processes in which to intervene. However, applying COR theory, we might also see that family conflict and overprotectiveness flow from a loss spiral starting with lack of parental support and thus put therapeutic energy toward recuperating that lack to help address these other processes. In short, COR may offer clinicians a unique way to conceptualize problematic dynamics as connected in loss spirals and thus see many new and different ways to systemically intervene in a loss-informed dynamic.

Ecological Theory

Ecological theory: An overview of tenets The Father of Ecological Theory is Urie Bronfenbrenner; his work is widely used today by researchers examining a variety of social science issues. The theory outlines structures around an individual, divided into rings, called the micro-, meso-, exo-, and macro-systems (Bronfenbrenner, 1979). The interaction of these systems becomes the substance of understanding experience and has many implications for potential intervention because it attends so deeply to the context of an individual. In other words, attending to ecology allows us to see and map the way that individuals are nested in families that are then nested in various layers of a larger community (Hobfoll, 1989). Ecological perspectives in disaster response are fruitful because they recognize the potential and ability and need for systems at various levels to respond because of the way the layered systems influence each other (Beaton et al., 2008). Ecological thinking also allows for flexibility in conceptualizing how change and even learning occur amid disaster (Preston et al., 2016).

Using an ecological framework is also effective to help researchers appreciate or better understand the context around natural disasters because it attends to diverse factors, namely, intraindividual factors and environmental and sociocultural forces. Specifically, an individual's behavior, development, and adaptation can be viewed as a process involving interaction between the individual and their environment – both proximal and distal influences which have an impact on disaster experience and response (Kytola, Cherry, Marks, & Hatch, 2015). This framework is especially effective to use when working to understand children and family's post-disaster experiences, looking particularly at the systems that impact them (Kilmer & Gil-Rivas, 2014; Weems & Overstreet, 2009). Taking a complex view of each child's ecosystem is important; generally, in disaster scholarship, children have often been lumped together into a subgroup that does not adequately capture the individual variation in personal and life characteristics in order to properly understand children's needs and vulnerabilities post-disaster (Peek, Abramson, Cox, Fothergill, & Tobin, 2018). However, some trends in the literature give hints as to where attention might be helpfully focused. For instance, children's needs have been demonstrated in some studies to be connected to their parent and caregivers' outcomes (Peek et al., 2018); also, the social disruption caused by disasters has been found to have bearing on educational processes affecting a variety of children's long-term learning outcomes (Gibbs et al., 2019).

Using ecological theory to conceptualize disaster research with families Ecological theory has been used widely in disaster scholarship, likely because disasters systemically affect the various layers between an individual and the broadest cultural context. Ecological thinking allows for scholars to consider about the complexities in how layers of context interact. In recent decades, researchers have also applied ecological theory with family-related samples and questions in disaster scholarship for several purposes. For instance, the use of the theory has allowed researchers to define what constitutes the ecology surrounding a family affected by disaster (e.g. Kilmer & Gil-Rivas, 2014). The theory has also been used to suggest how various layers of one's context interrelate for beneficial or deleterious outcomes post-disaster and in creating guidelines for where policy and intervention can be placed in the wider ecology to have a beneficial impact for individuals and families post-disaster (Danielson et al., 2017; Kamo et al., 2011; Somasundaram, 2007).

Kilmer and Gil-Rivas (2014) took an ecological approach to understanding children's post-disaster outcomes in the context of Hurricane Katrina, the most expensive natural disaster to strike in US history in 2005. They used ecological theory to designate variables as salient features to children's layered ecology for the purpose of enhancing our understanding of how the disaster had an impact, namely, they considered self-esteem, self-efficacy, future expectations, developmental level and maturity, cognitive capability, pre-disaster characteristics and adjustment, emotional regulation, and pre-disaster assumptive world. Along with these individual variables, they used ecological theory to examine another layer of context, specifically the children's caregiving systems (pre- and post-disaster) to assess for the protective potential of caregivers for children in the disaster context. Ecological theory suggests that caregiving systems for children are an important contextual layer to assess because traumatic experiences can affect both parenting behavior and quality.

In addition to using the theory to establish the salient layers of ecology within a disaster event, other researchers have used ecological theory to locate the interrelation between contextual layers. For instance, displacement is a critical consideration for disaster survivors because displacement from one's home is a frequent outcome of disaster exposure. Kamo et al., (2011) found that for older adults displaced from New Orleans by Hurricane Katrina, there was a strong association between social network and psychological well-being. This relationship is explained as a buffering effect where family functioning "may help an older person's mental health primarily through taking care of their physical needs" (Kamo et al., 2011, p. 1364). Through this example, we see how ecological theory can be applied to better understand how a contextual layer outside the individual can benefit or buffer effects of a disaster on an individual. Conversely, Danielson et al., (2017) found that survivors of a tornado who experienced loss of basic services and resources were at risk for alcohol use. This is an example of how ecological theory can help us organize and make sense of how losses or perturbations in outer layers of a person's ecology (e.g., community services) can leave individuals vulnerable post-disaster.

Ecological theory is also commonly used to make suggestions about how various layers of the ecological system can be mobilized to create beneficial outcomes post-disaster. For instance, Somasundaram (2007) detailed how changes in broader social systems in Sri Lanka amid the civil war crisis and tsunami trickled down through ecological layers to influence the family. The study also suggested that community approaches were helpful in some cases for families, specifically "creating public awareness, training of grass root workers, encouraging traditional practices and rituals, promoting positive family and community relationships and processes, rehabilitation and networking with other organizations" (Somasundaram, 2007, p. 24). Ecological theory in this case allowed for the analysis of how action at one level of a system can create influence at more proximal levels for individuals and families.

Implications for research and clinical practice: Ecological theory Perhaps the most obvious strength of ecological theory in research that lends itself so well to the study of disaster is its keen attention to context as it never fails to locate the individual, the family, and the community in a broader system bathed in cultural values, constraints, and opportunities. Ecological theory allows disaster scholars to cast variables of interest in their contextual position and make sense of how systems interact through reverberations of proximal and distal influence. The systemic nature of ecological theory makes it attractive for use in scholarship with family systems because it is flexible in its notions of nonlinear influence and allows for cultural influence and a variety of constraints and opportunities within and without systems to be accounted for.

Future research with families using ecological theory in the disaster setting could be infused with ever more complex questions that transcend reaction to immediate disaster scenarios and consider family embeddedness in their ecosystems accounting more overtly for the chronosystem surrounding disaster, for instance, research questions such as "How do interventions aimed at family from the exo- and macrosystems benefit families at various time points in relation to disaster?" and "Which systems outside the family are best positioned to aid in post-disaster recovery in a variety of settings with variable macro-system values?"

Clinicians working with families can certainly also benefit from ecological thinking in disaster intervention or preparation. Thorough assessment in understanding how the family is situated within broader layers of the ecosystem is likely the first beneficial step clinicians can take. From there, clinicians can actively position interventions that not only alter interactions, emotions, and cognitions within the family system but also potentially operate as a broker for the family in connection with the systems most proximate to them. While clinicians don't often have direct access to broad national and cultural systems or changing the family's status within those contexts, those doing disaster-related clinical work can be steeped in understanding of the family's need and connection to the exo- and meso-system. This can take the form of connecting them with information, services, community-related events, and even community-level thinking that aids the family in restoring a sense of belonging after disaster.

Attachment Theory

Attachment theory: An overview of tenets The genesis of attachment theory is found in John Bowlby's early writings (1977, 1978, 1982). Bowlby, a prominent developmental psychology scholar, offered the theory "as way of conceptualizing the propensity of human beings to make strong affectional bonds to particular others" (Bowlby, 1977, p. 201). Attachment theory also describes the adaptive role of fear and threat. Specifically, the process of unwilling separation and subsequent loss of affectional bonds was thought by Bowlby (1977) to lead to emotional distress and even personality disturbance(s). In short, a disruption of attachment elicits a fear response from individuals; similar to the responses, we might have to things like loud noises, new people, and darkness – things that are not intrinsically dangerous but that *could* be dangerous. Further examples of how we express fear, or a protest of disconnection, might include anxiety, depression, and mourning (Bowlby, 1977, 1982).

Bowlby (1978) defined and expanded the value of a secure base or patterned and reliable attachment relationships to facilitate emotional regulation. He also noted how the need for bonds is evident throughout the lifespan when he wrote, "attachment behavior is at its most obvious in early childhood, it can be observed through the life cycle, especially in emergencies" (Bowlby, 1982, p. 669). These essential tenets, the centrality of bonding for emotion regulation, the innate protest of separation from attachment figures, and the adaptive and hopefully connecting role of fear in emergencies, all have useful implications for research and clinical work in the context of natural disasters.

Susan Johnson's work through the last 15 years has proved influential in actively developing the tenets of attachment theory into clinical practice with couples. Primarily, her work has served to deepen understanding about the role of interdependence in the development and sustaining of autonomy and stress management and about the process of how secure bonds are developed through responsiveness (Johnson, 2004). Johnson (2004) has also suggested "that attachment is essentially a theory of trauma" (p. 32). Attachment theory essentially casts emotional reactions to traumatic events as predictable, adaptive, and revolving around the importance of human bonds to regulate distress. This approach has thus been used to inform and address research questions regarding families and natural disaster in the last two decades in several studies.

Using attachment to conceptualize disaster research with families The predictable and adaptive nature of stress and protest amid trauma and separation from affectional bonds and figures is designed to mobilize the attachment system and produce proximity to attachment figures (Johnson, 2004). However, traumatic events can trigger relational processes that thwart connection seeking, and thus the attachment system may be at once mobilized and yet impaired (Williams-Keeler & Johnson, 1998). Empirical examples utilizing an attachment frame have done so in several ways. For instance, attachment theory has been used to explain mental health distress amid disaster as a function of separation during disaster (Osofsky et al., 2015); it has also been used to explain child distress post-disaster as a function of lack of caregiver availability (Juth et al., 2015). Furthermore, attachment theory has been used to explore how the state of attachment bonds when disaster strikes connects with outcomes post-disaster (Gallagher et al., 2017) and finally to explain demographic changes after disaster as a function of bond preservation and creation (Cohan & Cole, 2002).

Osofsky et al. (2015) explored longitudinal patterns in symptoms post-Hurricane Katrina and conceptualized caregiver disruption as a risk factor for continued symptoms along with loss, other trauma prior to and following the storm, as well as general exposure to the disaster. The loss of connection for children comprised in caregiver disruption is positioned by attachment theory as a traumatic event in its own right and includes situations where a fear response intended to create proximity to an attachment figure is thwarted as in the case of death of a caregiver or separation from them during evacuations. In summary, attachment theory can help explain the chronicity of distress post-disasters in part through separation from key bonds.

Death and evacuation are not the only ways scholars have used attachment theory to conceptualize caregiver separation. Juth et al. (2015) found that in parents' post-traumatic stress symptoms after a major earthquake in Indonesia were associated with psychological distress in their children. They suggested, using attachment theory, that trauma symptoms experienced by parents inhibited their role as caregivers and thus dampened their responsiveness, which in turn may be determinant of a child's poor adjustment post-disaster. Both studies show how a key implication of using attachment theory in disaster research leads us to ask about how disaster events create or maintain physical or emotional separation, which predictably creates distress. Attachment theory also makes suggestions about how prior attachment bid outcomes results in styles of not only proximity seeking but also emotional awareness and expression in the form of attachment styles. Thus, attachment style is thought to have a direct bearing on how, behaviorally, the attachment system will be engaged upon distress in the context of close relationships (e.g., Johnson, 2004).

As such, disaster scholars who have also been interested in the role attachment style may play in explaining variation in mental health distress post-disaster. Gallagher et al. (2017) conducted research with heterosexual couples after the major bushfires in Australia; they found that attachment avoidance in male partners was correlated with heightened symptoms of depression and PTSD over time for both partners whereas attachment avoidance in female partners was correlated with heightened distress only in herself but not her partner. This finding has several implications; however, our main purpose in reviewing it here is to suggest how attachment theory can be used to understand dyadic patterns of distress following disaster. In other words, partners' responses to each other can be conceptualized as functions of emotion regulation arising in response to disaster that are driven by the quality of the attachment bond in one or both partners in adult relationships.

Another way attachment theory has been used to formulate questions and research findings in the context of disaster is to examine the formation of bonds in and of themselves. For example, after Hurricane Hugo, researchers found an increase in marriages and birth rates (Cohan & Cole, 2002). These researchers used attachment theory to suggest the salience of needing and attempting to activate bonds during times of loss, stress, separation, and even trauma. Thus, it was thought that disaster survivors may be prompted to take proximity, promoting action in their lives that may lead to all types of relationship transitions.

Implications for research and clinical practice: Attachment theory The role of attachment theory in disaster scholarship has served to underscore the value of being close and connected to loved ones over the life course from attachment bonds formed with caregivers up through pair bonds formed in adulthood. Attachment theory allows for a contextualization of why proximity and connection are desirable, even more so in disaster as it allows for emotion regulation. Attachment theory also lets us conjecture about why disaster-driven separation (or disengaged or insecure relationships before disaster) is likely to correlate with poor mental health outcomes over time.

Future disaster research framed within attachment theory could be directed toward investigations focused on resolution of the key dilemmas presented within this theory. That is, attachment theory itself underscores emotional dysregulation in the face of stress exacerbated by the potential separation of parents and children or partners, be it emotional or physical separation. We know that disasters can exacerbate preexisting styles of disconnection or even generate separation, be it physical and/or emotional. We could further ask programmatic questions to drive prevention and intervention efforts, which might include "Which patterns of connection or disconnection in families are most vulnerable to the disruptions following disaster?" or "What is the state of attachment relationships in geographical regions most vulnerable to disaster?" and "How is the state of attachment relationships in those regions affected by culture, oppression or lack of needed resources?"

Clinical practice with disaster-affected populations using an attachment lens leads us to consider perhaps a primary goal for intervention given the attachment-based findings we have reviewed, namely, to reduce physical emotional separation that is either emotional or physical within key attachment bonds, such as those between parents and children, or partners. This goal has several potential clinical inroads, including providing support for caregivers and partners to be accessible, responsive, and engaged (Sandberg, Busby, Johnson, & Yoshida, 2012) post-disaster. The other implication is that attachment bonds prior to a disaster are already in disrepair, and thus the security of the bond still needs to be improved, but prior to utilizing it as a holding space to process the potential fear and loss experienced in the disaster. Fortunately, attachment-based theories are widely available, evidenced-based, and adaptable for couples and families (Johnson, 2004; Siqueland, Rynn, & Diamond, 2005; Stavrianopoulos, Faller, & Furrow, 2014). Techniques within these models can be used to reduce distance and restructure bonds to allow for potentially better outcomes for couples and families post-disaster.

Family Stress Model

Family stress model: An overview of tenets The roots and history of family stress theory (FST) can be traced through work in psychobiology examining stimuli and psychological response to stress, sociology research of stress responses in the bereavement and grieving processes, psychiatric scholarship in looking at stressors and specific life events, and anthropological work paralleling psychiatry with the use of Hoff's Crisis Paradigm. Individual stress theory began to take form and develop through these trans-field iterations and lead in tandem to the development of family stress theory (Weber, 2011, p. 5).

The model of FST was first applied during the 1920s–1940s (Weber, 2011), a time in the history of the United States of America that is marked by the Great Depression. During the 1920s, American families suffered catastrophic financial losses which impacted every aspect of their day-to-day functioning. Angell (1936) described that family integration and adaptability impacted how families dealt with these crises [now called stressors]; results from the research indicate that successful adjustment came from high interdependence and flexibility. Likewise, during this time period, Cavan and Ranck (1938) examined what made a *well-organized* family. They suggested that families who had a higher degree of unity, reciprocal functioning, and definite function in the larger community could be described as *well-organized*. From these examples, the central tenets of FST is beginning to take shape.

As social scientists worked fervently to understand the impact of crises [stressors] on families, a terminology and body of research grew around FST. Hill's (1949, 1958) seminal development of the ABCX model is still a familiar model both in use and enjoying adaptation of researchers today. Because of his incredible contribution to the field, Reuben Hill is referred to as the Father of Family Stress Theory. This time period, from the 1940s to the early 1970s, is referred to as the "second era" of FST (Weber, 2011).

Up to this point in time, social science researchers focused on how families managed during a crisis by using the measurement of their *weaknesses*. The third era of FST, the late 1970s–1980s, was ushered in by a paradigm shift in the field toward examining how families managed during a crisis by using a measurement of their *strengths* (Weber, 2011). A direct result of this shift is the evolution of the ABCX formula into the double ABCX formula, presented by McCubbin and Patterson (1983). The main evolution of the formula includes their addition of coping and other strength-based variables. This strength-based view has been utilized in research with individuals to identify positive outcomes and resilience in the midst of stressful events, including disasters (Hatch, Cherry, Lu, & Marks, 2015).

By the mid-1980s and into the present time, FST continues to undergo refinement. The focus of researchers today is more post-modern, shifting to looking more at processes, shared meaning, context, cultures, and strengths rather than objective states of functioning. Throughout this fourth era, there has been an eruption in the work and development of models used to describe families coping with stressors/stress events. Examples of these include the Family Adjustment and Adaptation Response model (McCubbin & Patterson, 1983); Typology Double ABCX Model now referred to as the Typology Model of Family Adjustment and Adaptation (McCubbin and McCubbin (1987); Resilience Model of Family Stress, Adjustment, and Adaptation (McCubbin & McCubbin, 1991, 1993); an ecosystemic view (Burr, 1989); the Family Distress Model (Cornille & Boroto, 1992); and, more currently, the Contextual Model of Family Stress (Boss, 2002). Essentially, the family stress model (while an evolving canon of though) allows researchers to conceptualize systemic response to stress in a formulaic sense accounting for events that occur within and without the family. The structure of FST allows researchers to investigate mechanisms of influence, which may be direct or indirect in nature. In this way, FST conceptualization allows for complex process-oriented hypotheses to be tested.

Using the family stress model to conceptualize disaster research with families Reviewing the shifting growth and evolving history of FST helps better situate current research trends using stress models with families affected by disasters. Some researchers using FST with families affected by disasters focus on how the relationship between parents and children may be affected post-disaster. Felix et al. (2013) found that family environment factors may play a mediational role in adolescent's psychopathology post-disaster, whereas for children environmental factors play an additive role. They expand on this by explaining that hurricane exposure, from Hurricane Georges in Puerto Rico in 1998, had a direct relationship to internalizing psychopathology for children. For youth (11–17 years old), some parenting variables attenuated the relationship between exposure and internalizing psychopathology (Felix et al., 2013). In this study, FST was provided as a possible explanation for how external stressors affected parents – which in turn affect children and youth – ultimately finding that family environment and other family factors promoted resilience in children (Felix et al., 2013). Examining how economic hardships impact parents psychologically, which then increases problem behaviors in children, is a fundamental principle in FST. Scaramella et al. (2008) were able to replicate the family stress model (typically measures outcomes for adolescents) and apply the same principles to measure outcomes for low-income toddlers who were impacted by Hurricane Katrina. Again, we see the general pattern of FST use in disaster scholarship to be the lending of complex process-oriented hypotheses where one event or factor leads to another.

Other studies use FST to illustrate how family's experiences with post-disaster adjustment is dynamic within various populations in need of study and advocacy. One study used the family stress model to predict especially negative relationship outcomes for at-risk populations after Hurricane Katrina (Lowe et al., 2012). However, researchers also noted that disasters can sometimes pose new economic opportunities for low-income families that might lead them to relationship improvement (Lowe et al., 2012). In this case, FST was used to predict both relationship strain and growth, a bidirectional and dynamic capacity unique to FST and its derivatives. Another example of how FST has been used with disaster-affected families to capture dynamic processes is Peek and Marlatt's (2011) work, where they located functioning patterns representing downward and upward trajectories among both parents and children's adjustment.

Implications for research and clinical practice: The family stress model FST allows scholars and clinicians to place events, resources, losses, and even relationships in sequential, dynamic, and multidirectional paths to enhance complex understanding. In other words, FST easily lends itself to model testing that is complex and longitudinal and is adaptable using a range of qualitative and quantitative methods. Conceptually, FST requires an assessment of disasters that it inclusive of events and perceptions that tap into objective, perceived, internal, and systemic domains. As such, using the model to understand research findings and to pose questions assures that multiple facets of functioning can be accounted for in explaining outcomes of interest. Perhaps one way in which FST can be used in future research to deliberately advance the field of disaster scholarship with families is to consider the cultural context influencing the presentation of various concepts in the model. For instance, questions that could be addressed using the model might include understanding variants of coping and differential value of a variety of resources across culture.

Clinically, FST gives therapists working with families post-disaster a host of ways to help families understand chaos they may be experiencing internally and externally. The value of information that is culturally sensitive, offered in the context of a supportive relationship and given on a timely basis, has been noted in the trauma literature (Whitworth, 2016). One way in which conceptualizing family response from an FST perspective may be helpful is that it provides a way for clinicians to inform and empower families about what they are experiencing or might later experience. This perspective also allows therapists to visualize several potential interventions based on a family's position in the model. Perhaps the family needs help stimulating resources or may benefit from one or more challenges to their perception about events that have happened and their meaning assigned to such events. Clinicians may use FST to pinpoint areas in the conceptual sequelae, wherein the family could use help and then intervene in a targeted way.

Conclusion

The purpose of this chapter was to review the use, benefits, and potential future applications of theories already being used in disaster scholarship with families to spur continued theory-driven inquiry. The present analysis has allowed for an examination of how theory is commonly used to ask and answer questions and even how the theories relate on a broader scale as mechanisms of knowledge development regarding disaster events. For instance, COR theory suggests that loss, above all, is operative in the inherent nature of stress, while ecological theory allows for a bird's eye view of how human domains are nested and interrelated and organize interaction on multiple levels across time. Attachment theory then draws our focus to the salience of the loss of bonds. Family stress theory finally allows for empirical work incorporating these various elements (losses, contexts, bonds) in complex sequences, suggesting a host of variation in a general sequela: event, response, and outcome. Taken together, these theoretical lenses have given disaster scholarship an organization to help researchers make meaning of the effects of disasters for families. Furthermore, the continued use of theory applied in less reactive and more intentional ways to drive empirical inquiry with families affected by disaster has numerous potential benefits to advance the field and improve relational and mental health outcomes.

Finally, the use of these theories is likely to be enhanced with the advent of greater cross-disciplinary research. The usefulness of research is improved when scholars from multiple domains collaborate in efforts to translate findings across fields, despite barriers which may present challenges to interdisciplinary scholarship, such as determining the unit of analysis be that individual, family, or community. However, continued efforts to transcend traditional field boundaries with the intentional use of theory are likely to broaden the impact of disaster research. For instance, public health scholars are uniquely positioned to apply psychological research to broad audiences, and family therapy scholars are poised to consider systemic and relational outcomes. In summary, collaboration and programmatic inquiry are needed to address the existing and forthcoming challenges of disaster, which will confront individuals, families, and communities across the globe.

References

Alexander, D. E. (2001). *Natural disasters*. New York: Routeledge.

Angell, R. C. (1936). *The family encounters the depression*. Gloucester, MA: Peter Smith; (Reprinted in 1965).

Banford Witting, A., & Busby, D. (2018). The long arm of trauma in childhood: Associations with resources in couple relationships. *Journal of Marital and Family Therapy*. https://doi.org/10.1111/jmft.12354

Banford Witting, A., Bagley, L., Nelson, K., & Lindsay, T. (Under review). Natural disasters and the relational study of the family: A 2-Decade Scoping Review.

Beaton, R., Bridges, E., Salazar, M. K., Oberle, M. W., Stergachis, A., Thompson, J., et al. (2008). Ecological model of disaster management. *Business and Leadership, 56*(11), 471–478.

Benight, C. C., & McFarlane, A. C. (2007). Challenges for disaster research: Recommendations for planning and implementing disaster mental health studies. *Journal of Loss and Trauma, 12*(5), 419–434.

Bokszczanin, A. (2008). Parental support, family conflict, and overprotectiveness: Predicting PTSD symptom levels of adolescents 28 months after a natural disaster. *Anxiety, Stress & Coping: An International Journal, 21*(4), 325–335. https://doi.org/10.1080/10615800801950584

Bonanno, G. A., & Mancini, A. D. (2012). Beyond resilience and PTSD: Mapping the heterogeneity of responses to potential trauma. *Psychological Trauma: Theory, Research, Practice, and Policy, 4*(1), 74.

Bonanno, G. A., Westphal, M., & Mancini, A. D. (2011). Resilience to loss and potential trauma. *Annual Review of Clinical Psychology, 7*, 511–535.

Boss, P. (2002). *Family stress management: A contextual approach* (2nd ed.). Thousand Oaks, CA: Sage.

Botzen, W. J. W., & Van Den Bergh, J. C. J. M. (2009). Managing natural disaster risks in a changing climate. *Environmental Hazards, 8*(3), 209–225.

Bowlby, J. (1977). The making and breaking of affectional bonds: I. Aetiology and psychopathology in the light of attachment theory. *The British Journal of Psychiatry, 130*(3), 201–210.

Bowlby, J. (1978). Attachment theory and its therapeutic implications. *Adolescent Psychiatry, 6*, 5–33.

Bowlby, J. (1982). Attachment and loss: Retrospect and prospect. *American Journal of Orthopsychiatry, 52*(4), 664–678. https://doi.org/10.1111/j.1939-0025.1982.tb01456.x

Bronfenbrenner, U. (1979). The ecology of human development: Experiments by nature and design. Cambridge, MA: Harvard University Press. psycnet.apa.org/psycinfo/1962-00777-001

Brock, R. L., O'Hara, M. W., Hart, K. J., McCabe, J. E., Williamson, J. A., Laplante, D. P., et al. (2014). Partner support and maternal depression in the context of the Iowa floods. *Journal of Family Psychology, 28*(6), 832–843. https://doi.org/10.1037/fam0000027.supp

Burr, R. G. (1989). *Reframing family stress theory: From the ABC-X model to a family ecosystemic model*. Unpublished master's thesis, Brigham Young University, Provo.

Cavan, R. S., & Ranck, K. H. (1938). *The family and the depression: A study of one hundred Chicago families*. Chicago: The University of Chicago Press.

Clayton, S., Devine-Wright, P., Stern, P. C., Whitmarsh, L., Carrico, A., Steg, L., et al. (2015). Psychological research and global climate change. *Nature Climate Change, 5*(7), 640.

Cohan, C. L., & Cole, S. W. (2002). Life course transitions and natural disaster: Marriage, birth, and divorce following hurricane Hugo. *Journal of Family Psychology, 16*(1), 14–25. https://doi.org/10.1037//0893-3200.16.1.14

Cornille, T. A., & Boroto, D. R. (1992). The family distress model: A conceptual and clinical application of Reiss' strong bonds finding. *Contemporary Family Therapy, 14*(3), 181–198.
Danielson, C. K., Sumner, J. A., Adams, Z. W., McCauley, J. L., Carpenter, M., Amstadter, A. B. & Ruggiero, K. J. (2017). Adolescent substance use following a deadly U.S. tornado outbreak: A population-based study of 2,000 families. *Journal of Clinical Child & Adolescent Psychology, 46*(5), 732–745, https://doi.org/10.1080/15374416.2015.1079780
Doherty, T. J., & Clayton, S. (2011). The psychological impacts of global climate change. *American Psychologist, 66*(4), 265.
Felix, E., You, S., Vernberg, E., & Canino, G. (2013). Family influences on the long term post-disaster recovery of Puerto Rican youth. *Journal of Abnormal Child Psychology, 41*(1), 111–124. https://doi.org/10.1007/s10802-012-9654-3
Gallagher, H. C., Lusher, D., Gibbs, L., Pattison, P., Forbes, D., Block, K., et al. (2017). Dyadic effects of attachment on mental health: Couples in a post disaster context. *Journal of Family Psychology, 31*(2), 192–202. https://doi.org/10.1037/fam0000256
Gerhart, J. I., Canetti, D., & Hobfoll, S. E. (2015). Traumatic stress in overview: Definition, context, scope, and long-term outcomes. In K. E. Cherry (Ed.), *Traumatic stress and long-term recovery: Coping with disasters and other negative life events* (pp. 3–24). New York: Springer International Publishing Switzerland.
Gibbs, L., Nursey, J., Cook, J., Ireton, G., Alkemade, N., Roberts, M., et al. (2019). Delayed disaster impacts on academic performance of primary school children. *Child Development, 90*, 1–11. https://doi.org/10.1111/cdev.13200
Hatch, T. G., Cherry, K. E., Lu, Y., & Marks, L. D. (2015). Seeing silver linings after catastrophic loss: Personal growth, positive adaption, and relationships that matter. In K. E. Cherry (Ed.), *Traumatic stress and long-term recovery: Coping with disasters and other negative life events* (pp. 389–402). New York: Springer International Publishing Switzerland.
Hill, R. (1949). *Families under stress*. New York: Harper & Brothers.
Hill, R. (1958). Generic features of families under stress. *Social Casework, 49*, 139–150.
Hobfoll, S. E. (1989). Conservation of resources: A new attempt at conceptualizing stress. *American Psychologist, 44*(3), 513–524.
Hobfoll, S. E., Watson, P., Bell, C. C., Bryant, R. A., Brymer, M. J., Friedman, M. J., et al. (2007). Five essential elements of immediate and mid-term mass trauma intervention: Empirical evidence. *Psychiatry, 70*(4), 283–315. https://doi.org/10.1521/psyc.2007.70.4.283
Hyndman, J. (2008). Feminism, conflict and disasters in post-tsunami Sri Lanka. *Gender, Technology and Development, 12*(1), 101–121.
Javidi, H., & Yadollahie, M. (2012). Post-traumatic stress disorder. *International Journal of Occupational and Environmental Medicine, 3*, 2.
Johnson, S. M. (2004). *The practice of emotionally focused couple therapy: Creating connection* (2nd ed.). New York: Brunner/Mazel.
Jones, R. T., Ribbe, D. P., Cunningham, P. B., Weddle, J. D., & Langley, A. K. (2002). Psychological impact of fire disaster on children and their parents. *Behavior Modification, 26*(2), 163–186. https://doi.org/10.1177/0145445502026002003
Juth, V., Silver, R. C., Seyle, D. C., Widyatmoko, C. S., & Tan, E. T. (2015). Post-disaster mental health among parent–child dyads after a major earthquake in Indonesia. *Journal of Abnormal Child Psychology, 43*(7), 1309–1318. https://doi.org/10.1007/s10802-015-0009-8
Kahn, M. E. (2005). The death toll from natural disasters: The role of income, geography, and institutions. *The Review of Economics and Statistics, 82*(2), 271–284.
Kamo, Y., Henderson, T. L., & Roberto, K. A. (2011). Displaced older adults' reactions to and coping with the aftermath of Hurricane Katrina. *Journal of Family Issues, 32*(10), 1346–1370. "https://psycnet.apa.org/doi/10.1177/0192513X11412495" https://doi.org/10.1177/0192513X11412495
Kenny, D. A., Kashy, D. A., & Cook, W. L. (2006). Methodology in the social sciences (David A. Kenny, series editor). In *Dyadic data analysis*. New York: Guilford Press.
Kilmer, R. P., & Gil-Rivas, V. (2014). Introduction: Attending to ecology. In R. P. Kilmer, V. Gil-Rivas, R. G. Tedeschi, & L. G. Calhoun (Eds.), *Helping families and communities recover from disaster: Lessons learned from Hurricane Katrina and its aftermath*. Washington, DC: American Psychological Association.

Kirmayer, L. J., Kienzler, H., Afana, A. H., & Pedersen, D. (2010). Trauma and disasters in social and cultural context. In *Principles of social psychiatry* (2nd ed., pp. 155–177). Hoboken, NJ: Wiley-Blackwell.

Krausmann, E., Cozzani, V., Salzano, E., & Renni, E. (2011). Industrial accidents triggered by natural hazards: An emerging risk issue. *Natural Hazards and Earth System Sciences, 11*(3), 921–929.

Kristensen, P., Weisæth, L., & Heir, T. (2010). Predictors of complicated grief after a natural disaster: A population study two years after the 2004 south-east Asian tsunami. *Death Studies, 34*(2), 137–150.

Kytola, K. L., Cherry, K. E., Marks, L. D., & Hatch, T. G. (2015). When neighborhoods are destroyed by disaster: Relocate or return and rebuild? In K. E. Cherry (Ed.), *Traumatic stress and long-term recovery: Coping with disasters and other negative life events* (pp. 211–229). New York: Springer International Publishing Switzerland.

Lowe, S. R., Rhodes, J. E., & Scoglio, A. J. (2012). Changes in marital and partner relationships in the aftermath of hurricane Katrina: An analysis with low-income women. *Psychology of Women Quarterly, 36*(3), 286–300. https://doi.org/10.1177/0361684311434307

Margolin, G., Ramos, M. C., & Guran, E. L. (2010). Earthquakes and children: The role of psychologists with families and communities. *Professional Psychology: Research & Practice, 41*(1), 1–9. https://doi.org/10.1037/a0018103

McCubbin, H. I., & Patterson, J. M. (1983). The family stress process: The double ABCX model of family adjustment and adaptation. In H. I. McCubbin, M. Sussman, & J. M. Patterson (Eds.), *Social stress and the family: Advances and developments in family stress theory and research* (pp. 7–37). New York: Haworth.

McCubbin, M. A., & McCubbin, H. I. (1987). Family stress theory and assessment: The T-double ABCX model of family adjustment and adaptation. In H. I. McCubbin & A. Thompson (Eds.), *Family assessment inventories for research and practice* (pp. 3–32). Madison, WI: University of Wisconsin.

McCubbin, M. A., & McCubbin, H. I. (1991). Family stress theory and assessment: The resiliency model of family stress adjustment and adaptation. In H. I. McCubbin & A. Thompson (Eds.), *Family assessment inventories for research and practice* (pp. 3–31). Madison, WI: University of Wisconsin.

McCubbin, M. A., & McCubbin, H. I. (1993). Families coping with illness: The resiliency model of family stress adjustment and adaptation. In C. Danielson, B. Hamel-Bissell, & P. Winstead-Fry (Eds.), *Families, health & illness: Perspectives on coping and intervention* (pp. 21–64). St. Louis, MO: Mosby.

McDermott, B. M., & Cobham, V. E. (2012). Family functioning in the aftermath of a natural disaster. *BMC Psychiatry, 12*, 1. https://doi.org/10.1186/1471-244X-12-55

Morrow, B. H. (1999). Identifying and mapping community vulnerability. *Disasters, 23*(1), 1–18.

Newman, E., Pfefferbaum, B., Kirlic, N., Tett, R., Nelson, S., & Liles, B. (2014). Meta-analytic review of psychological interventions for children survivors of natural and man-made disasters. *Current Psychiatry Reports, 16*(9), 462.

Nygaard, E., & Heir, T. (2012). World assumptions, posttraumatic stress and quality of life after a natural disaster: A longitudinal study. *Health and Quality of Life Outcomes, 10*(1), 76.

Osofsky, J., Kronenberg, M., Bocknek, E., & Cross Hansel, T. (2015). Longitudinal impact of attachment-related risk and exposure to trauma among young children after Hurricane Katrina. *Child & Youth Care Forum, 44*(4), 493–510. https://doi.org/10.1007/s10566-015-9300-7

Padli, J., Habibullah, M. S., & Baharom, A. H. (2010). Economic impact of natural disasters' fatalities. *International Journal of Social Economics, 37*(6), 429–441. https://doi.org/10.1108/03068291011042319

Peek, L., Abramson, D. M., Cox, R. S., Fothergill, A., & Tobin, J. (2018). Children and disasters. In H. Rodriguez et al. (Eds.), *Handbook of disaster research* (pp. 243–262). New York: Springer International Publishing. https://doi.org/10.1007/978-3-319-63254-4_13

Peek, L. B., & Marlatt, H. (2011). Disaster hits home: A model of displaced family adjustment after Hurricane Katrina. *Journal of Family Issues, 32*(10), 1371–1396. https://doi.org/10.1177/0192513X11412496

Preston, J., Chadderton, C., Kitagawa, K., & Edmonds, C. (2016). Community response in disasters: An ecological learning framework. *International Journal of Lifelong Education, 34*(6), 727–753. https://doi.org/10.1080/02601370.2015.1116116

Rahman, M. S. (2013). Climate change, disaster and gender vulnerability: A study on two divisions of Bangladesh. *American Journal of Human Ecology, 2*(2), 72–82.

Raschky, P. A. (2008). Institutions and the losses from natural disasters. *Natural Hazards and Earth System Sciences, 8*, 627–634.

Reid, M. (2013). Disasters and social inequalities. *Sociology Compass, 7*(11), 984–997.

Rivera, F. I. (2012). Cultural mechanisms in the exchange of social support among Puerto Ricans after a natural disaster. *Qualitative Health Research, 22*(6), 801–809. https://doi.org/10.1177/1049732311432719

Robertson, A. A., Morse, D. T., & Baird-Thomas, C. (2009). Hurricane Katrina's impact on the mental health of adolescent female offenders. *Anxiety, Stress & Coping: An International Journal, 22*(4), 433–448. https://doi.org/10.1080/10615800802290634

Sandberg, J. G., Busby, D. M., Johnson, S. M., & Yoshida, K. (2012). The brief accessibility, responsiveness, and engagement (BARE) scale: A tool for measuring attachment behavior in couple relationships. *Family Process, 51*(4), 512. https://doi.org/10.1111/j.1545-5300.2012.01422.x

Sattler, D. N. (2006). Family resources, family strains, and stress following the Northridge earthquake. *Stress, Trauma and Crisis: An International Journal, 9*(3–4), 187–202. https://doi.org/10.1080/15434610600854038

Scaramella, L. V., Sohr-Preston, S. L., Callahan, K. L., & Mirabile, S. P. (2008). A test of the family stress model on toddler-aged children's adjustment among hurricane Katrina impacted and non-impacted low-income families. *Journal of Clinical Child & Adolescent Psychology, 37*(3), 530–541. https://doi.org/10.1080/15374410802148202

Siqueland, L., Rynn, M., & Diamond, G. S. (2005). Cognitive behavioral and attachment based family therapy for anxious adolescents: Phase I and II studies. *Journal of Anxiety Disorders, 19*(4), 361–381. https://doi.org/10.1016/j.janxdis.2004.04.006

Somasundaram, D. (2007). Collective trauma in northern Sri Lanka: A qualitative psychosocial-ecological study. International Journal of Mental Health Systems, 1, Article 5. https://doi.org/10.1186/1752-4458-1-5

Stavrianopoulos, K., Faller, G., & Furrow, J. L. (2014). Emotionally focused family therapy: Facilitating change within a family system. *Journal of Couple and Relationship Therapy, 13*(1), 25–43. https://doi.org/10.1080/15332691.2014.865976

Suar, D., Sekhar Das, S., Alat, P., & Kumar, R. (2017). Exposure, loss, and support predicting the dimensions of posttsunami trauma. *Journal of Loss & Trauma, 22*(5), 427–439. https://doi.org/10.1080/15325024.2017.1310499

Tang, B., Liu, X., Liu, Y., Xue, C., & Zhang, L. (2014). A meta-analysis of risk factors for depression in adults and children after natural disasters. *BMC Public Health, 14*(1), 623.

Walsh, J., Gibson, A., & Brown, L. M. (2016). Peace of mind's price tag: The psychological costs of financial stressors on older adults postdisaster. *Translational Issues in Psychological Science, 2*(4), 408–417. https://doi.org/10.1037/tps0000089

Walters, V., & Gaillard, J. C. (2014). Disaster risk at the margins: Homelessness, vulnerability and hazards. *Habitat International, 44*, 211–219.

Weber, J. G. (2011). *Individual and family stress and crises*. Thousand Oaks, CA: SAGE Publications.

Weems, C. B., & Overstreet, S. (2009). An ecological-needs-based perspective of adolescent and youth emotional development in the context of disaster: Lessons from Katrina. In K. E. Cherry (Ed.), *Lifespan perspectives on natural disasters: Coping with Katrina, Rita and other storms* (pp. 27–44). New York: Springer Science+Business Media.

Whitworth, J. D. (2016). The role of psychoeducation in trauma recovery: Recommendations for content and delivery. *Journal of Evidence-Informed Social Work, 13*(5), 442–451. https://doi.org/10.1080/23761407.2016.1166852

Williams-Keeler, L., & Johnson, S. M. (1998). Creating healing relationships for couples dealing with trauma: The use of emotionally focused marital therapy. *Journal of Marital and Family Therapy, 24*, 25–40. https://doi.org/10.1111/j.1752-0606.1998.tb01061.x

Chapter 6
Psychological Effects of Natural Disaster: Traumatic Events and Losses at Different Disaster Stages

Shuei Kozu and Allison Gibson

Introduction

The International Federation of Red Cross and Red Crescent Societies (IFRC) divide natural disasters into four categories: geophysical (e.g., earthquakes, tsunamis), meteorological (e.g., cyclones), hydrological (e.g., avalanches and floods), and climatological (e.g., drought and wildfires) (IFRC, 2016). Data show there have been significant increases in all types of natural disasters worldwide during the last several decades. For example, the combination of all four types of disasters in 1980 numbered approximately 300 events, while this number almost tripled by 2014, showing a steady escalation in the years from 1980 to 2014 (Münchener Rückversicherungs-Gesellschaft, 2015). According to the United Nations Office for Disaster Risk Reduction (2019), between the years of 2005 and 2014, approximately 1.7 billion people have been affected by natural disasters, with 700,000 experiencing loss of life. Scientists predict that extreme weather linked to climate change is on the rise and will likely cause more natural disasters (Shepherd et al., 2013). Given these data and predictions, it is important for psychosocial providers to understand the biopsychosocial impacts of natural disasters, because more people will likely suffer physical and mental health problems following disasters.

When multiple losses and traumatic events hit in succession, it can lead to complicated or traumatic grief, making it difficult for some people to recover from psychological distress without timely and appropriate identification of their conditions (Neria & Litz, 2004). Cohen and Mannarino (2004) define the childhood traumatic grief as a "condition in which a child or adolescent has lost a loved one in

S. Kozu (✉)
Department of Social Work, Southern Connecticut State University, New Haven, CT, USA
e-mail: kozus1@southernct.edu

A. Gibson
College of Social Work, University of Kentucky, Lexington, KY, USA

© Springer Nature Switzerland AG 2021

K. E. Cherry, A. Gibson (eds.), *The Intersection of Trauma and Disaster Behavioral Health*, https://doi.org/10.1007/978-3-030-51525-6_6

circumstances that are objectively or subjectively traumatic and in which trauma symptoms impinge on the child's ability to negotiate the normal grieving process" (p. 819). Adult complicated bereavement refers to "a condition in which bereavement is complicated due to the nature of losing a relationship that provided significant security and dependency for the bereaved" (p. 159, Cohen & Mannarino, 2004, cited in Edgar-Bailey & Kress, 2010). In the first section of this chapter, we review traumatic grief in children, including intervention strategies. The second section of the chapter will focus on adults, followed by the last section that describes community-based services utilized in disaster areas after the Great East Japan Earthquakes in 2011.

People involved in a natural disaster are highly likely to be exposed to traumatic events, with many people exposed to multiple events during a single disaster. For example, during the Great East Japan Earthquakes in 2011, residents of the Tohoku area in Japan experienced such traumas as the strong and prolonged shaking of the earth, being sucked into the tsunami water, witnessing someone else drowning, and seeing massive fires in the coastal area (Kozu & Homma, 2014). These traumatic events, unfortunately, are not singular, but may occur repeatedly, and can lead to the snowball effect of other catastrophic events (Kukihara, Yamawaki, Uchiyama, Arai, & Horikawa, 2014). For example, after one episode of violent shaking, earthquakes often bring on aftershocks for many days, weeks, and months, delaying the rescue effort due to unsafe conditions. As seen in Fukushima, Japan, damages from the tsunami water induced the explosion of two nuclear plants (Maeda & Oe, 2015). Thus, people residing in the disaster area may experience multiple life-threatening events for a prolonged period of time. However, it is extremely important to keep in mind that most people who experience traumatic events do not develop a psychiatric disorder (Cherry, Allen, & Galea, 2010). Rather, at an early stage of a disaster, most people may experience hypervigilance and have acute stress reactions, or they can be overwhelmed by shock and become numb. Intervention strategies in this stage of disaster should focus on triaging people with acute symptoms, providing psychological first aid, providing information and resources to meet the basic needs for survival, and locating loved ones.

In addition to the psychological impact on people residing in the area where the disaster hits, family members and friends living away can be equally distressed. Disasters often cause disruption in communication, as cell phone connections and utilities such as electricity services may be interrupted. While it may not be life threatening, it still is a traumatic experience for loved ones who live away from the disaster area, knowing their family and friends live where the disaster is occurring, but not being able to communicate and confirm the well-being of the persons for a prolonged period of time. One of the criteria for post-traumatic stress disorders in Diagnostic and Statistical Manual of Mental Disorders, fifth edition (American Psychiatric Association [DSM-5], 2013), includes the threat to the life of a close family member or close friend. This is especially true for someone who has unresolved loss in the past, and this may lead to complicated traumatic grief (Neria & Litz, 2004). When a large-scale disaster occurs, the media focuses on the event, and there are public outcries and willingness to support those who are in the area where the disaster hits. While the number may be small, there are people in communities far away from the disaster who are severely impacted by the disaster and suffering silently. It is important to keep in mind that victims of a disaster are not always in the immediate vicinity of the geographical area.

During the immediate aftermath of a disaster, chaos continues to ensue. People find that their loved ones are missing, killed, or severely injured. Nearby hospitals may not be open or able to accommodate an overwhelming number of patients who seek medical attentions as the result of the disaster. People may also discover that their property has been damaged or destroyed, that they lost their jobs, or that they are trapped in their homes without food or water, with transportation means destroyed by the disaster. In short, people may suffer multiple layers of loss and negative changes in their lives. There are many factors that influence psychological impacts on people during this stage. Studies found early unification with family members and access to resources and support, for example, to be protective factors (Overstreet, Salloum, Burch, & West, 2011). On the other hand, the level and frequency of exposure is strongly associated with the severity of the post-traumatic stress reaction (Pynoos, Goenjian, & Steinberg, 1998). An ecological perspective by Bronfenbrenner, which examined the influence of individual's relationships with his/her families, communities, and the wider society, should be utilized when assessing the survivors of a disaster, because merely focusing on the traumatic aspect may risk misdiagnosis or overlooking critical factors that impact the survivors' overall functions (e.g., Beaton et al., 2008; Gibson, Kusmaul, Leedahl, & Rizzo, 2020; Modica & Zoboli, 2016).

A natural disaster is often followed by secondary adversities (Overstreet et al., 2011). These include relocation, change of jobs or schools, secondary health problems, changes in family dynamics and roles in the family, changes in family finances or socioeconomic status, and alterations in school or job performance. Pynoos et al. (1998) stated, "Secondary stresses increase the risk of co-morbidity, complicate efforts at adjustment, initiate maladaptive coping, interfere with the availability of social support, family functioning and reintegration with peers" (p. 130). During the early stages of a disaster, the immediate realization of loss tends to be of concrete entities, such as property, or physical concerns, such as the death or injury of loved ones. During the later stages of disaster recovery, people start to experience psychological loss. This might take the form of losing their sense of safety or losing community connection at familiar places they had been affiliated with, like school, work, or clubs. Another loss to consider in this stage is the withdrawal of disaster emergency response teams as well as volunteers. During the early stages of disasters, especially those that receive extensive media coverage, an overwhelming number of people express their willingness to support the area. After several weeks or months, this support dwindles, leaving the local residents to move forward on their own. Survivors of disasters have voiced that they felt they were left behind or abandoned (Dembert & Simmer, 2000; Mitchell, Griffin, Stewart, & Loba, 2004). A coordination in psychosocial provider response would be helpful: instead of numerous clinicians responding during the immediate aftermath, there would be ongoing psychosocial support at a later stage, when the psychological distress may increase and psychopathologic states might begin to develop. Access to psychosocial providers in the disaster area will ensure accurate diagnoses and referral to treatment centers in a timely manner.

While a natural disaster can hit anywhere in the world, regardless of the race, gender, or socioeconomic status of the residents of the community, these factors greatly affect the community's capacity to recover from the disasters. Preexisting conditions, such as poverty and discrimination, hinder the individuals and community's ability to receive external support, which might delay rebuilding the community. It is important for psychosocial providers to understand the culture of the community and assess, if possible, if there are people who are isolated and not able to receive or utilize any support. Discrimination and social isolation are factors known to negatively impact one's health and could exaggerate mental health issues after experiencing multiple losses (Smith, Jaurique, & Ryan, 2016; Wang et al., 2017).

Traumatic Grief in Children

Grief is a natural and healthy response to loss, which involves physical, emotional, cognitive, and spiritual processes (Andrews & Marotta, 2005). People need to grieve in their own ways, and it is important for adults who are around children to allow them to go through the grief process at their own pace. When children suffer from traumatic grief, remembering the deceased functions as a trigger for the traumatic event. In this case, not only is the healthy grief process interrupted but also the children suffer from symptoms of trauma, triggered by the memory.

According to Mannarino and Cohen (2011), children with traumatic grief are "unable to complete the tasks of reconciliation because remembering the loved ones typically serves as a trauma reminder, with the subsequent development of trauma symptoms" (p. 24). Even when children remember happy memories of loved ones, this triggers the traumatic event in which the loved ones died, and subsequently, they suffer from traumatic symptoms, and they may even try to avoid remembering the loved ones (Cohen & Mannarino, 2010). Pynoos et al. (1998) state that children's subjective experience during a disaster, as well as appraisals of their own actions and those of their parents, influences the course of children's post-traumatic distress and recovery. Because they are still in the process of developing, children may not able to accurately appraise the situation or integrate cognitive knowledge with their emotions. For example, knowing they could not have predicted an earthquake, children tend to think, "If only I had not asked to go to back inside the house, my father would not have been buried under the debris when the earthquake hit." Thus, it is common for children to blame themselves for the death of their loved ones, and in turn, they experience intense feelings of shame and guilt, in addition to trauma symptoms (Jones et al., 2009).

Interventions for Traumatic Grief in Children

Traumatic grief cognitive behavioral therapy (TG-CBT) One of the most studied and empirically supported intervention methods to treat childhood traumatic grief is traumatic grief cognitive behavioral therapy (TG-CBT). This method is practiced in a therapeutic setting and helps to challenge and address children's distorted notions that they are responsible for the deaths of their loved ones. The intent of this intervention is to reduce the traumatic symptoms, such as recurrent intrusive thoughts, sleep disturbance, irritability, and hypervigilance, and assist children in getting "unstuck" from traumatic memories of the loved ones. Inviting caregivers into the treatment is crucial, as they would benefit from psychoeducation about what children are experiencing at their particular developmental stages and learn how to manage challenging behaviors that are associated with trauma. During the therapy sessions, children can discuss healthy coping strategies with the therapist, and parents can assist children in utilizing these coping strategies or model healthy coping at home. As children continue to appraise caregivers' behaviors and emotions, developing plans to cope effectively with painful memories can help both of them. It is well-documented that children do better when their caregivers' mental health improves (La Greca & Prinstein, 2012; Overstreet et al., 2011). To reiterate, traumatic grief is not a typical or expected reaction (Edgar-Bailey & Kress, 2010). Further, over-pathologizing or misdiagnosing, when children are exhibiting acute stress or going through a normal grieving process, could lead to delay in receiving appropriate intervention (Currier, Holland, & Neimeyer, 2007).

Helping Children with Grief

Once the trauma-oriented aspect of treatment is completed and children are no longer triggered by memories of the loved ones' deaths, they can begin the process of mourning. In other words, treatment does not end simply by addressing traumatic reactions and feelings associated with the trauma. As stated earlier, grief is a natural and healthy process that involves physical, emotional, cognitive, and spiritual functions (Andrews & Marotta, 2005). The goal is for integration or reconciliation, which includes reestablishing one's sense of self without the deceased, finding meaning in the death, and engaging with other caregivers (Wolfelt, 1996; Worden, 1996). Treatment of children requires different skills and strategies, based on their age; adolescents may have different needs than younger children. Yet, the following recommendations apply to all children and adolescents, regardless of their age (Lowenstein, 2006)

Be aware of the impact of your personal experience This is especially true if the clinician is also impacted by the same disaster. It is common for local providers to play the multiple roles of being a survivor and a provider. Seyle, Widyatmoko, and Silver stated, "In small communities, the teachers and staff from the schools are likely to be living in the area and hence likely to be directly exposed to the same dangers as others in the area" (2013, p. 390). It is extremely important for clinicians to be aware of their own health status and triggers for trauma.

Be present and authentic While it is important for children to receive age-appropriate information to process their experiences, it is equally important for them to be heard and validated. Therefore, clinicians can listen more and talk less, allowing children to express a full range of emotions rather than having the clinician interpret their emotions. Again, if the clinician experienced the same trauma, this may become challenging, because the children's stories might trigger one's own experience, causing the clinician to become emotionally distracted. Grieving children require the full presence, physically and emotionally, of someone who cares and listens to them.

Avoid trying to "cheer up" children As stated earlier, the children may have been exposed to multiple traumatic events as well as losses and changes in their lives. Children also have a tendency to hide their emotions, such as sadness, so as not to worry adults around them. Children benefit greatly by being given permission to be sad, angry, or any other feelings they might consider "negative," as well as knowing they do not have to have positive attitudes all the time.

Involvement in meaningful rituals Children respond well to routine and rituals, as they provide structure, predictability, and familiarity. Connecting to religious beliefs can be helpful, as can be community-based or personal or familial rituals. At times, after a natural disaster, practicing an existing local custom or religious practice can be challenging. For example, when a body was not found for days, the family was prevented from following the religious guidelines for a proper burial. Involving children, if age appropriate, in discussing how best to respectfully send off the loved ones, or inviting the religious leader for guidance, can be a positive activity.

Ongoing connection with person(s) who died The sudden onset of the death of the loved ones can be hard for children, especially if the deceased are the caregivers who would have been there for the children. It is important to communicate with children that they are allowed to "maintain connections" with loved ones even after the death, to reduce their sense of abandonment (Stein et al., 2018). It also gives time for children to reestablish a sense of self in the new world that does not have the physical presence of their loved ones. For children, a concrete prop, such as a disconnected telephone, may be helpful to allow "communication" with loved ones who have passed on. The "wind phone" provides an excellent example and will be discussed more fully in the community-based programs section later on.

Fun and play Encourage children to play and have fun. Children tend to be more reactive and changeable in their affect expression than adults (Mannarino & Cohen, 2011). They may need permission that it is okay to have fun and laugh while simultaneously mourning. Play is an important learning tool and process for children. Play can be incorporated in individual therapy sessions, such as using play therapy techniques, or in a group setting, when multiple children of the same situation attend a support group. It could also be in the form of creative arts therapy (Edgar-Bailey & Kress, 2010).

Keep children in their routine Following a disaster, children are likely to experience a sense of danger, sometimes referred to as one's "protective shield" (Pynoos, Steinberg, & Weaith, 1995). This can lead to a variety of negative impacts for a child and their family. Keeping children in their routine (e.g., waking up at a regular time, engaging in their typical morning/getting ready routine) can help rebuild a sense of safety following a traumatic event. It allows children to anticipate what is going to happen next. Rebuilding a sense of safety promotes a sense of calmness, which has been found to be a protective factor for those recovering from a disaster and ultimately reducing risk of developing PTSD post-disaster (McNally, Bryant, & Ehlers, 2003; Shalev & Freedman, 2005).

Supporting caregivers As stated earlier, often, caregivers are also the survivors of the disaster and sustained the same loss as the child. If a child lost a father, it means the mother lost her spouse (or a former partner). The surviving adults have numerous and frequently tedious responsibilities to take care, such as filling out paperwork for insurance coverage, funeral arrangements, and everyday tasks. In order for caregivers to be present and authentic to their children during their grief process, it is important for caregivers to acknowledge their own losses and take time for themselves to grieve and exercise self-care. This includes ensuring adequate sleep, exercising, and eating healthy meals. Developing supportive rituals or engaging in support groups may be another way for adult caregivers to cope with the loss while modeling for their children.

Traumatic Grief in Adults

Similar to children, adults also experience traumatic grief post-disaster. Traumatic grief is different from post-traumatic stress disorder (PTSD). With PTSD, a person is often anxious and fearful that the traumatic event that caused the loss will occur again. In traumatic grief, anxiety results because the person is searching and yearning for their loved one. It is common for adults experiencing traumatic grief to go on to develop PTSD and/or depression (Norris, Friedman, & Watson, 2002; O'Connor, 2010).

Traumatic grief has been observed following non-bereavement loss related to disasters (Shear et al., 2011; Shear & Delaney, 2015). There is evidence for elevated risk of traumatic grief among women (Lobb et al., 2010), persons with lower levels of education (Lobb et al., 2010), and personality traits including insecure attachment and neuroticism (Meij et al., 2007). The nature of the relationship with the deceased is also important: the death of a partner or child is associated with traumatic grief (Lobb et al., 2010) as well as losing a person with whom one had a close and supportive relationship increases the risk (Bonanno et al., 2002; Prigerson, Maciejewski, & Rosenheck, 2000).

Sometimes also referred to as separation trauma, prolonged-acute grief, or complicated grief, traumatic grief has been rarely assessed in epidemiological studies of disasters. In Shear and colleagues' study of the community following Hurricane Katrina, a brief screen for grief was explored (Shear et al., 2011). Of 3088 residential respondents, 26.1% had significant loss with possible complicated grief and 7.0% had moderate-severe complicated grief. Death of loved one was associated with the highest conditional probability of moderate-severe complicated grief (18.5%) but accounted for only 16.5% of moderate-severe complicated grief due to its comparatively low prevalence. Most moderate-severe complicated grief was due to tangible (52.9%) or interpersonal (24.0%) losses. Significant predictors of complicated grief were mostly unique to either bereavement (e.g., social support) or other losses (e.g., pre-hurricane history of psychopathology, social competence).

Interventions for Traumatic Grief in Adults

For adults with traumatic grief, acute grief symptoms (e.g., strong feelings of longing-yearning, preoccupation with thoughts-memories of the deceased, withdrawal, loss of interest) persist longer (Ghaffari-Nejad, Ahmadi-Mousavi, Gandomkar, & Reihani-Kermani, 2007; Johannesson et al., 2009; Kristensen, Weisæth, & Heir, 2010). Research has documented that traumatic grief is distinguishable from major depression and post-traumatic stress disorder (Horowitz et al., 1997; Prigerson et al., 1995, 1996; Shear, Jackson, Essock, Donahue, & Felton, 2006; Simon et al., 2007) and is impairing to one's daily activities (Prigerson et al., 1995, 1996; Shear et al., 2006).

For older populations, they often have experienced multiple losses, such as jobs, health, independence, social roles, familiar living surroundings, and loved ones earlier along in their lives, thereby building resilience in one's ability to cope with disaster (Norris & Murrell, 1988). Although loss is often expected during this age, unexpected losses of a child and/or partner during a disaster or crisis event can be detrimental. Loss of a spouse or partner might result in feeling more dependent on others. Often the elderly adult lacks the needed social support post-disaster (Kaniasty & Norris, 1995).

Psychological Interventions

One of the best treatments for traumatic grief is a targeted psychotherapy (Wittouck, Van Autreve, De Jaegere, Portzky, & van Heeringen, 2011). Complicated grief therapy emphasizes loss processing and restoration of life without the deceased, using a range of techniques including cognitive behavioral therapy, interpersonal therapy, and motivational interviewing. Teaching about what grief is and a psychological model of how it develops, termed psychoeducation, is often also included (Barbosa, Sá, & Carlos Rocha, 2014; Papa, Sewell, Garrison-Diehn, & Rummel, 2013; Wagner, Knaevelsrud, & Maercker, 2006).

Growing evidence supports interventions that include repeatedly telling the story of the death is helpful as well (Barbosa et al., 2014; Papa et al., 2013; Wagner et al., 2006). One study had participants underwent 10 weeks of group therapy and 4 weeks of individual cognitive behavioral therapy for complicated grief. This intervention included psychoeducation, discussing positive and negative memories of the deceased, telling the story of the death, addressing errors in thoughts—cognitive restructuring, a communication with the deceased exercise—and goal and pleasant event setting (Wagner et al., 2006), which also found beneficial effects. While this intervention is appropriate for adults experiencing traumatic grief, it may not be appropriate for those that suffer from PTSD symptoms, as talking about death or loved ones could trigger the traumatic symptoms.

Community or peer-based bereavement support groups enhance social support, which may be particularly helpful for older adults experiencing traumatic grief. Such groups have demonstrated being a very helpful intervention for bereavement-related distress. Although the content of such groups may not be standardized and efficacy data for complicated grief are lacking, bereaved patients should be connected to available, local support groups. Further, there are still relatively few controlled studies examining psychological treatments for traumatic grief. Recommended psychological therapies, tested in at least two independent and controlled studies, include "complicated grief treatment" (encompassing elements of exposure, cognitive restructuring, and interpersonal therapy; Supiano & Luptak, 2014; Shear et al., 2016; Shear, Frank, Houck, & Reynolds, 2005) cognitive behavioral therapy (combining exposure and cognitive interventions; Boelen, de Keijser, van den Hout, & van den Bout, 2007; Bryant et al., 2014) and therapist-assisted online cognitive behavioral therapy (encompassing exposure, cognitive interventions, and behavioral activation applied using writing assignments; Wagner et al., 2006; Eisma et al., 2015).

Promising interventions tested in single controlled studies include behavioral activation (Papa et al., 2013), cognitive narrative therapy (Barbosa et al., 2014) and integrated cognitive behavioral therapy (Rosner, Pfoh, Kotoučová, & Hagl, 2014). Interpretative and supportive time-limited group therapies (Piper, McCallum, Joyce, Rosie, & Ogrodniczuk, 2001; Piper, Ogrodniczuk, Joyce, Weideman, & Rosie, 2007) are promising interventions not compared with alternative treatment but yielding moderate effects on grief reduction.

Psychological First Aid

Psychological First Aid (PFA) is a disaster response intervention many organizations utilize (Speier, Osofsky, & Osofsky, 2009). PFA is taught to all disaster responders as part of disaster response. While its testing with traumatic grief is limited (Boelen & Smid, 2017), PFA's ultimate goal is to mitigate stress and to bolster adaptive coping by offering direct assistance, support, and referral. Adults, and especially older adults, have been found to be particularly responsive to natural spontaneous discussions amongst peers, reporting accompanying feelings of normalcy, validation, and calm (Langan & Palmer, 2012). More broadly, social support has been found to be a critical resource in coping with trauma (Norris et al., 2002), especially for older persons given that they are at heightened risk for social isolation and are more likely than other age groups to live alone. Reminders of resiliency having survived other traumatic and taxing events (e.g., the Vietnam War, the Holocaust; 9/11) may also be beneficial (Lewis, 2003).

Adults who are identified post-disaster as needing more assistance with psychological recovery are frequently offered crisis counseling. Crisis counselors can play a key role in helping survivors beyond initial disaster response (Norris, Hamblen, & Rosen, 2009). If the problems appear to be more closely related to lack of coping skills with increasing or persistent symptoms of depression, anxiety, and substance abuse, then some direct training in developing such skills would be offered during crisis counseling. In cases where the need for crisis counseling and additional treatment is identified from PFA by disaster responders, it is important to have such clients referred to and assessed by psychosocial providers with appropriate education and credentials.

While programs such as PFA and crisis counseling are designed to mitigate the immediate impact of trauma-related grief, they are just the first and second states of effective psychological care and provide a critical foundation on which to lay subsequent treatments. If these two interventions do not provide sufficient relief, formal psychological treatment is offered. However, professionals should bear in mind that while psychological first aid and crisis counseling programs are evidence-informed, evidence-based interventions have been slower to develop (see Cherry, Calamia, & Elliott, Chap. 8, this volume, for a related discussion). Therefore, any psychological intervention for persons following disasters should be delivered with caution and, most importantly, individualized to the unique needs of the patient (Gibson, Walsh, & Brown, 2018).

Studies on pharmacotherapy for traumatic grief and depression related to bereavement provide evidence for the efficacy and safety of antidepressant medications (Boelen & Smid, 2017). In most studies, pharmacotherapy was used to target depression rather than traumatic grief and, thus, saw reduction of depressive symptoms (Shear et al., 2016). In some situations, medication should be offered in addition to (rather than instead of) psychosocial interventions and attempts to optimize social support (Boelen & Smid, 2017).

Community-Based Programs to Support Grieving Children, Adults, and Families

In addition to helping children and adults individually, psychosocial providers can look into community-based organizations for additional support and resources. When a disaster hits, local community-based organizations may extend their services to those whom they might not otherwise offer services. Also, while some agencies may not be able to take new clients at the time of chaos and recovery, new organizations may start to meet the needs of the local residents. The followings are some organizations and activities made available to Japanese residents in Tohoku after the Great East Japan Earthquakes in 2011. These services are helpful for both adults and children.

Rainbow House

Rainbow House was established by the philanthropic foundation Ashinaga after the earthquake in Kobe, Japan, in 1995. The house is dedicated to children who are mourning the loss of loved ones, especially orphaned children. Their services include group activities for children and support for caregivers, among many others. Please refer to their website to see the full range of service they offer: https://en.ashinaga.org/what-we-do/japan/rainbow-houses/.

After the Great East Japan Earthquakes in 2011, a Rainbow House was built in three locations in the Tohoku area to provide a safe place for children to grieve, while also providing support to caretakers. The director of the establishment explained that many young adults from Kobe came to the area to assist the mission of the organization who were survivors of the Kobe earthquakes in 1995 and had utilized the services at the Rainbow House in Kobe. When parents in the Tohoku area learned the background of the volunteers, they shared how relieved and comforted they were by the presence of the volunteers who had grown up as caring and thoughtful adults. They also shared they were especially encouraged by the young adults who would come to the disaster area to give back, despite the loss of their own parents. This gave them hope that their own children could get through this challenging time and become healthy and well-adjusted adults, as they saw the volunteers to be. These real-life role models perhaps served as the best support for parents who had anxieties around raising their children on their own.

Because children's recovery and healing largely align with parental mental health (La Greca & Prinstein, 2012; Overstreet et al., 2011), by supporting parents and reassuring them that their children may be able to get through challenges in their lives, both parents and children can regain confidence and move toward reconciliation.

Wind Phone

In 2010, Itaru Sasaki, a resident of Otsuchi, Japan, built a glass-paneled phone booth in his garden after his cousin passed away (Atlas Obscura, 2014). Inside the booth, there was a disconnected old phone he would use to communicate with his cousin. When the Great East Japan Earthquakes hit his hometown of Otsuchi a year later, with an estimated loss of 10% of the residents due to the tsunami, he shared his mourning ritual with the residents of the town. Known as the "wind phone," or "kaze no denwa," reportedly over 10,000 visitors, including children, utilized the devise. They would dial the phone number that had belonged to their loved ones to ask where they had gone, to report their daily activities, and to communicate their feelings, including sadness and confusion. It was clear that these were one-way conversations, but people—adults and children alike—felt the need to keep connections with their loved ones. As time went on, the content of the conversations changed, indicating that most people were able to move on with their lives (Atlas Obscura, 2014; BBC, 2019).

Shared Tragedies of Disastrous Events

There are other examples of how human relationships and connections can serve as important support for grieving families across various events. For instance, in 2001, the United States experienced a terrorist attack in New York City. Later in 2011, Japan saw the Fukushima nuclear plant explosion. Families from the 9/11 Foundation visited to the Tohoku area, and vice versa, with survivors of the Fukushima nuclear plant explosion to the 9/11 museum in New York City Survivors and families of 9/11 victims (9/11 Tribute Museum, 2017a). These survivors of previous disasters shared their stories of trauma and resilience. This provided meaningful experiences to the survivors of 9/11, as well as inspiring the residents of Tohoku, who might never be able to return back to their ancestral home (9/11 Tribute Museum, 2017b). While a natural disaster destroys communities and lives, by reaching out to the local residents, new connections and human relationships can be achieved through shared experiences, establishing a new community across the world. Through the process of new community building, healing can begin to take place.

Conclusion and Future Directions

Natural disasters have been happening more frequently and with increased severity worldwide. Psychosocial providers are required to have a wide range of knowledge and skills to be able to provide the needed assistance to those who are affected, at different stages of recovery. While rare, traumatic grief has been seen in children and adults after natural disasters. As this is a very complicated condition, those who are suffering from the condition need to be evaluated and treated by providers who are trained both in trauma treatment and grief counseling. Trauma treatment should take precedent to reduce the likelihood of affected individuals reexperiencing traumatic memories. The treatment should not end before the person can go through the grief process and achieve cognitive, emotional, and spiritual integration at the end of the treatment.

As mentioned earlier, there are certain populations, or communities, that are underserved or left out from disaster support services. Women who are not allowed to step outside their homes without male companions due to their religious beliefs may be isolated when their husbands or male relatives die, for example. Material support may not be evenly distributed, particularly in the area known to have a high rate of community violence. It is important for psychosocial providers to make outreach to communities or populations who may be underserved. Further, additional research on traumatic grief post-disaster across the lifespan, as well as appropriate interventions, is warranted. However, this is especially true for older adult populations where PTSD, acute stress disorder, and traumatic grief are gravely understudied (Gibson et al., 2018).

When natural disasters alter and destroy communities, communities often experience shared grief. Clinical professions can provide critical support for persons affected by traumatic grief post-disaster. In addition to assisting survivors directly, clinicians and health providers can partner with local providers (i.e., school administrators and educators, senior center staff) to educate them on what signs of traumatic grief to look for and how to make referrals to mental health services. Clinicians can also develop or bolster community resources for transition of care.

References

9/11 Tribute Museum. (2017a, August 7). *Goldman Sachs hosts as volunteers connect with students from Fukushima*. Retrieved from https://911tributemuseum.org/tag/tohoku/

9/11 Tribute Museum. (2017b). *9/11 community's connections to Japan continue to grow*. Retrieved from https://911tributemuseum.org/911-communitys-connections-to-japan-continue-to-grow/

American Psychiatric Association. (2013). *Diagnostic and statistical manual of mental disorders* (5th ed.). Washington, DC: Author.

Andrews, C. R., & Marotta, S. A. (2005). Spirituality and coping among grieving children: A preliminary study. *Counseling & Values, 50*(1), 38–50. https://doi.org/10.1002/j.2161-007X.2005.tb00039.x

Ashinaga. (2017). *Rainbow houses*. Retrieved from https://en.ashinaga.org/what-we-do/japan/rainbow-houses/

Atlas Obscura. (2014). *Wind telephone*. Retrieved from https://www.atlasobscura.com/places/wind-telephone

Barbosa, V., Sá, M., & Carlos Rocha, J. (2014). Randomised controlled trial of a cognitive narrative intervention for complicated grief in widowhood. *Aging & Mental Health, 18*(3), 354–362. https://doi.org/10.1080/13607863.2013.833164

BBC. (2019, June 9). *Japan's phone to the dead*. Retrieved from https://www.bcc.co.uk/programmes/p07cj0h3

Beaton, R., Bridges, E., Salazar, M. K., Oberle, M. W., Stergachis, A., Thompson, J., et al. (2008). Ecological model of disaster management. *Journal of American Association of Occupational Health Nurses (AAOHN), 56*(11), 471–478. https://www.ncbi.nlm.nih.gov/pubmed/19051571

Boelen, P. A., de Keijser, J., van den Hout, M. A., & van den Bout, J. (2007). Treatment of complicated grief: A comparison between cognitive-behavioral therapy and supportive counseling. *Journal of Consulting and Clinical Psychology, 75*(2), 277–284. https://doi.org/10.1037/0022-006X.75.2.277

Boelen, P. A., & Smid, G. E. (2017). Disturbed grief: Prolonged grief disorder and persistent complex bereavement disorder. *The BMJ, 357*, j2016. https://doi.org/10.1136/bmj.j2016

Bonanno, G. A., Wortman, C. B., Lehman, D. R., Tweed, R. G., Haring, M., Sonnega, J., et al. (2002). Resilience to loss and chronic grief: A prospective study from pre-loss to 18-months post-loss. *Journal of Personality and Social Psychology, 83*(5), 1150–1164. https://doi.org/10.1037/0022-3514.83.5.1150

Bryant, R. A., Kenny, L., Joscelyne, A., Rawson, N., Maccallum, F., Cahill, C., et al. (2014). Treating prolonged grief disorder: A randomized clinical trial. *JAMA Psychiatry, 71*(12), 1332–1339. https://doi.org/10.1001/jamapsychiatry.2014.1600

Cherry, K. E., Allen, P. D., & Galea, S. (2010). Older adults and natural disasters: Lessons learned from Hurricanes Katrina and Rita. In P. Dass-Brailsford (Ed.), *Crisis and disaster counseling: Lessons learned from Hurricane Katrina and other disasters* (pp. 115–130). Thousand Oaks, CA: Sage.

Cohen, J. A., & Mannarino, A. P. (2004). Treatment of Childhood Traumatic Grief. *Journal of Clinical Child and Adolescent Psychology, 33*(4), 819–831. https://doi.org/10.1207/s15374424jccp3304_17

Cohen, J. A., & Mannarino, A. P. (2010). Childhood traumatic grief. In M. D. Dulcan (Ed.), *Textbook of child and adolescent psychiatry* (pp. 509–516). Washington, DC: American Psychiatric Publishing.

Currier, J. M., Holland, J. M., & Neimeyer, R. A. (2007). The effectiveness of bereavement interventions with children: A meta-analytic review of controlled outcome research. *Journal of Clinical Child and Adolescent Psychology, 36*, 253–259.

Dembert, M. L., & Simmer, E. (2000). When trauma affects a community: Group interventions and support after a disaster. In R. H. Klein & V. L. Schermer (Eds.), *Group psychotherapy for psychological trauma* (pp. 239–264). New York: The Guilford Press.

Edgar-Bailey, M., & Kress, V. E. (2010). Resolving child and adolescent traumatic grief: Creative techniques and interventions. *Journal of Creativity in Mental Health, 5*(2), 158–176. https://doi.org/10.1080/15401383.2010.485090

Eisma, M. C., Boelen, P. A., van den Bout, J., Stroebe, W., Schut, H. A., Lancee, J., et al. (2015). Internet-based exposure and behavioral activation for complicated grief and rumination: A randomized controlled trial. *Behavior Therapy, 46*(6), 729–748. https://doi.org/10.1016/j.beth.2015.05.007

Ghaffari-Nejad, A., Ahmadi-Mousavi, M., Gandomkar, M., & Reihani-Kermani, H. (2007). The prevalence of complicated grief among Bam earthquake survivors in Iran. *Archives of Iran Medicine, 10*(4), 525–528. doi: 07104/AIM.0019.

Gibson, A., Kusmaul, N., Leedahl, S., & Rizzo, V. M. (2020). Create social responses to a changing environment. In S. Sanders, S. Kolomer, C. W. Spellman, & V. M. Rizzo (Eds.), *Gerontological social work and the grand challenges: Focusing on policy and practice.* (pp. 93-110). New York: Springer.

Gibson, A., Walsh, J., & Brown, L. M. (2018). Disaster mental health services review of care for older persons following disasters. *Disaster Medicine and Public Health Preparedness, 12*(3), 366–372.

Horowitz, M. J., Siegel, B., Holen, A., Bonanno, G. A., Milbrath, C., & Stinson, C. H. (1997). Diagnostic criteria for complicated grief disorder. *American Journal of Psychiatry, 154*(7), 904–910. https://doi.org/10.1176/ajp.154.7.904

International Federation of Red Cross and Red Crescent. (2016). *Types of disasters: Definition of hazard.* Retrieved from https://www.ifrc.org/en/what-we-do/disaster-management/about-disasters/definition-of-hazard/

Johannesson, K. B., Lundin, T., Hultman, C. M., Lindam, A., Dyster-Aas, J., Arnberg, F., et al. (2009). The effect of traumatic bereavement on tsunami-exposed survivors. *Journal of Trauma Stress, 22*(6), 497–504. https://doi.org/10.1002/jts.20467

Jones, R. T., Burns, K. D., Immel, C. S., Moore, R. M., Schwartz-Goel, K., & Culpepper, B. (2009). The impact of Hurricane Katina on children and adolescents: Conceptual and methodological implications for assessment and intervention. In K. E. Cherry (Ed.), *Lifespan perspectives on natural disasters* (pp. 65–94). New York: Springer.

Kaniasty, K., & Norris, F. H. (1995). Mobilization and deterioration of social support following natural disasters. *Current Directions in Psychological Science, 4*(3), 94–98. https://doi.org/10.1111/1467-8721.ep10772341

Kozu, S., & Homma, H. (2014). Lessons learned from the Great East Japan Earthquake: The need for disaster preparedness in the area of disaster mental health for children. *Journal of Emergency Management, 12*(6), 431–439. https://doi.org/10.5055/jem.2014.0206

Kristensen, P., Weisæth, L., & Heir, T. (2010). Predictors of complicated grief after a natural disaster: A population study two years after the 2004 South-East Asian Tsunami. *Death Studies, 34*(2), 137–150. https://doi.org/10.1080/07481180903492455

Kukihara, H., Yamawaki, N., Uchiyama, K., Arai, S., & Horikawa, E. (2014). Trauma, depression, and resilience of earthquake/tsunami/nuclear disaster survivors of Hirono, Fukushima, Japan. *Psychiatry & Clinical Neurosciences, 68*(7), 524–533. https://doi.org/10.1111/pcn.12159

La Greca, A. M., & Prinstein, M. J. (2012). Hurricanes and earthquakes. In A. M. La Greca, W. K. Silverman, E. M. Vernberg, & M. C. Roberts (Eds.), *Helping children cope with disasters and terrorism* (pp. 107–138). Washington, DC: American Psychological Association.

Langan, J. C., & Palmer, J. L. (2012). Listening to and learning from older adult Hurricane Katrina survivors. *Public Health Nursing, 29*, 126–135.

Lewis, M. (2003). *The frail and the hardy seniors of 9/11: The needs and contributions of older Americans*. The Johns Hopkins University website. http://www.upmcbiosecurity.org

Lobb, E. A., Kristjanson, L. J., Aoun, S. M., Monterosso, L., Halkett, G. K., & Davies, A. (2010). Predictors of complicated grief: a systematic review of empirical studies. *Death Studies, 34*, 673–698. https://doi.org/10.1080/07481187.2010.496686. pmid:24482845.

Lowenstein, L. (2006). *Creative interventions for bereaved children*. Toronto, ON: Champion Press.

Maeda, M., & Oe, M. (2015). The Great East Japan Earthquake: Tsunami and nuclear disaster. In K. E. Cherry (Ed.), *Traumatic stress and long-term recovery: Coping with disasters and other negative life events* (pp. 71–90). New York: Springer.

Mannarino, A. P., & Cohen, J. A. (2011). Traumatic loss in children and adolescents. *Journal of Child & Adolescent Trauma, 4*(1), 22–33. https://doi.org/10.1080/19361521.2011.545048

McNally, R. J., Bryant, R. A., & Ehlers, A. (2003). Does early psychological intervention promote recovery from posttraumatic stress? *Psychological Science in the Public Interest, 42*(2), 45–79.

Meij, L. W.-D., Stroebe, M., Schut, H., Stroebe, W., van den Bout, J., van der Heijden, P., et al. (2007). Neuroticism and attachment insecurity as predictors of bereavement outcome. *Journal of Research in Personality, 41*(2), 498–505. https://doi.org/10.1016/j.jrp.2006.06.001

Mitchell, T. L., Griffin, K., Stewart, S. H., & Loba, P. (2004). 'We will never ever forget.': The Swissair flight 111 disaster and its impact on volunteers and communities. *Journal of Health Psychology, 9*(2), 245–262. https://doi.org/10.1177/1359105304040890

Modica, M., & Zoboli, R. (2016). Vulnerability, resilience, hazard, risk, damage, and loss: A socio-ecological framework for natural disaster analysis. *Web Ecology, 16*, 59–62. https://doi.org/10.5194/we-16-59-2016

Münchener Rückversicherungs-Gesellschaft. (2015). *Loss events worldwide 1980-2014*. Retrieved from https://www.munichre.com/site/touch-naturalhazards/get/documents_E2080665585/mr/assetpool.shared/Documents/5_Touch/_NatCatService/Focus_analyses/1980-2014-Loss-events-worldwide.pdf

Neria, Y., & Litz, B. (2004). Bereavement by traumatic means: The complex synergy of trauma and grief. *Journal of Loss & Trauma, 9*(1), 73–87. https://doi.org/10.1080/15325020490255322

Norris, F. H., Friedman, M. J., & Watson, P. J. (2002). 60,000 disaster victims speak: Part II. Summary and implications of the disaster mental health research. *Psychiatry, 65*(3), 240–260. https://www.ncbi.nlm.nih.gov/pubmed/12405080.

Norris, F. H., Hamblen, J. L., & Rosen, C. S. (2009). Service characteristics and counseling outcomes: Lessons from a cross-site evaluation of crisis counseling after Hurricanes Katrina, Rita and Wilma. *Administration and Policy in Mental Health and Mental Health Services Research, 36*, 176–185.

Norris, F. H., & Murrell, S. A. (1988). Prior experience as a moderator of disaster impact on anxiety symptoms in older adults. *American Journal of Community Psychology, 16*(5), 665–683. https://doi.org/10.1007/BF00930020

O'Connor, M. (2010). PTSD in older bereaved people. *Aging & Mental Health, 14*(6), 670–678. https://doi.org/10.1080/13607860903311725

Overstreet, S., Salloum, A., Burch, B., & West, J. (2011). Challenges associated with childhood exposure to severe natural disasters: Research review and clinical implications. *Journal of Child & Adolescent Trauma, 4*(1), 52–68. https://doi.org/10.1080/19361521.2011.545103

Papa, A., Sewell, M. T., Garrison-Diehn, C., & Rummel, C. (2013). A randomized open trial assessing the feasibility of behavioral activation for pathological grief responding. *Behavior Therapy, 44*(4), 639–650. https://doi.org/10.1016/j.beth.2013.04.009

Piper, W. E., McCallum, M., Joyce, A. S., Rosie, J. S., & Ogrodniczuk, J. S. (2001). Patient personality and time-limited group psychotherapy for complicated grief. *International Journal of Group Psychotherapy, 51*(4), 525–552. https://doi.org/10.1521/ijgp.51.4.525.51307

Piper, W. E., Ogrodniczuk, J. S., Joyce, A. S., Weideman, R., & Rosie, J. S. (2007). Group composition and group therapy for complicated grief. *Journal of Consulting and Clinical Psychology, 75*(1), 116–125. https://doi.org/10.1037/0022-006X.75.1.116

Prigerson, H., Maciejewski, P., & Rosenheck, R. (2000). The interactive effects of marital harmony and widowhood on health, health service utilization and costs. *Gerontologist, 40*(3), 349–357. https://doi.org/10.1093/geront/40.3.349

Prigerson, H. G., Bierhals, A. J., Kasl, S. V., Reynolds, C. F., Shear, M. K., Newsom, J. T., et al. (1996). Complicated grief as a disorder distinct from bereavement-related depression and anxiety: A replication study. *American Journal of Psychiatry, 153*(11), 1484–1486. https://doi.org/10.1176/ajp.153.11.1484

Prigerson, H. G., Frank, E., Kasl, S. V., Reynolds, C. F., Anderson, B., Zubenko, G. S., et al. (1995). Complicated grief and bereavement-related depression as distinct disorders: Preliminary empirical validation in elderly bereaved spouses. *American Journal of Psychiatry, 152*(1), 22–30. https://www.ncbi.nlm.nih.gov/m/pubmed/7802116/.

Pynoos, R. S., Goenjian, A. K., & Steinberg, A. M. (1998). Children and disasters: A developmental approach to posttraumatic stress disorder in children and adolescents. *Psychiatry & Clinical Neurosciences, 52*, S129–S138. https://doi.org/10.1046/j.1440-1819.1998.0520s5S129.x

Pynoos, R. S., Steinberg, A. M., & Weaith, R. (1995). A developmental model of childhood traumatic stress. In D. Cicchetti & D. J. Cohen (Eds.), *Developmental psychopathology, vol 2: Risk, disorder, and adaption* (pp. 72–95). Oxford, UK: Wiley.

Rosner, R., Pfoh, G., Kotoučová, M., & Hagl, M. (2014). Efficacy of an outpatient treatment for prolonged grief disorder: A randomized controlled clinical trial. *Journal of Affective Disorders, 167*, 56–63. https://doi.org/10.1016/j.jad.2014.05.035

Seyle, D. C., Widyatmoko, C. S., & Silver, R. C. (2013). Coping with natural disasters in Yogyakarta, Indonesia: A study of elementary school teachers. *School Psychology International, 34*(4), 387–404. https://doi.org/10.1177/0143034312446889

Shalev, A. Y., & Freedman, S. (2005). PTSD following terrorist attacks: A prospective evaluation. *American Journal of Psychiatry, 162*, 1188–1191.

Shear, K., Frank, E., Houck, P. R., & Reynolds, C. F. (2005). Treatment of complicated grief: A randomized controlled trial. *JAMA, 293*(21), 2601–2608. https://doi.org/10.1001/jama.293.21.2601

Shear, K. M., Jackson, C. T., Essock, S. M., Donahue, S. A., & Felton, C. J. (2006). Screening for complicated grief among Project Liberty service recipients 18 months after September 11, 2001. *Psychiatric Services, 57*(9), 1291–1297. https://doi.org/10.1176/ps.2006.57.9.1291

Shear, M. K., & Delaney, S. (2015). On bereavement and grief: A therapeutic approach to healing. In K. E. Cherry (Ed.), *Traumatic stress and long-term recovery: Coping with disasters and other negative life events* (pp. 403–418). New York: Springer.

Shear, M. K., McLaughlin, K. A., Ghesquiere, A., Gruber, M. J., Sampson, N. A., & Kessler, R. C. (2011). Complicated grief associated with hurricane Katrina. *Depression & Anxiety, 28*(8), 648–657. https://doi.org/10.1002/da.20865. pmid:21796740.

Shear, M. K., Reynolds, C. F. I., Simon, N. M., Zisook, S., Wang, Y., Mauro, C., et al. (2016). Optimizing treatment of complicated grief: A randomized clinical trial. *JAMA Psychiatry, 73*(7), 685–694. https://doi.org/10.1001/jamapsychiatry.2016.0892

Shepherd, A., Mitchell, T., Lewis, K., Lenhardt, A., Jones, L., Scott, L., et al. (2013). *The geography of poverty, disasters and climate extremes in 2030. Executive summary.* Retrieved from www.odi.org/poverty-disasters-2030

Simon, N. M., Shear, K. M., Thompson, E. H., Zalta, A. K., Perlman, C., Reynolds, C. F., et al. (2007). The prevalence and correlates of psychiatric comorbidity in individuals with complicated grief. *Comprehensive Psychiatry, 48*(5), 395–399. https://doi.org/10.1016/j.comppsych.2007.05.002

Smith, H., Jaurique, A., & Ryan, D. (2016). The mistreatment of others: Discrimination can undermine university identification, student health, and engagement. *Social Justice Research, 29*(4), 355–374. https://doi.org/10.1007/s11211-016-0274-x

Speier, A. H., Osofsky, J. D., & Osofsky, H. J. (2009). Building a disaster mental health response to a catastrophic event: Louisiana and Hurricane Katrina. In K. E. Cherry (Ed.), *Lifespan perspectives on natural disasters: Coping with Katrina, Rita and Other storms* (pp. 241–260). New York: Springer.

Stein, C. H., Petrowski, C. E., Gonzales, S. M., Mattei, G. M., Hartl Majcher, J., Froemming, M. W., et al. (2018). A matter of life and death: Understanding continuing bonds and posttraumatic growth when young adults experience the loss of a close friend. *Journal of Child and Family Studies, 27*(3), 725–738. https://doi.org/10.1007/s10826-017-0943-x

Supiano, K. P., & Luptak, M. (2014). Complicated grief in older adults: A randomized controlled trial of complicated grief group therapy. *Gerontologist, 54*(1), 840–856. https://doi.org/10.1093/geront/gnt076

United Nations Office for Disaster Risk Reduction. (2019). *Disaster statistics.* Retrieved from https://www.unisdr.org/we/inform/disaster-statistics

Wagner, B., Knaevelsrud, C., & Maercker, A. (2006). Internet-based cognitive-behavioral therapy for complicated grief: A randomized controlled trial. *Death Studies, 30*(5), 429–453. https://doi.org/10.1080/07481180600614385

Wang, J., Lloyd-Evans, B., Giacco, D., Forsyth, R., Nebo, C., Mann, F., et al. (2017). Social isolation in mental health: A conceptual and methodological review. *Social Psychiatry & Psychiatric Epidemiology, 52*(12), 1451–1461.

Wittouck, C., Van Autreve, S., De Jaegere, E., Portzky, G., & van Heeringen, K. (2011). The prevention and treatment of complicated grief: A meta-analysis. *Clinical Psychological Review, 31*(1), 69–78. https://doi.org/10.1016/j.cpr.2010.09.005

Wolfelt, A. D. (1996). *Healing the bereaved child: Grief gardening, growth through grief, and other touchstones for caregivers.* Fort Collins, CO: Companion Press.

Worden, J. W. (1996). *Children and grief: When a parent dies.* New York: Guilford.

Part II
Direct Practice

Chapter 7
Cultural Competence and Disaster Mental Health

Mandana Mostofi and Lisa M. Brown

Introduction

Disaster is a broad term that captures a diverse range of circumstances from natural (e.g., earthquakes and tsunamis) to human-made disasters (e.g., wars, acts of terrorism, technological accidents) (Pourhosseini, Ardalan, & Mehrolhassani, 2015). Although the effects of some disasters can be mitigated or prevented, most people will be affected by a disaster during the course of their life. Disasters have the potential to destroy the physical environment and adversely impact the health of large numbers of people in a very short period of time (Math, Nirmala, Moirangthem, & Kumar, 2015). Because most disasters occur with little advance warning and result in widespread damage, post-event communities require urgent support that often exceeds local resources. Disaster responders and recovery teams have to be quickly mobilized from unaffected regions to provide needed aid and assistance to affected populations (Wells, Springgate, Lizaola, Jones, & Plough, 2013).

In December 2004, a 9.2 Richter scale earthquake, followed by a three-wave tsunami, destroyed many areas in the Aceh province of Indonesia. The Médecins Sans Frontières' (MSF) mental health team undertook the responsibility of assessing the residents and responding to their mental health needs (Eggertson, 2006). The MSF team initially faced multiple challenges that they attributed to the mismatch between the health assessment tools and interventions that were developed in a Western context and the cultural setting where the services were to be delivered. To successfully meet local needs, the MSF team shifted their focus and reallocated their resources by obtaining qualitative data from focus groups and informant interviews to guide their activities (de Jong, Prosser, & Ford, 2005).

This example illustrates the importance of cultural competence in providing disaster mental health services. Disaster response teams need some familiarity with

M. Mostofi · L. M. Brown (✉)
Palo Alto University, Palo Alto, CA, USA
e-mail: lbrown@paloaltou.edu

© Springer Nature Switzerland AG 2021

K. E. Cherry, A. Gibson (eds.), *The Intersection of Trauma and Disaster Behavioral Health*, https://doi.org/10.1007/978-3-030-51525-6_7

the local culture prior to deployment as survivors' interpretation of catastrophic events is shaped by cultural beliefs. For instance, in Indonesia, mental health problems are understood as shameful psychiatric illnesses. Furthermore, residents who completed the assessments indicated that they felt the tsunami was a punishment or a warning for being "immoral." From this experience, the team learned that it was best to avoid using psychiatric terms and making diagnoses. The need for caution when applying Western constructs and terminology highlights the importance of learning about the local cultural and religious beliefs prior to starting work in another country. Culture not only shapes people's understanding of why a disaster occurred, it also influences their preferences in using disaster mental health services. Notably, due to the long-term oppressive nature of political conflict between Aceh province and the central government of Indonesia, the mere presence of Western humanitarian assistance in the region had increased people's level of hopefulness. Many of the local individuals expressed the hope that the government would stay more engaged and that the ongoing conflict would cease simply due to Western involvement and attention in the region.

Disaster Mental Health

Disasters, whether natural or human-made, can be life-threatening, can be extremely disruptive to daily functioning, and cause a number of secondary stressors in people's lives (e.g., loss of property, reduction of social support networks, destruction of community structures) (Goldmann & Galea, 2014). As a result, disaster survivors can experience increased anxiety, depression, or anger, which may, in turn, interfere with adaptive functioning capacity (Math et al., 2015; Weissbecker, 2009). Disaster mental health aims to support survivors in coping with psychological stressors in the aftermath of disasters (SRA International, 2008; Stebnicki, 2017). Disaster mental health services are intended to alleviate additional psychological stressors, support adaptive coping strategies, and restore survivors' abilities to a level of satisfactory functioning (SRA International, 2008; Stebnicki, 2017).

Cultural Competence and Disaster Mental Health

Culture

Culture is defined in various ways depending on the discipline or field of study (Gopalkrishnan, 2018). In general, it refers to shared characteristics of a group of people, such as social values, norms, customs, and beliefs that a group of people follows, as well as forms of expression that they engage in (i.e., traditions, rituals, art, etc.) (Brown, 2008; Department of Health and Human Services (DHHS), 2003; Butler et al., 2016). Culture is multilayered and dynamic and involves a range of group identities that include, but are not limited to, age, gender, sexual orientation, disability status, race, ethnicity, religion, language, and socioeconomic status (Hays, 2016; Sue, 2001). It is also closely connected to people's conceptualization of human life and natural events, health beliefs and behaviors, communication style, and help-seeking behaviors (Hall & Theriot, 2016; Plotkin, 2011). Thus, people's perceptions of a disaster as well as their methods for coping with threat, response, and recovery will differ significantly based on their cultural identities, life experiences, and previous exposure to traumatic events (Mendenhall, 2006).

It is key to remember that disaster mental health is based on studies that examined the mental health consequences of wars during the first half of the twentieth century (Goldmann & Galea, 2014; Murthy & Lakshminarayana, 2006; Raphael & Maguire, 2009). However, it was not until 1974 that mental health services became a part of the US Federal Disaster Relief Fund (Adler, 1943; Morris, 2011; Raphael & Maguire, 2009; Tyhurst, 1951). These studies provided evidence for the notion that experiencing extreme stress could cause enduring psychological consequences. However, some disaster relief interventions may have unintentionally exacerbated the psychological distress of the survivors because they were either not pilot tested or developed to be culturally relevant (Adler, 1943; Tyhurst, 1951). As a result, a new office in the National Institute of Mental Health was formed to specifically address the mental health needs of impacted populations post-disaster (Morris, 2011).

Cross-Cultural Practices: Needs and Challenges

Disaster relief work can potentially bring service providers to any town or country in the world (Gopalkrishnan, 2018). Moreover, because the US population is rapidly becoming more ethnically and culturally diverse, US-based relief workers are likely to encounter people from diverse backgrounds at different stages of acculturation (US Census Bureau, 2018). For example, large numbers of Vietnamese refugees who had settled in Louisiana during the 1970s were ill-equipped for preparing and recovering from Hurricane Katrina (Vu & Van Landingham, 2012). Working with individuals, families, and communities with various cultural identities and backgrounds is now the norm in the field of disaster behavioral health (Bouye, McCleary, & Williams, 2016).

In the past, counseling and disaster mental health services were generic and considered culturally neutral, with little or no attention to implications for working with diverse populations (Gopalkrishnan, 2018). Since the majority of mental health theories, measures, and tools were developed in Western cultures, Eurocentricity and monoculturalism in these practices went unnoticed for many years (Gopalkrishnan, 2018). Though these practices were indeed beneficial in addressing mental distress in some settings, reevaluation of past events have revealed that some practices or approaches were problematic or even harmful when applied without cultural diversity considerations (Schnyder et al., 2016).

To effectively work cross-culturally, individuals and organizations that provide disaster health care must collaborate with local communities (DHHS, 2003; King, Burkle, Walsh, & North, 2015). This requires cultural competency at both individual and institutional levels (Rosen, Greene, Young, & Norris, 2010). Cultural competence is one of the key constructs of a larger cultural framework for practitioners to use when engaging in multicultural contexts (Danso, 2018).

Program development and service provision in the absence of cultural competence can lead to missteps in the process, or worse, such as exacerbating distress of recipients (Brown, 2008). Conversely, cultural competence provides continuity for healing following disasters by supporting the survivors' recovery in the context of their own life experiences and values (Brown, 2008; DHHS, 2003). It enhances service utilization by improving alliance and engagement (Anderson, Bautista, & Hope, 2019). It can also speed the healing process as it promotes resilience by connecting the survivors to their resources and strengthening social bonds (Brown, 2008; DHHS, 2003). Cultural competence is essential in the development and delivery of effective and efficacious disaster behavioral health services.

Though critical, integrating cultural competence into disaster health programs is quite challenging. Major barriers include the dynamic essence of both culture and disasters which demands ongoing assessment and adaptation in design and implementation of disaster health operations (DHHS, 2003). Another major barrier is the unpredictable timing of disasters, which shapes the urgent, high-intensity, and temporary makeup of disaster health programs, while not allowing adequate time for preparation (Dominici, Levy, & Louis, 2005). The two overarching goals of this chapter are to provide guiding principles and resources to support providers in overcoming the challenges just described and to offer suggestions for effectively planning and delivering trauma-informed and culturally competent mental health services. Behavioral health providers can utilize and adapt these guidelines to address the unique needs of culturally diverse individuals and communities.

Trauma-Informed Care in Disaster Behavioral Health

According to Substance Abuse and Health Services Administration (SAMHSA), trauma-informed care (TIC) offers awareness and information about the prevalence, nature, and the pervasive impact of trauma to behavioral health providers (see Kusmaul, Chap. 9, this volume). The TIC model highlights that trauma results from serious disruption, damage, and loss that overwhelms people's normal coping capacities, subsequently resulting in physical and psychological stress reactions. Exposure to trauma places people at increased risk of developing mental illness, substance misuse, and other health problems. This model emphasizes the need for development and utilization of science-informed and trauma-sensitive assessment and interventions strategies across settings.

A core principle of TIC is a collaborative and strength-based approach that acknowledges the resilience of trauma survivors and facilitates healing by promoting a culture of safety, trust, transparency, and empowerment. TIC recognizes that people's response to disaster is based on their personal history, their coping capacity, and the availability of accessible resources which is impacted by cultural, historical, and other contextual issues. Therefore, this model leverages the healing value of authentic cultural connections, emphasizes the importance of cultural competence and culture-responsive services, and addresses historical trauma (SAMHSA, 2014b). For future information on TIC, see Key Ingredients for Successful Trauma-Informed Care Implementation by Menschner and Maul (2016) and SAMHSA's Concept of Trauma and Guidance on Trauma-Informed Approach (SAMHSA, 2014a).

Cultural Competence

Cultural competence has been promoted as one approach to address and reduce mental health disparities (Butler et al., 2016). It goes beyond having textbook knowledge about the particular demographics of a subgroup or the population, being politically correct, or merely working through translation and interpretation issues (Mendenhall, 2006). It requires the ability to be respectful of and responsive to diversity. To do so involves an ongoing process of adapting services to the unique characteristics and current needs of individuals and communities (Sue, Zane, Nagayama Hall, & Berger, 2009).

At its core, cultural competence requires the ability of providers to acknowledge and embrace human diversity and differences (Nassar-McMillan, 2014). It also encompasses humility toward local wisdom of people, knowing and acknowledging that one does not know, and the openness and flexibility to be educated (Brown, 2008; Danso, 2018). This attitude toward diversity allows professionals to get a better grasp of the context through which local people make sense of their experience (Mendenhall, 2006).

At the individual level, cultural competence is comprised of three principals: (1) awareness, (2) knowledge, and (3) skills (Danso, 2018; Yamada, 2015). Cultural awareness emphasizes the need to be aware of one's assumptions, beliefs, attitudes, and biases and how they may differ from those of other cultural groups (Danso, 2018; Yamada, 2015). This includes asking difficult, yet critical, questions about assumptions that are taken as truth by the dominant culture. It also challenges the providers to reflect on their social location, power, and privilege (Brown, 2008; Danso, 2018).

Furthermore, cultural competence involves acquiring adequate knowledge of other cultural groups, including their norms, values, and beliefs as well as the unique historical and sociopolitical factors that have shaped their experiences of power or oppression (Brown, 2008; West-Olatunji, Henesy, & Varney, 2015). Finally, to be culturally competent, one should demonstrate the skills to communicate, assess, plan, intervene, and advocate in a culturally appropriate and relevant manner (Danso, 2018; King et al., 2015; SAMHSA, 2014b; Yamada, 2015).

At the institutional level, cultural competence is often presented as a set of consistent behaviors, attitudes, and policies that enable an organization to work effectively in a multicultural setting (Cross, 1989; SAMHSA, 2014b). It requires a rational and inclusive plan that encompasses policy-making and program design, evaluation, and implementation (DHHS, 2003). A chief obligation at the institutional level includes the commitment to allocate funds and resources to provide ongoing training and culturally competent supervision to direct-care workers (SAMHSA, 2014b).

Design, implementation, monitoring, and evaluation (DIME) is one great example of models that successfully addresses cultural competence at an institutional level. DIME is a four-step model that utilizes evidence-based approaches for programming and combines it with rigorous monitoring and evaluation. This model highlights the essential role of culturally sensitive training and ongoing supervision and outlines some of the specific ways to select, train, and monitor consultant, trainers, supervisors, and direct-care providers to ensure effectiveness and sustainability (Applied Mental Health Research (AMHR) Group, 2013).

The complexity and fluidity of culture in conjunction with existing diversities between and within cultures limits the ability of providers to arrive at a point where it can be said that he or she is fully competent in another culture (Saunders, Haskins, & Vasquez, 2015). Therefore, it is more accurate to conceptualize cultural competence as a dynamic and continuous developmental process (Cross, 1989; Danso, 2018). It is best to describe it as a commitment to engage in an ongoing process of learning about oneself and others and to build upon experiences over time (Danso, 2018).

In order to better understand the complexity of diversity, it is helpful to utilize models such as the ADDRESSING (Age, Disability-development, Disability-acquired, Religion, Ethnicity, Socioeconomic status, Sexual orientation, Indigenous heritage, National origin, and Gender) or the RESPECTFUL (Religion, Economic class, Sexuality, Psychological health, Ethnicity, Chronological, Trauma, Family history, Unique physical characteristics, and Location and Language) model of cultural competence (D'Andrea & Daniels, 2001; Hays, 2016). The ADDRESSING and RESPECTFUL acronyms both depict social locations related to lifespan, disability, religion, ethnicity and race, social class, sexual orientation, and gender (D'Andrea & Daniels, 2001; Hays, 2016). Additionally, the RESPECTFUL acronym acknowledges a unique factor in identity development: trauma and threats to well-being which references prior lifetime trauma that predate the disaster event (D'Andrea & Daniels, 2001). Trauma is highly relevant when discussing disaster mental health and should be considered as it not only influences an individuals' health but also their responsiveness to health interventions (Shimmin, Wittmeier, Lavoie, Wicklund, & Sibley, 2017).

These models consider human identity as an interwoven web that is shaped by the intersection of various social and personal categories to which people belong. They also provide a foundational algorithm for identifying these multiple identities and an explanation for how they are socially constructed, meaning that their development was informed by the historical, social, and interpersonal contexts that people belong to (i.e., social location) (D'Andrea & Daniels, 2001; Hays, 2016). They also allow mental health workers to understand the complexities of cultural influences and form strategic considerations when working cross-culturally. Though incredibly helpful, these lists are not exhaustive and all-inclusive. There are other social locations such as legal status, geographic location, and state of health that impact individuals' identity (Brown, 2008; Shimmin et al., 2017). Readers are encouraged to be sensitive to identifying other social locations that can inform services. Below, the ADDRESSING model is presented in greater detail.

The ADDRESSING Model

To understand how the ADDRESSING model can assist providers when working with disaster-affected individuals, it is useful to discuss each of these social locations within the framework of trauma-informed disaster behavioral health. Because the most vulnerable survivors of disasters are arguably those who are already disadvantaged along a variety of group statuses, it is essential to discuss power and privilege while exploring each social location (Baker & Cormier, 2015). Moreover, people belong to an interwoven mixture of social groups with varying degrees of power, privilege, and influence (i.e., intersectionality) and can move from one category to another throughout their lifetime.

Because the purpose of this chapter is to provide an introduction to the topic of cultural competence in the context of disasters, only the eight variables in Hays' model will be discussed in more depth in the following sections. Indeed, other social locations and identities can be identified in addition to variables highlighted through Hays' ADDRESSING model such as partnership status. Readers are invited to think critically about these domains and consider them in their work with disaster survivors.

Age

As a biological construct, chronological age is the length of time someone has existed (Age, n.d.), which has implications for one's developmental needs and capacities. As a social construct, age informs the cultural context, social norms, expectations, and meanings that are associated with ones' age group (Field & Syrett, 2015). Disasters happen to people from all ages and disrupt the social and interpersonal context and routines in which they function (Brown, 2008; Freeman, Nairn, & Gollop, 2015; Heath, Nickerson, Annandale, Kemple, & Dean, 2009). As a result, disasters can alter the course of development and even undermine the achievement of succeeding developmental milestones (Brown, 2008;Freeman et al., 2015 ; Heath et al., 2009). Although all age groups are vulnerable in unique ways, individuals' needs and their access to resources vary depending on their age (SAMHSA, 2014c).

Moreover, age can be a risk factor for certain kinds of oppression, such as ageism, which may look different across different cultures (Brown, 2008; Löckenhoff et al., 2009; Vauclair, Hanke, Huang, & Abrams, 2017).

Children may lack sufficient psychological and material resources to independently cope in the face of disasters (Freeman et al., 2015; Heath et al., 2009). Children's ability to grasp and cope with disasters depends on their psychological maturity and their caregivers' adjustment following disasters (Brown, 2008; Heath et al., 2009; Juth, Silver, Seyle, Widyatmoko, & Tan, 2015). Unfortunately, disasters also impact caregivers and often compromise their abilities to provide necessary resources and support to their children (Heath et al., 2009; Pfefferbaum & North, 2008). Age awareness in mental health services should be targeted by assessing for children's symptoms and developmental capacities in relation to normative developmental milestones in their cultural context (Brown, 2008).

Older adults (i.e., adults age 65 or older) are another age group that is at great risk during and after disasters (Brown, 2008; Field & Syrett, 2015; SAMHSA, 2014c). Certain aspects of aging may exacerbate mental health vulnerabilities in older adults. For instance, for some people old age may bring loss of functional capacities and increase dependence on others (Brown, 2008; Colón-Emeric, Whitson, Pavon, & Hoenig, 2013; Shih et al., 2018). Moreover, older adults often have limited financial resources and decreasing social support networks that may be further depleted by disasters (Malik et al., 2017; Shih et al., 2018). Age-related cultural competence can be increased by integrating these dynamic considerations into trauma-informed disaster behavioral health.

Understanding age cohort effects is another important component of age identity. Cohort effects are differences caused by experiences and circumstances unique to the generation to which one belongs (Bell & Jones, 2015). People exposed to certain historical events (e.g., previous disasters, war, financial depression) or social forces (e.g., discrimination, bias based on sex or skin color, microaggressions) may have different perceptions of and responses to disasters (Brown, 2008).

Disability

It is difficult to universally define disability because disability is viewed differently across cultures (Brown, 2008; Bunning, Gona, Newton, & Hartley, 2017). The International Classification of Functioning, Disability, and Health (ICF) describes disability as an impairment or a limitation that restricts an individual from participating in certain functions of life (World Health Organization (WHO), 2002). Disability is on a continuum with health and is the product of a dynamic interaction between a person, at the level of body, and their social context (Goering, 2015; Kostanjsek, 2011; WHO, 2002).

Disability can be developmental or acquired (Hays, 2016). Developmental disabilities are those that are congenital in nature, meaning individuals were either born with the disabling condition or displayed signs of disability early in life (Krahn,

Walker, & Correa-De-Araujo, 2015). For these individuals, disability is a long-lasting and core component of their life experience and their identity (Brown, 2008). Disability may also be acquired. Acquired disabilities are those that occur or develop later in life and result in a shift in functioning that requires adaptation (Bogart, 2014).

Cultural competency in work with people with disabilities requires awareness and knowledge regarding locally held beliefs about causes of disability, the oppression of individuals and communities with disabilities (i.e., ableism), and personal and institutional ableist biases (Bunning et al., 2017; Butler et al., 2016; Goering, 2015). Moreover, there should be awareness and sensitivity that like any other minority group, individuals with disability may identify with the disability culture at various degrees (Bogart, 2014; Butler et al., 2016). For example, the hallmark of deaf culture is that sign language is the primary means of communication for literary traditions, art, and history. However, deaf culture extends beyond sign language and includes social beliefs, traditions, and values that are shaped by deafness.

It is also important to highlight the importance of using person-first language (i.e., "person with disability") as more accurate and appropriate by arguing that individuals should not be labeled or defined by their diagnosis (Brown, 2008; Gernsbacher, 2017). On the other hand, some, including members of disability groups, argue that identity-first language ("disabled person") is less stigmatizing and more validating of this intersection of identity as it is phrased similar to other social locations (e.g., Irish person) (Collier, 2012; Gernsbacher, 2017). Furthermore, providers must consider that not all disabilities are visible (e.g., auditory or visual impairment) and ensure to explore this intersection of identity regardless of a person's appearance (Butler et al., 2016). The best course of action is to ask the person what their preference is and not to assume.

In the framework of trauma-informed disaster behavioral health, it is crucial to explore the intersection of disability and disasters. Research suggests that disability status amplifies the challenges and barriers faced during and after disasters (Baker & Cormier, 2015; Stough, Sharp, Resch, Decker, & Wilker, 2016). For instance, disability may limit a person's ability to respond to disaster the way others do and hinder their capacity to complete tasks that they were previously able to accomplish independently (Baker & Cormier, 2015). It is also possible that the disability was acquired as a result of disaster (Stough et al., 2016).

Religion and Spirituality

From a mental health perspective, religion and spirituality, as a social location, address the belief system through which individuals make meaning about the world (Behere, Das, Yadav, & Behere, 2013; Hays, 2016). Even atheists or religiously unaffiliated individuals have a belief system or an existential theory that allows them to assign meaning to their lives (Brown, 2008; Speed, Coleman, & Langston, 2018). Meaning-making, as a process and social practice, directly impacts survivors' narratives of disaster and their strategies for coping with its repercussions

(Brown, 2008; Ha, 2015; Joakim & White, 2015). The presence of meaning-making systems is generally found to serve as a protective factor in adjustment to adverse events (Brown, 2008; Henslee et al., 2015; Park, 2016).

In many instances, exposure to adverse life events, such as disasters, has a transformative effect on the survivors' meaning-making processes (Aten, O'Grady, Milstein, Boan, & Schruba, 2014; Brown, 2008). Shifts in belief system following disaster exposure and recovery are extremely informative regarding survivors' coping trajectory (Aten et al., 2014; Brown, 2008). Positive transformations (i.e., benevolent framing of disaster) have been associated with resilience, while transformations toward decline (i.e., punitive reappraisal of disaster) have been linked to increased distress and risk (Aten et al., 2014; Ha, 2015; Henslee et al., 2015).

Considering religion-related oppression, discrimination, and bias is of utmost importance when working with religious minorities in the context of disasters and trauma. Having one's faith marginalized can constrain the very means through which individuals and communities cope with adversities and can lead to feelings of exclusion, increase distress, and decrease trust of professionals (Marks, Dollahite, & Young, 2018). Culturally competent disaster services demand inquiry about survivors' belief systems (i.e., religious, spiritual, and existential) and should reinforce recovery or consolidation of meaning-making (Aten et al., 2014).

Ethnicity and Race

Race and ethnicity are two separate yet closely related macro-level categories of culture that have long been employed to describe human ancestry (Mersha & Abebe, 2015). Race refers to a selective set of visually obvious phenotypic characteristics such as bone structure and skin, eye, and hair color that have a high degree of variation and often reflect natural selection to specific geographic regions (Brown, 2008; Feliciano, 2015; Relethford, 2009). Ethnicity is defined as shared cultural heritage or national ancestry of a group of people that includes common history, language, and preferences for food and music (Brown, 2008; DHHS, 2003; Ford & Harawa, 2010; Hays, 2008). Although empirical evidence has confirmed that race does not have any biological meaning, it continues to be a commonly used social construct (Brown, 2008; Smedley & Smedley, 2005). Both race and ethnicity are fundamental elements of sociocultural identity and often profoundly impact the ways in which people respond to and cope with disasters (DHHS, 2003).

In fact, ethnicity is a major approach to differentiate cultural identification (Brown, 2008). The role of racial and ethnic cultural identity is so profound that it impacts the emotional meaning that people attribute to events and their expressions of affective distress (Paniagua & Yamada, 2013). For instance, many people from Asian cultures do not associate as strongly with the Western philosophies of mind and body dualism and thus tend to experience and communicate their emotional distress or social discord as physical or somatic symptoms (i.e., somatization) (Bauer, Chen, & Alegría, 2012; Ferrari et al., 2015). However, it must be noted that people may not closely identify with their racial or ethnic ancestry. One possible reason can be stronger identification with a culture that an individual has been raised in rather than their familial and ancestral culture. Therefore, it is important to directly inquire about this social identity without assuming that one necessarily identifies with their phenotype.

The major considerations for ethnocultural competency are tailoring services to linguistic, communication, and cultural norms and values of the local community (Butler et al., 2016). It is perhaps unnecessary to explain why healthcare delivery would suffer in the absence of effective communication. To demonstrate cultural competency and sensitivity, institutions and mental health providers must reach beyond mere translation of material to ensure consideration of the nuances that impact the effectiveness of communication and adherence to services. This includes intercultural communication competency (e.g., culture's propriety or etiquette, verbal and non-verbal differences), modifying cultural variables, and tailoring services to health beliefs and values of individuals (Butler et al., 2016; Delvecchio Good & Hannah, 2015; Taylor & Lurie, 2004).

Dynamics of power, privilege, and oppression against racial and ethnic minorities are necessary to understand and break the cycle of health disparities. Globally, ethnic and racial minority status is associated with increased risk following disasters (Bethel, Burke, & Britt, 2013; Smid, Drogendijk, Knipscheer, Boelen, & Kleber, 2018). For generations, many groups of people have endured bias, prejudice, discrimination, and violence on the basis of their phenotype (Goosby & Heidbrink, 2013; Ortiz & Telles, 2012). Ethnic minorities often experience increased interpersonal stress due to reasons such as marginalization, discrimination, immigration status, limited social support, language difficulties, and acculturation issues (Cleveland, Rousseau, & Guzder, 2014; Drogendijk, van der Velden, Gersons, & Kleber, 2011; Drogendijk, van der Velden, & Kleber, 2012; Smid et al., 2018). Engaging racial and ethnic minorities in interventions that promote increased social connectedness and restoration of cultural meaning can reduce mental health disparities (Smid et al., 2018).

Socioeconomic Status

Socioeconomic status (i.e., social class or class) is defined as a location in the social and economic hierarchy that reflects a combination of income status, educational attainment, and occupational prestige (Kraus, Park, & Tan, 2017). It also involves the presence of social capital, which refers to resources (e.g., information, influence, financial capital, emotional support, etc.) available through a person's social networks (Manstead, 2018; Pitkin Derose & Varda, 2009). Though individuals may fall in multiple social classes during their lifetime, socioeconomic status often has long-lasting effects on people's personal and social identity. It influences thoughts, emotions, and behaviors in response to social environment (Dorius, 2018; DuPont, Noy, Okuyama, & Sawada, 2015; Kraus et al., 2017; Manstead,

2018). All too often, disasters cause acute decline of individual and communal socioeconomic status, which can be extremely distressing (Brown, 2008; Joseph, Matthews, & Myers, 2014).

Social class plays a significant role in people's ability to prepare for, respond to, and recover from disasters (Brown, 2008; SAMHSA, 2017). Research suggests that in many instances lack of resources causes people of low socioeconomic status to be significantly less prepared for disasters, unable to respond to official warnings about disasters, and more vulnerable to adversity during the recovery phase (Al-Rousan, Rubenstein, & Wallace, 2014; Fothergill & Peek, 2004; SAMHSA, 2017). For instance, individuals and families who lack financial resources and are unable to afford sufficient insurance will likely be unable to rebuild their homes, which in turn increases their risk of homelessness or dislocation (Brown, 2008; SAMHSA, 2017). Losses that cannot be replaced can cause individuals and communities extreme distress and feelings of hopelessness (DHHS, 2003; SAMHSA, 2017).

Cultural competency in disaster mental health requires awareness about classism (i.e., bias, prejudice, discrimination, and stigmatization based on social class) (Fuller-Rowell, Evans, & Ong, 2012; Simons, Koster, Groffen, & Bosma, 2017). Classism denies the presence of systemic barriers to economic equality and blames the person for lacking class privilege (Manstead, 2018; Simons et al., 2017). Moreover, cultural competence demands awareness and knowledge about the intersection of social class with other social locations, including gender, disability, age, race and ethnicity, geographic location, etc., in the context of the local culture (Brown, 2008; Weber & Hilfinger Messias, 2012). Addressing individuals' subjective experiences of classism (i.e., perceived or internalized classism) with attention to intersectionality of their social identities can promote empowerment and resiliency (Schwarzbaum & Thomas, 2008).

Sexual Orientation and Gender

The following paragraphs discuss cultural competency in relation to the two separate, yet closely related, social locations of sexual orientation and gender. Sexual orientation can be described as either fixed or fluid patterns of an individual's direction and expression of sexual and romantic thoughts, affection, attraction, or desires toward members of same, other, both, or neither sexes (Brown, 2008; Moleiro & Pinto, 2015; Morgan, 2013). Gender is a social construct that was originally conceptualized as a person's personification of the biological makeup of their body (i.e., sex), which itself is indicated through morphological variations (i.e., genomics, reproductive organs, and external genitals) (Brown, 2008). While sex is often sorted using the binary categories of male and female, the concept of gender has evolved to emphasize the socially constructed variations in individuals' self-concept on a spectrum of masculinity and femininity, which may be static or fluid over time (Moleiro & Pinto, 2015; Short, Yang, & Jenkins, 2013).

Sexual minority cultural competency demands a certain degree of open-mindedness and humility to embrace individual and cultural variations in sexual orientation and gender identity (Berke, Maples-Keller, & Richards, 2016; Brown, 2008). It also requires that providers have awareness and knowledge about systemic inequalities benefiting heterosexual and cisgender (i.e., when gender identity matches sex at birth) ways of being and relating. Some examples include homophobia (i.e., fear or hatred of gay, lesbian, or bisexual individuals) and/or transphobia (i.e., dislike of or prejudice against transgender people). Provider's own personal biases and prejudices are also crucial to reflect on (Moore, 2017). In addition, disaster mental health services should communicate a culture of competency by considering inclusion throughout practices (e.g., familiarity with the most recent terms used among local sexual minority groups (Berke et al., 2016). An essential consideration is to engage in routine inquiry about gender identity and sexual orientation and never to assume regardless of seemingly obvious cues (Brown, 2008).

Indigenous Heritage and Colonization

People with indigenous heritage (i.e., aboriginal, native) are the ethnic group known as the original inhabitants of a region prior to its occupation or colonization (Currie, Wheat, & Wess, 2018). Their shared histories of colonization often involved driving away or enslaving indigenous people and even acts of genocide (Brown, 2008). Unfortunately, because indigenous people continue to be marginalized today, they experience a number of adverse consequences, such as lower incomes, lower educational attainment, and high rates of substance use (Burns, 2013; Currie et al., 2018). Trauma has been a ubiquitous reality in the life and heritage of native people.

Models of trauma-informed and culturally competent indigenous services acknowledge the historical and contemporary societal factors that perpetuate oppression and make efforts to transform systems of care by reasserting indigenous knowledge into practices (Brown, 2008; Hill, Lau, & Sue, 2010). Some general objectives include obtaining knowledge and understanding about the history of colonization, indigenous and nonindigenous relations, the transgenerational impact of colonization, and contemporary indigenous communities (Burns, 2013). Moreover, considering the intersection of indigenous heritage and spirituality is critical in the context of culturally appropriate and empowering health practices (Dudgeon, Bray, D'Costa, & Walker, 2017).

National Origin

In Hays' ADDRESSING model, national origin mainly emphasizes the issue of migration, which is closely related to ethnicity (Hays, 2016). Migration refers to a change of residence and can be self-determined (i.e., voluntarily) or with little choice and mainly determined by external circumstances (i.e., involuntary) (Nirmala, Kumar, & Virupaksha, 2014; Verkuyten, Mepham, & Kros, 2018). Moreover, migration may be within or across national borders (Zimmerman, Kiss, & Hossain, 2011). Though the primary goal of migration is to improve living conditions, it poses multiple challenges and life stressors, including adjustment difficulties, language barriers, cultural disparities, etc. (Nirmala et al., 2014; Zimmerman et al., 2011). Subsequently, migration can negatively impact the health of migrant populations (Nirmala et al., 2014; Zimmerman et al., 2011).

Migration as a broad concept has been separated into various categories based on reasons for migration as well as the legal processes involved (Nirmala et al., 2014). These categories often entail useful information about the risks associated with the migration process and status of the migrant and therefore useful to discuss. Immigrant refers to those migrants who freely make the decision to leave their homeland with the intention of permanently settling in a foreign country (Nirmala et al., 2014). On the other hand, refugees are those who fled from their homeland due to legitimate fear for their lives and safety and cannot return unless conditions improve (Nirmala et al., 2014). While refugees have been approved for protection under international laws, asylum seekers are yet in the quest of this protection from the country to which they have fled (Crawley & Skleparis, 2018).

Trauma-informed and culturally competent disaster behavioral health care for migrants requires a thorough exploration of their cultural beliefs about mental health as well as their access to resources locally and globally (Wylie et al., 2018). Providers should consider the motives and processes involved in one's migration and assess for encounters with trauma along this journey accordingly (Brown, 2008). Moreover, the course of resettlement can be distressing or even traumatic in and of itself and may intersect with low socioeconomic status, limited social capital, unauthorized or illegal immigration status, and other risk factors that increase vulnerability at times of disaster (Andersson, Edling, & Rydgren, 2018; Bhugra & Becker, 2005; Fussell, Delp, Riley, Chávez, & Valenzuela Jr., 2018).

Conclusion and Implications

Disasters often result in physical and mental suffering due to their profound and enduring effects on the quality of life of those who are affected. Consequently, disaster-affected individuals and communities are at higher risk for developing mental and behavioral health disorders that can be debilitating and disruptive to their ability to adaptively recover. This is particularly true among those who have already endured other traumas, especially those were already disadvantaged prior to the disaster. Disaster mental and behavioral health services aim to protect and promote the mental well-being of all who were affected by disasters and reduce access and use disparities. Just as people's understanding of and response to a disaster differs significantly based on their values, beliefs, and life experiences, so does their comfort level and willingness to use available disaster mental health services. People who would benefit from disaster mental health services may be reluctant to use available treatment, even if offered for free, because of stigma and their impressions concerning the value of behavioral health treatment in general. To provide accessible and effective care, disaster mental and behavioral health providers must accurately assess the needs of the local communities and successfully engage them in services that are deemed as acceptable and culturally sensitive. Cultural competency is a construct that was developed as a framework to assist professionals in providing tailored and culturally appropriate mental health care. This framework is equally useful when developing and implementing disaster behavioral health programs.

Cultural competence refers to a set of consistent values and practices at the level of policy, administration, and direct-care providers that honors, respects, and responds to the range of beliefs, attitudes, behaviors, and interpersonal styles of individuals from culturally diverse backgrounds. These values and practices are essential for accurate assessments of local needs as well as facilitating access to and improving standards of care. Eliminating disaster mental health service disparities requires culturally competent involvement at all levels to overcome barriers, stigma, and social adversities and to encourage positive identity and resilience. Professionals should always acknowledge the resilience of the survivors and validate the strategies that they used to cope in the face of adversity. These strategies are often reflective of one or more social contexts in which individuals are situated. Therefore, it is crucial to address and attentively explore the various social locations in relation to the local context.

To assist professionals in their efforts to effectively serve diverse individuals and communities post-disaster, this chapter describes the Hays' ADDRESSING model of diversity and reviews its list of social categories in relation to trauma-informed disaster mental health (2016). Social categories such as age, disability, religion and spirituality, ethnicity and race, socioeconomic statue, sexual orientation and gender, indigenous heritage and colonization, and national origin were discussed in detail and are examined through the systemic lens of social power, privilege, and oppression. In each section, specific considerations were made for direct-care providers.

In summary, cultural competence is not an end goal that can be achieved as a product or deliverable; it is rather an ongoing process where professionals at all levels can grow in their ability to respond respectfully and effectively to people from all cultures. The goal of this chapter was to illustrate the complexity of cultural competency and provide methods and strategies for professionals to navigate the care of diverse populations seen after disasters. The hope is for readers to think beyond this useful framework by continuing to understand that people are shaped by the intersection of their social identities.

References

Adler, A. (1943). Neuropsychiatric complications in victims of Boston's cocoanut grove disaster. *Journal of the American Medical Association, 123*, 1098–1101. https://doi.org/10.1001/jama.1943.02840520014004

Age [Def. 1.c]. (n.d.). *In Merriam Webster online dictionary*. Retrieved January 16, 2019, from https://www.merriam-webster.com/dictionary/age

Al-Rousan, T. M., Rubenstein, L. M., & Wallace, R. B. (2014). Preparedness for natural disasters among older US adults: A nationwide survey. *American Journal of Public Health, 104*(3), 506–511.

Anderson, K. N., Bautista, C. L., & Hope, D. A. (2019). Therapeutic alliance, cultural competence and minority status in premature termination of psychotherapy. *American Journal of Orthopsychiatry, 89*(1), 104–114. https://doi.org/10.1037/ort0000342

Andersson, A., Edling, C., & Rydgren, J. (2018). The intersection of class origin and immigration background in structuring social capital: The role of transnational ties. *British Journal of Sociology, 69*(1), 99–123. https://doi.org/10.1111/1468-4446.12289

Applied Mental Health Research (AMHR) Group. (2013). Design, implementation, monitoring, and evaluation of mental health and psychological assistance programs for trauma survivors in low resource countries: A user's manual for researchers and program implementers (Adult version). In *Module 5: Intervention selection, adaptation, and implementation*. Baltimore, MD: Johns Hopkins University.

Aten, J. D., O'Grady, K. A., Milstein, G., Boan, D., & Schruba, A. (2014). Spiritually oriented disaster psychology. *Spirituality in Clinical Practice, 1*(1), 20–28. https://doi.org/10.1037/scp0000008

Baker, L. R., & Cormier, L. A. (2015). *Disasters and vulnerable populations: Evidence-based practice for the helping professions*. New York: Springer Publishing Company.

Bauer, A. M., Chen, C. N., & Alegría, M. (2012). Prevalence of physical symptoms and their association with race/ethnicity and acculturation in the United States. *General Hospital Psychiatry, 34*(4), 323–331. https://doi.org/10.1016/j.genhosppsych.2012.02.007

Behere, P. B., Das, A., Yadav, R., & Behere, A. P. (2013). Religion and mental health. *Indian Journal of Psychiatry, 55*(Suppl 2), S187–S194.

Bell, A., & Jones, K. (2015). Age, period and cohort processes in longitudinal and life course analysis: A multilevel perspective. In C. Burton-Jeangros, S. Cullati, A. Sacker, et al. (Eds.), *A life course perspective on health trajectories and transitions [Internet]*. Cham, Switzerland: Springer. Retrieved from: https://www.ncbi.nlm.nih.gov/books/NBK385364/

Berke, D. S., Maples-Keller, J. L., & Richards, P. (2016). LGBTQ perceptions of psychotherapy: A consensual qualitative analysis. *Professional Psychology: Research and Practice, 47*(6), 373–382. https://doi.org/10.1037/pro0000099

Bethel, J. W., Burke, S. C., & Britt, A. F. (2013). Disparity in disaster preparedness between racial/ethnic groups. *Disaster Health, 1*(2), 110–116. https://doi.org/10.4161/dish.27085

Bhugra, D., & Becker, M. A. (2005). Migration, cultural bereavement and cultural identity. *World Psychiatry, 4*(1), 18–24. https://doi.org/10.1002/(ISSN)2051-5545

Bogart, K. R. (2014). The role of disability self-concept in adaptation to congenital or acquired disability. *Rehabilitation Psychology, 59*(1), 107–115. https://doi.org/10.1037/a0035800

Bouye, K. E., McCleary, K. J., & Williams, K. B. (2016). Increasing diversity in the health professions: Reflections on student pipeline programs. *Journal of Healthcare, Science and the Humanities, 6*(1), 67–79. Retrieved from http://www.ncbi.nlm.nih.gov/pubmed/29644118; http://www.pubmedcentral.nih.gov/articlerender.fcgi?artid=PMC5890504

Brown, L. S. (2008). *Cultural competence in trauma therapy: Beyond the flashback*. Washington, DC: American Psychological Association.

Bunning, K., Gona, J. K., Newton, C. R., & Hartley, S. (2017). The perception of disability by community groups: Stories of local understanding, beliefs and challenges in a rural part of Kenya. *PLoS One, 12*(8), 1–20. https://doi.org/10.1371/journal.pone.0182214

Burns, M. (2013). Towards growing Indigenous culturally competent legal professionals in Australia. *The International Education Journal: Comparative Perspectives, 12*(1), 226–248.

Butler, M., McCreedy, E., Schwer, N., Burgess, D., Call, K., Przedworski, J. M., et al. (2016). *Improving cultural competence to reduce health disparities (Comperative effectiveness reviews, no. 170)*. Rockville, MD: Agency for Healthcare Research and Quality (US). Retrieved from: https://www.ncbi.nlm.nih.gov/books/NBK361126/

Cleveland, J., Rousseau, C., & Guzder, J. (2014). Cultural consultation for refugees. In L. J. Kirmayer, J. Guzder, & C. Rousseau (Eds.), *Cultural consultation* (pp. 245–268). New York: Springer.

Collier, R. (2012). Person-first language: Noble intent but to what effect? *CMAJ: Canadian Medical Association Journal = Journal de l'Association Medicale Canadienne, 184*(18), 1977–1978. https://doi.org/10.1503/cmaj.109-4319

Colón-Emeric, C. S., Whitson, H. E., Pavon, J., & Hoenig, H. (2013). Functional decline in older adults. *American Family Physician, 88*(6), 388–394. https://doi.org/10.1097/01.NAJ.0000277831.12796.84

Crawley, H., & Skleparis, D. (2018). Refugees, migrants, neither, both: Categorical fetishism and the politics of bounding in Europe's 'migration crisis'. *Journal of Ethnic and Migration Studies, 44*(1), 48–64. https://doi.org/10.1080/1369183X.2017.1348224

Cross, T. L. (1989). *Towards a culturally competent system of care: A monograph of effective services for children who are severely emotionally disturbed*. Washington, DC: Georgetown University Child Development Center.

Currie, G., Wheat, J., & Wess, T. (2018). Building foundations for indigenous cultural competence: An institution's journey toward "closing the gap". *Journal of Medical Imaging and Radiation Sciences, 49*(1), 6–10. https://doi.org/10.1016/j.jmir.2017.12.003

D'Andrea, M., & Daniels, J. (2001). RESPECTFUL counseling: An integrative model for counselors. In D. Pope-Davis & H. Coleman (Eds.), *The interface of class, culture and gender in counseling* (pp. 417–466). Thousand Oaks, CA: Sage.

Danso, R. (2018). Cultural competence and cultural humility: A critical reflection on key cultural diversity concepts. *Journal of Social Work, 18*(4), 410–430. https://doi.org/10.1177/1468017316654341

de Jong, K., Prosser, S., & Ford, N. (2005). Addressing psychosocial needs in the aftermath of the tsunami. *PLoS Medicine, 2*(6), e179. https://doi.org/10.1371/journal.pmed.0020179

Delvecchio Good, M. J., & Hannah, S. D. (2015). "Shattering culture": Perspectives on cultural competence and evidence-based practice in mental health services. *Transcultural Psychiatry, 52*(2), 198–221. https://doi.org/10.1177/1363461514557348

Department of Health and Human Services (US). (2003). *Developing cultural competence in disaster mental health programs: Guiding principles and recommendations. DHHS Pub. No. SMA 3828*. Rockville, MD: Center for Mental Health Services, Substance Abuse and Mental Health Services Administration.

Dominici, F., Levy, J. I., & Louis, T. A. (2005). Methodological challenges and contributions in disaster epidemiology. *Epidemiologic Reviews, 1*(1), 9–12. https://doi.org/10.1093/epirev/mxi009

Dorius, S. F. (2018). Socioeconomic status mobility in the modern world system: Growth and allocation effects. *Social Indicators Research, 138*(3), 1049–1074. https://doi.org/10.1007/s11205-017-1688-z

Drogendijk, A. N., van der Velden, P. G., Gersons, B. P. R., & Kleber, R. J. (2011). Lack of perceived social support among immigrants after a disaster: A comparative study. *The British Journal of Psychiatry, 198*, 317–322.

Drogendijk, A. N., van der Velden, P. G., & Kleber, R. J. (2012). Acculturation and post-disaster mental health problems among affected and non-affected immigrants: A comparative study. *Journal of Affective Disorders, 138*, 485–489. https://doi.org/10.1016/j.jad.2012.01.037

Dudgeon, P., Bray, A., D'Costa, B., & Walker, R. (2017). Decolonising psychology: Validating social and emotional wellbeing. *Australian Psychologist, 52*(4), 316–325. https://doi.org/10.1111/ap.12294

DuPont, W., Noy, I., Okuyama, Y., & Sawada, Y. (2015). The long-run socio-economic consequences of a large disaster: The 1995 earthquake in Kobe. *PLoS One, 10*(10), 1–17. https://doi.org/10.1371/journal.pone.0138714

Eggertson, L. (2006). Tsunami donations help worldwide. *CMAJ: Canadian Medical Association Journal = Journal de l'Association Medicale Canadienne, 174*(3), 299. https://doi.org/10.1503/cmaj.051642

Feliciano, C. (2015). Shades of race: How phenotype and observer characteristics shape racial classification. *American Behavioral Scientist, 60*(4), 390–419. https://doi.org/10.1177/0002764215613401

Ferrari, S., Burian, R., Hahn, E., Chaudhry, N., Chaudhry, I. B., Husain, N., et al. (2015). Somatization among ethnic minorities and immigrants: Why does it matter to consultation liaison psychiatry? *Journal of Psychosomatic Research, 79*(1), 85–86. https://doi.org/10.1016/j.jpsychores.2015.02.011

Field, C. T., & Syrett, N. L. (2015). *Age in America: The colonial era to the present*. New York: New York University Press.

Ford, C. L., & Harawa, N. T. (2010). A new conceptualization of ethnicity for social epidemiologic and health equity research. *Social Science and Medicine, 71*(2), 251–258. https://doi.org/10.1016/j.socscimed.2010.04.008

Fothergill, A., & Peek, L. A. (2004). Poverty and disasters in the United States: A review of recent sociological findings. *Natural Hazards, 32*(1), 89–110. https://doi.org/10.1023/B:NHAZ.0000026792.76181.d9

Freeman, C., Nairn, K., & Gollop, M. (2015). Disaster impact and recovery: What children and young people can tell us. *Kotuitui: New Zealand Journal of Social Sciences Online, 10*(2), 103–115. https://doi.org/10.1080/1177083X.2015.1066400

Fuller-Rowell, T. E., Evans, G. W., & Ong, A. D. (2012). Poverty and health: The mediating role of perceived discrimination. *Psychological Science, 23*(7), 734–739. https://doi.org/10.1177/0956797612439720

Fussell, E., Delp, L., Riley, K., Chávez, S., & Valenzuela Jr., A. (2018). Implications of social and legal status on immigrants' health in disaster zones. *American Journal of Public Health, 108*(12), 1617–1620. https://doi.org/10.2105/AJPH.2018.304554

Gernsbacher, M. A. (2017). Editorial perspective: The use of person-first language in scholarly writing may accentuate stigma. *Journal of Child Psychology and Psychiatry and Allied Disciplines, 58*(7), 859–861. https://doi.org/10.1111/jcpp.12706

Goering, S. (2015). Rethinking disability: The social model of disability and chronic disease. *Current Reviews in Musculoskeletal Medicine, 8*(2), 134–138. https://doi.org/10.1007/s12178-015-9273-z

Goldmann, E., & Galea, S. (2014). Mental health consequences of disasters. *Annual Review of Public Health, 35*(1), 169–183. https://doi.org/10.1146/annurev-publhealth-032013-182435

Goosby, B. J., & Heidbrink, C. (2013). The transgenerational consequences of discrimination on African-American health outcomes. *Sociology Compass, 7*(8), 630–643. https://doi.org/10.1111/soc4.12054

Gopalkrishnan, N. (2018). Cultural diversity and mental health: Considerations for policy and practice. *Frontiers in Public Health, 6*, 179. https://doi.org/10.3389/fpubh.2018.00179

Ha, K.-M. (2015). The role of religious beliefs and institutions in disaster management: A case study. *Religions, 6*(4), 1314–1329. https://doi.org/10.3390/rel6041314

Hall, J. C., & Theriot, M. T. (2016). Developing multicultural awareness, knowledge, and skills: Diversity training makes a difference? *Multicultural Perspectives, 18*(1), 35–41. https://doi.org/10.1080/15210960.2016.1125742

Hays, P. A. (2008). *Addressing cultural complexities in practice : assessment, diagnosis, and therapy* (2nd ed.). Washington, DC: American Psychological Association.

Hays, P. A. (2016). *Addressing cultural complexities in practice: Assessment, diagnosis, and therapy* (3rd ed.). Washington, DC: American Psychological Association.

Heath, M. A., Nickerson, A. B., Annandale, N., Kemple, A., & Dean, B. (2009). Strengthening cultural sensitivity in children's disaster mental health services. *School Psychology International, 30*(4), 347–373. https://doi.org/10.1177/0143034309106944

Henslee, A. M., Coffey, S. F., Schumacher, J. A., Tracy, M., Norris, F. H., & Galea, S. (2015). Religious coping and psychological and behavioral adjustment after Hurricane Katrina. *Journal of Psychology: Interdisciplinary and Applied, 149*(6), 630–642. https://doi.org/10.1080/00223980.2014.953441

Hill, J. S., Lau, M. Y., & Sue, D. (2010). Integrating trauma psychology and cultural psychology: Indigenous perspectives on theory, research, and practice. *Traumatology, 16*(4), 39–47. https://doi.org/10.1177/1534765610388303

Joakim, E. P., & White, R. S. (2015). Exploring the impact of religious beliefs, leadership, and networks on response and recovery of disaster-affected populations: A case study from indonesia. *Journal of Contemporary Religion, 30*(2), 193–192. https://doi.org/10.1080/13537903.2015.1025538

Joseph, N. T., Matthews, K. A., & Myers, H. F. (2014). Conceptualizing health consequences of Hurricane Katrina from the perspective of socioeconomic status decline. *Health Psychology: Official Journal of the Division of Health Psychology, American Psychological Association, 33*(2), 139–146. https://doi.org/10.1037/a0031661

Juth, V., Silver, R. C., Seyle, D. C., Widyatmoko, C. S., & Tan, E. T. (2015). Post-disaster mental health among parent–child dyads after a major earthquake in Indonesia. *Journal of Abnormal Child Psychology, 43*(7), 1309–1318. https://doi.org/10.1007/s10802-015-0009-8

King, R. V., Burkle, F. M., Walsh, L. E., & North, C. S. (2015). Competencies for disaster mental health. *Current Psychiatry Reports, 17*(3), 548. https://doi.org/10.1007/s11920-015-0548-2

Kostanjsek, N. (2011). Use of the International Classification of Functioning, Disability and Health (ICF) as a conceptual framework and common language for disability statistics and health information systems. *BMC Public Health, 11*(Suppl4), S3. https://doi.org/10.1186/1471-2458-11-S4-S3

Krahn, G. L., Walker, D. K., & Correa-De-Araujo, R. (2015). Persons with disabilities as an unrecognized health disparity population. *American Journal of Public Health, 105*(S2), S198–S206. https://doi.org/10.2105/AJPH.2014.302182

Kraus, M. W., Park, J. W., & Tan, J. J. X. (2017). Signs of social class: The experience of economic inequality in everyday life. *Perspectives on Psychological Science, 12*(3), 422–435. https://doi.org/10.1177/1745691616673192

Löckenhoff, C. E., De Fruyt, F., Terracciano, A., McCrae, R. R., De Bolle, M., Costa, P. T., et al. (2009). Perceptions of aging across 26 cultures and their culture-level associates. *Psychology and Aging, 24*(4), 941–954. https://doi.org/10.1037/a0016901

Malik, S., Lee, D. C., Doran, K. M., Grudzen, C. R., Worthing, J., Portelli, I., et al. (2017). Vulnerability of older adults in disasters: Emergency department utilization by geriatric patients after Hurricane Sandy. *Disaster Medicine and Public Health Preparedness, 12*(2), 184–193. https://doi.org/10.1017/dmp.2017.44

Manstead, A. S. (2018). The psychology of social class: How socioeconomic status impacts thought, feelings, and behaviour. *British Journal of Social Psychology, 57*(2), 267–291. https://doi.org/10.1111/bjso.12251

Marks, L. D., Dollahite, D. C., & Young, K. P. (2018). Struggles experienced by religious minority families in the United States. *Psychology of Religion and Spirituality. Advance Online Publication, 11*, 247. https://doi.org/10.1037/rel0000214

Math, S., Nirmala, M., Moirangthem, S., & Kumar, N. (2015). Disaster management: Mental health perspective. *Indian Journal of Psychological Medicine, 37*(3), 261. https://doi.org/10.4103/0253-7176.162915

Mendenhall, T. J. (2006). Trauma-response teams: Inherent challenges and practical strategies in interdisciplinary fieldwork. *Families, Systems & Health, 24*(3), 357–362. https://doi.org/10.1037/1091-7527.24.3.357

Menschner, C., & Maul, A. (2016). *Key ingredients for successful trauma-informed care implementation*. Trenton, NJ: Center for Health Strategies, Inc.

Mersha, T. B., & Abebe, T. (2015). Self-reported race/ethnicity in the age of genomic research: Its potential impact on understanding health disparities. *Human Genomics, 9*(1), 1–34. https://doi.org/10.1186/s40246-014-0023-x

Moleiro, C., & Pinto, N. (2015). Sexual orientation and gender identity: Review of concepts, controversies and their relation to psychopathology classification systems. *Frontiers in Psychology, 1*(1), 52–66. https://doi.org/10.3389/fpsyg.2015.01511

Moore, L. M. (2017). Heterosexism and homophobia. In the Wiley-Blackwell encyclopedia of social theory, B. S. Turner (Ed.). doi:https://doi.org/10.1002/9781118430873.est0162.

Morgan, E. M. (2013). Contemporary issues in sexual orientation and identity development in emerging adulthood. *Emerging Adulthood, 52*, 1. https://doi.org/10.1177/2167696812469187

Morris, A. J. (2011). Psychic aftershocks: Crisis counseling and disaster relief policy. *History of Psychology, 14*(3), 264–286. https://doi.org/10.1037/a0024169

Murthy, R. S., & Lakshminarayana, R. (2006). Mental health consequences of war: A brief review of research findings. *World Psychiatry: Official Journal of the World Psychiatric Association (WPA), 5*(1), 25–30. https://doi.org/10.1080/13518040701205365

Nassar-McMillan, S. C. (2014). A framework for cultural competence, advocacy, and social justice: Applications for global multiculturalism and diversity. *International Journal for Educational and Vocational Guidance, 14*(1), 103–118. https://doi.org/10.1007/s10775-014-9265-3

Nirmala, B., Kumar, A., & Virupaksha, H. (2014). Migration and mental health: An interface. *Journal of Natural Science, Biology and Medicine, 5*(2), 233–239. https://doi.org/10.4103/0976-9668.136141

Ortiz, V., & Telles, E. (2012). Racial identity and racial treatment of Mexican Americans. *Race and Social Problems, 4*(1), 41–56. https://doi.org/10.1007/s12552-012-9064-8

Paniagua, F. A., & Yamada, A. M. (2013). *Handbook of multicultural mental health: Assessment and treatment of diverse populations* (2nd ed.). San Diego, CA: Elsevier Academic Press. https://doi.org/10.1016/C2011-0-06838-2

Park, C. L. (2016). Meaning making in the context of disasters. *Journal of Clinical Psychology, 72*(12), 1234–1246. https://doi.org/10.1002/jclp.22270

Pfefferbaum, B., & North, C. S. (2008). Children and families in the context of disasters: Implications for preparedness and response. *The Family Psychologist: Bulletin of the Division of Family Psychology, 24*(2), 6–10.

Pitkin Derose, K., & Varda, D. M. (2009). Social capital and health care access: A systematic review. *Medical Care Research and Review, 66*(3), 272–306. https://doi.org/10.1177/1077558708330428

Plotkin, H. (2011). Human nature, cultural diversity and evolutionary theory. *Philosophical Transactions of the Royal Society B: Biological Sciences, 366*(1563), 454–463. https://doi.org/10.1098/rstb.2010.0160

Pourhosseini, S. S., Ardalan, A., & Mehrolhassani, M. H. (2015). Key aspects of providing healthcare services in disaster response stage. *Iranian Journal of Public Health, 44*(1), 111–118. Retrieved from https://www.ncbi.nlm.nih.gov/pubmed/26060782

Raphael, B., & Maguire, P. (2009). Disaster mental health research: Past, present, and future. In Y. Neria, S. Galea, & F. H. Norris (Eds.), *Mental health and disasters* (pp. 7–28). New York: Cambridge University Press. https://doi.org/10.1017/CBO9780511730030.002

Relethford, J. H. (2009). Race and global patterns of phenotypic variation. *American Journal of Physical Anthropology, 139*(1), 16–22. https://doi.org/10.1002/ajpa.20900

Rosen, C. S., Greene, C. J., Young, H. E., & Norris, F. H. (2010). Tailoring disaster mental health services to diverse needs: An analysis of 36 crisis counseling projects. *Health and Social Work, 35*(3), 211–220. https://doi.org/10.1093/hsw/35.3.211

Saunders, J. A., Haskins, M., & Vasquez, M. (2015). Cultural competence: A journey to an elusive goal. *Journal of Social Work Education, 51*(1), 19–34. https://doi.org/10.1080/10437797.2015.977124

Schnyder, U., Bryant, R. A., Ehlers, A., Foa, E. B., Hasan, A., Mwiti, G., et al. (2016). Culture-sensitive psychotraumatology. *European Journal of Psychotraumatology, 7*(1), 31179. https://doi.org/10.3402/ejpt.v7.31179

Schwarzbaum, S. E., & Thomas, A. J. (2008). Oppression and resilience. In *Dimensions of multicultural counseling: A life story approach* (pp. 13–44). Thousand Oaks, CA: SAGE Publications, Inc.

Shih, R. A., Acosta, J. D., Chen, E. K., Carbone, E. G., Xenakis, L., Adamson, D. M., et al. (2018). Improving disaster resilience among older adults: Insights from Public Health Departments and aging-in-place efforts. *Rand Health Quarterly, 8*(1), 3.

Shimmin, C., Wittmeier, K. D. M., Lavoie, J. G., Wicklund, E. D., & Sibley, K. M. (2017). Moving towards a more inclusive patient and public involvement in health research paradigm: The incorporation of a trauma-informed intersectional analysis. *BMC Health Services Research, 17*(1), 539–548. https://doi.org/10.1186/s12913-017-2463-1

Short, S. E., Yang, Y. C., & Jenkins, T. M. (2013). Sex, gender, genetics, and health. *American Journal of Public Health, 103*(S1), S93–S101. https://doi.org/10.2105/AJPH.2013.301229

Simons, A. M. W., Koster, A., Groffen, D. A. I., & Bosma, H. (2017). Perceived classism and its relation with socioeconomic status, health, health behaviours and perceived inferiority: The Dutch Longitudinal Internet Studies for the Social Sciences (LISS) panel. *International Journal of Public Health, 62*(4), 433–440. https://doi.org/10.1007/s00038-016-0880-2

Smedley, A., & Smedley, B. D. (2005). Race as biology is fiction, racism as a social problem is real: Anthropological and historical perspectives on the social construction of race. *American Psychologist, 60*(1), 16–26. https://doi.org/10.1037/0003-066X.60.1.16

Smid, G. E., Drogendijk, A. N., Knipscheer, J., Boelen, P. A., & Kleber, R. J. (2018). Loss of loved ones or home due to a disaster: Effects over time on distress in immigrant ethnic minorities. *Transcultural Psychiatry, 55*(5), 648–668. https://doi.org/10.1177/1363461518784355

Speed, D., Coleman, T. J., & Langston, J. (2018). What do you mean, "what does it all mean?" atheism, nonreligion, and life meaning. *SAGE Open, 8*(1), 1–13. https://doi.org/10.1177/2158244017754238

SRA International, Inc. (2008). *Cultural competency in disaster response: A review of current concepts, policies, and practices*. Rockville, MD: SRA International, Inc.

Stebnicki, M. A. (2017). *Disaster mental health counseling: Responding to trauma in a multicultural context*. New York: Springer Publishing Company.

Stough, L. M., Sharp, A. N., Resch, J. A., Decker, C., & Wilker, N. (2016). Barriers to the long-term recovery of individuals with disabilities following a disaster. *Disasters, 40*(3), 387–410. https://doi.org/10.1111/disa.12161

Substance Abuse and Mental Health Services Administration, (US). (2014a). *SAMHSA's concept of trauma and guidance for a trauma-informed approach. HHS Publication*. Rockville, MD: Substance Abuse and Mental Health Services Administration.

Substance Abuse and Mental Health Services Administration, (US). (2014b). *Treatment Improvement Protocol (TIP) Series: No. 59. Improving cultural competence. HHS Publication No. (SMA) 14-4849*. Rockville, MD: Substance Abuse and Mental Health Services Administration.

Substance Abuse and Mental Health Services Administration, (US). (2014c). *Treatment Improvement Protocol (TIP) Series: No. 57. Trauma-informed care in behavioural health services. HHS Publication No. (SMA) 13-4801*. Rockville, MD: Substance Abuse and Mental Health Services Administration.

Substance Abuse and Mental Health Services Administration, (US). (2017). *Greater impact: How disasters affect people of low socioeconomic status*. SAMHSA Disaster Technical Assistance Center Supplemental Research Bulletin.

Sue, D. W. (2001). Multidimensional facets of cultural competence. *The Counseling Psychologist, 29*(6), 790–821. https://doi.org/10.1177/0011000001296002

Sue, S., Zane, N., Nagayama Hall, G. C., & Berger, L. K. (2009). The case for cultural competency in psychotherapeutic interventions. *Annual Review of Psychology, 60*(1), 525–548. https://doi.org/10.1146/annurev.psych.60.110707.163651

Taylor, S. L., & Lurie, N. (2004). The role of culturally competent communication in reducing ethnic and racial healthcare disparities. *The American Journal of Managed Care, 10*(Special Issue), SP1–SP4. https://doi.org/http://www.ncbi.nlm.nih.gov/pubmed/15481430

Tyhurst, J. S. (1951). Individual reactions to community disaster: The natural history of psychiatric phenomena. *The American Journal of Psychiatry, 107*(10), 764–769. https://doi.org/10.1176/ajp.107.10.764

US Census Bureau. (2018). *Older people projected to outnumber children*. Retrieved from https://www.census.gov/newsroom/press-releases/2018/cb18-41-population-projections.html

Vauclair, C. M., Hanke, K., Huang, L. L., & Abrams, D. (2017). Are Asian cultures really less ageist than Western ones? It depends on the questions asked. *International Journal of Psychology, 52*(2), 136–144. https://doi.org/10.1002/ijop.12292

Verkuyten, M., Mepham, K., & Kros, M. (2018). Public attitudes towards support for migrants: The importance of perceived voluntary and involuntary migration. *Ethnic and Racial Studies, 41*(5), 901–918. https://doi.org/10.1080/01419870.2017.1367021

Vu, L., & Van Landingham, M. J. (2012). Physical and mental health consequences of Katrina on Vietnamese immigrants in New Orleans: A pre-and post-disaster assessment. *Journal of Immigrant and Minority Health, 14*(3), 386–394.

Weber, L., & Hilfinger Messias, D. A. K. (2012). Mississippi front-line recovery work after Hurricane Katrina: An analysis of the intersections of gender, race, and class in advocacy, power relations, and health. *Social Science and Medicine, 74*(11), 1833–1841. https://doi.org/10.1016/j.socscimed.2011.08.034

Weissbecker, I. (2009). Mental health as a human right in the context of recovery after disaster and conflict. *Counselling Psychology Quarterly, 22*(1), 77–84. https://doi.org/10.1080/09515070902761065

Wells, K. B., Springgate, B. F., Lizaola, E., Jones, F., & Plough, A. (2013). Community engagement in disaster preparedness and recovery: A tale of two cities – Los Angeles and New Orleans. *Psychiatric Clinics of North America, 36*(3), 451–466. https://doi.org/10.1016/j.psc.2013.05.002

West-Olatunji, C., Henesy, R., & Varney, M. (2015). Group work during international disaster outreach projects: A model to advance cultural competence. *Journal for Specialists in Group Work, 40*(1), 38–54. https://doi.org/10.1080/01933922.2014.992504

World Health Organization. (2002). *Towards a common language for functioning, disability and health*. Geneva, Swtizerland: ICF/WHO. https://doi.org/WHO/EIP/GPE/CAS/01.3

Wylie, L., Van Meyel, R., Harder, H., Sukhera, J., Luc, C., Ganjavi, H., et al. (2018). Assessing trauma in a transcultural context: Challenges in mental health care with immigrants and refugees. *Public Health Reviews, 39*, 22. https://doi.org/10.1186/s40985-018-0102-y

Yamada, A. M. (2015). Applications of the Cultural Awareness, Skill, & Knowledge (C-ASK) interview tool: Cultivating cultural competence. *Hawai'i Journal of Medicine & Public Health, 74*(7 Suppl 1), 19.

Zimmerman, C., Kiss, L., & Hossain, M. (2011). Migration and health: A framework for 21st century policy-making. *PLoS Medicine, 8*(5), e1001034. https://doi.org/10.1371/journal.pmed.1001034

Chapter 8
When Disasters Strike: Navigating the Challenges of "Sudden Science"

Katie E. Cherry, Matthew R. Calamia, and Emily M. Elliott

Introduction

Natural disasters happen worldwide. Disaster exposure is often quite stressful for survivors, with threats to health and well-being. Understanding how disaster-related stressors affect behavioral health outcomes in the immediate aftermath and in the years after calamitous disasters is a looming public health concern, which calls for continued funding for research to measure disaster impacts (Bonnano, Brewin, Kaniasty, & La Greca, 2010). Despite the importance of this topic and need for basic research, carrying out well-designed studies in disaster contexts remains a daunting challenge for behavioral health researchers in the twenty-first century.

The purpose of this chapter is to examine the practice of conducting behavioral research in the immediate aftermath of a natural disaster based on previous experience doing this type of work (see Cherry, 2009, 2015). For expository convenience, the term, *sudden science*, is offered to capture situations where a devastating environmental event prompts a scholarly response. Social scientists must design and quickly implement a research protocol to capture a disaster's immediate impact on directly affected samples of survivors. To be clear, quickly in this sense does not mean in haste. Rather, quickly means carried out within a tight timeline, though well done nevertheless. The key to the success of disaster-prompted research projects is to stay within the parameters of well-designed and carefully executed research as discussed in this chapter and elsewhere (see Banford-Witting & Bagley, Chap. 5, this volume).

Throughout this chapter, a conversational tone is employed, together with a lessons-learned approach to enhance relevance and applicability for future research endeavors. While a lessons learned approach might seem cliché, the surprises, setbacks, serendipity, and rewards of sudden science may be best communicated as

K. E. Cherry (✉) · M. R. Calamia · E. M. Elliott
Department of Psychology, Louisiana State University, Baton Rouge, LA, USA
e-mail: pskatie@lsu.edu

© Springer Nature Switzerland AG 2021

K. E. Cherry, A. Gibson (eds.), *The Intersection of Trauma and Disaster Behavioral Health*, https://doi.org/10.1007/978-3-030-51525-6_8

lessons learned (see also Richardson, Plummer, Barthelemy, & Cain, 2009). This chapter is organized as follows. First, the concept of sudden science is introduced, followed by three real-world examples for illustration. Next, we cover three, broad categories of concerns that may emerge while conducting a sudden science project. These reflect ethical, methodological, and pragmatic concerns for researchers to keep in mind while engaged in sudden science (Knack, Chen, Williams, & Jensen-Campbell, 2006). In closing, we point out post-disaster vulnerabilities that may exist among survivors who volunteer to participate in a sudden science research project. We also note post-disaster resilience that is sometimes apparent alongside of a disaster's devastation. Evidence of vulnerabilities and resilience intertwined can be interpreted to indicate that disaster-related losses and setbacks can be overcome as life goes on after a natural disaster (Cherry, 2020).

What Is "Sudden Science"?

Disasters are disruptive events that can be frightening and a threat to safety. Some disasters happen quickly with little to no advanced warning, as is often the case with tornados and cyclones. In other instances, the National Weather Service provides valuable forecasting information concerning the intensity and tracking of severe weather systems, as is the case with hurricanes and tropical storms. Volcanic eruptions and wildfires are other events of nature where potential paths of destruction can be anticipated over wide territorial expanses. Whether sudden or slow moving, one remarkable thing disasters have in common is that they showcase in a very real way the frailty of human life when juxtaposed against the powerful and destructive forces of nature.

Disasters are likely to set in motion a chain of predictable events. In the United States, a presidential declaration may happen, sometimes before a catastrophic, environmental event has taken place, to release federally-based disaster relief funding. At the same time, individuals, families, and communities may engage in preparatory responses before a natural disaster strikes, such as securing homes and businesses and evacuating to physically remove self and family from harm's way. A variety of different disaster responses may also occur after a disaster has taken place. For instance, governmental officials at the federal, state, and local levels are catapulted into action when a disaster has occurred. Community-wide, coordinated disaster responses often begin immediately as first responders are mobilized and deployed. The American Red Cross, among other disaster relief organizations, move in, establishing shelters and distributing food and water to disaster-impacted communities. Faith-based organizations provide assistance at the local level in an effort to meet the immediate needs of disaster survivors. Public libraries may also become a second line of response, serving as a community-wide source of information and networking opportunities.

With this maelstrom of activity underway, search and rescue operations are at the forefront. Saving lives is the rallying cry, though a somber reality exists as well.

That is, many may not survive the environmental event. Sadly, Disaster Mortuary Operational Response Team (DMORT) services are part of a disaster response. With the emphasis on saving lives and recovering bodies in the immediate aftermath of a disaster, one might reasonably ask whether conducting scholarly research is even appropriate to do, let alone feasible to carry out after a disaster has taken place?

To anticipate, the authors are of the opinion that a resounding "yes" is in order. Throughout this chapter, we will make the point that among the many disaster responses set in motion by devastating environmental events, behavioral research designed to yield scholarly new evidence informing the human impact of catastrophic events – in effect, sudden science – does have its rightful place among the various responses to a disaster.

And one may ask just why this is the case? Provided that ethical safeguards are established and applied throughout the duration of the project,[1] a carefully designed and well-executed research project can yield a trove of valuable new information. Importantly, the majority of research participants in trauma-focused studies, including studies of natural disasters, report that the benefits of participation outweigh any negative consequences. This is true even among those who initially experience some emotional distress as part of their participation (Newman & Kaloupek, 2004). Participants may derive both direct benefits (e.g., learning about available resources and being able to talk about the trauma in a supportive environment) and indirect benefits related to novel findings that may help others in their community. Knowledge generation, which happens when sound research studies are carried out, is key for the development of evidence-based interventions to mitigate adversity in the wake of a disaster. New research findings can also dispel misconceptions that may negatively impact disaster survivors' well-being. In short, knowledge generation, via the efforts of sudden science, has its rightful place in disaster response.

Three Examples of Sudden Science

For the first author, sudden science began as a research exercise in the fall of 2005. The purpose of this exercise was to confirm the integrity of the Louisiana Healthy Aging Study (LHAS) sample after Hurricanes Katrina and Rita, two back-to-back disasters, happened in rapid succession. The LHAS was an interinstitutional and multidisciplinary research project on the determinants of longevity and healthy aging in the Louisiana population. Funded by a National Institute on Aging (NIH) program project grant, data collection for the LHAS had begun in 2002. On August 29, 2005, Hurricane Katrina pummeled the US Gulf Coast region, which includes the states of Texas, Louisiana, Mississippi, Alabama, and Florida. A catastrophic

[1] As is the case for all research with human subjects, researchers must obtain IRB oversight to carry out a sudden science project. Sometimes an existing research program is already underway when a disaster strikes, as has been our experience with the LHAS. Even still, an ongoing project would need to have the existing protocol amended to cover a disaster assessment.

event, Katrina's 90,000 square miles swath of destruction was similar in size to Great Britain (Weisler, Barbee, & Townsend, 2006). A few weeks later, Hurricane Rita, a second Category 3 hurricane at landfall, struck the western side of Louisiana and southeast Texas on September 24, 2005. There is a voluminous literature documenting the psychosocial consequences of these disasters (e.g., Cherry, 2009; Galea, Tracy, Norris, & Coffey, 2008b; Kessler et al., 2009). Hurricane Katrina and Rita's devastating impact on commercial fishers and workers in the shrimping industry along the US Gulf Coast is also well known (Cherry, Marks, Adamek, & Lyon, 2015b; Cutter et al., 2006).

Looking back to the fall of 2005, Cherry and her colleagues were in the third year of LHAS data collection when Katrina struck. Research participants, many of whom were nonagenarians (90 years of age and older), lived in Baton Rouge and surrounding communities, which were at least 85 miles inland from the coast. Consequently, Katrina's impact was indirect, and damages were minimal for the LHAS participants compared to coastal residents who lost their homes and communities. Nevertheless, there was urgency at the time to determine how Hurricane Katrina may have impacted LHAS nonagenarians and their younger counterparts. With pre-Katrina data available for these participants, the research team was also in a unique position to generate new knowledge about disaster impacts on very old adults based on multiple cognitive and psychosocial health indicators and a prospective design (Cherry et al., 2011; Cherry, Galea, & Silva, 2008; Cherry, Silva, & Galea, 2009). In the paragraphs that follow, we outline the LHAS hurricane assessment, highlighting the methodological features of this work, along with some first findings that emerged from this early excursion into the realm of sudden science.

Katrina and Rita in 2005: The LHAS Hurricane Assessment

When Katrina struck on August 29, 2005, all of the New Orleans–based LHAS collaborators had evacuated to distant cities and different states. Baton Rouge, LA, was a receiving city for countless displaced coastal residents, becoming the largest parish (county) in the state virtually overnight. On the Louisiana State University campus, the Maddox Fieldhouse was used as a temporary field hospital to treat and triage medically challenged evacuees. The Pete Maravich Assembly Center was also used as a temporary shelter for evacuees. Interested readers are referred elsewhere for coverage of LSU's remarkable disaster response in the wake of Katrina (see Bacher, Devlin, Calongne, Duplechain, & Pertuit, 2005; Cherry, Allen, & Galea, 2010a).

Fortunately, the Institutional Review Board at LSU was fully operational during this chaotic time. After a brief research meeting to conceptualize the scope of work for LHAS hurricane assessment, an amendment was made to the existing IRB protocol to permit recontacting and retesting a subset of LHAS participants for whom pre-Katrina data were available. To enhance the scholarship and rigor, we sought guidance from a social epidemiologist with expertise in post-disaster assessment. The LHAS protocol was further amended to include an assessment of storm impact; symptoms of depression, anxiety, and post-traumatic stress; as well as open-ended questions that were new to the LHAS protocol. Following IRB approval, those participants who had completed LHAS testing up to 8 months before Katrina were identified and contacted for participation in the hurricane assessment.

In brief, three age groups comprising middle-aged (47–64 years), older (65–89 years), and the oldest-old adults (90 years and older) were tested during the immediate impact period, defined as October 2005 to January 2006 (Wave 1 testing). All completed a subset of LHAS health indicators for which we had pre-Katrina data, a structured storm impact questionnaire, and the three new mental health measures. We also included a short set of open-ended questions to capture participants' stories and reflections on this historic environmental event. Narrative responses to the open-ended questions were recorded and later subjected to qualitative analyses based on grounded theory methodology (Marks, 2015). To measure potential changes in responses over time, a follow-up hurricane assessment was conducted, which took place from March to November 2006 (Wave 2 testing). We controlled for the amount of time between Wave 1 and Wave 2 assessments by retesting LHAS participants within a uniform (8 ± 3 months) interval. At Wave 2, we added a group of younger controls (26–54 years), a religiosity questionnaire developed for this study, and two more open-ended questions designed to capture religious beliefs and practices as a form of coping with post-disaster stressors and the role of faith-based communities in the disaster response (Marks, Cherry, & Silva, 2009; Tausch et al., 2011).

In short, the LHAS hurricane assessment was conceptualized quickly and carried out speedily – an exercise in sudden science that fulfilled its purpose. The research team accomplished what we had set out to do, which was to gather evidence to clarify Katrina and Rita's impact on the LHAS sample. Findings confirmed that LHAS participants' scores were no different than their pre-Katrina baseline on nearly all of the principle dependent measures assessed during the immediate aftermath (Cherry, Galea, et al., 2010b) and at Wave 2 follow-up (Cherry et al., 2011; Silva Brown et al., 2010). The interesting exception to this cache of null outcomes concerned the cognitive performance data, where middle-aged and older participants' scores on a measure of working memory had dipped during the immediate impact period (Wave 1) relative to pre-Katrina baseline but returned to normal within 2 years post-event (Wave 2; see Cherry et al., 2011). A noteworthy result, this finding implies that disaster stressors may negatively impact cognition (especially for middle-aged and older adults) during a natural disaster's immediate aftermath, as discussed more fully later on. Lastly, the inclusion of open-ended questions in the hurricane assessment during both waves of testing proved to be propitious, yielding a variety of new findings concerning volunteerism after a natural disaster (Silva, Marks, & Cherry, 2009), religiosity as a coping mechanism (Marks et al., 2009; Tausch et al., 2011), and perceived silver linings associated with the hurricanes (Stanko et al., 2015).

Five Years Later: What Happened Next?

In anticipation of the 5-year anniversary of Hurricanes Katrina and Rita, we were motivated to address the question of what happened to the evacuees who had lost their homes in the 2005 storms (Hatch, Cherry, Kytola, Lu, & Marks, 2015a). By 2010, some had returned to their devastated coastal parish communities to rebuild homes and recover former lifestyles, while others had relocated permanently inland to new homes and cities in south Louisiana and elsewhere. The decision to return versus relocate after a natural disaster is certainly not a trivial one, having numerous consequences on many levels, yet there was surprisingly little research at that time to address this question empirically (Kytola, Cherry, Marks, & Hatch, 2015). There was also scant data available on older adults' health and well-being 5 years (or more) post-disaster. Building on the earlier LHAS Hurricane Assessment, the next logical step was to look at risk and resilience characteristics among directly affected adults who had lost homes and property in these two back-to-back disasters.

Resilience is a rich concept with a long history, whose origins can be traced to the field of developmental psychology (Cherry & Galea, 2015). For the present discussion, resilience pertains to one's ability to "bounce back" after adversity (Bonanno, 2004; Bonanno, Galea, Bucciarelli, & Vlahov, 2007; Bonnano & Gupta, 2009). To study post-Katrina resilience, Cherry and her colleagues used a mixed method design with quantitative (surveys, questionnaires) and qualitative (open-ended questions) measures. Two groups of adults who had lost homes in Katrina were selected for comparison: former coastal parish residents (relocated permanently) and current coastal residents (returned to rebuild). We used the same structured storm impact questionnaire and mental health measures (depression, post-traumatic stress, anxiety) from the earlier LHAS Hurricane Assessment. A resilience questionnaire was added to the protocol. In addition, seven open-ended questions were administered, two of which were the same religiosity questions used in the earlier LHAS Hurricane Assessment.

By January of 2010, new grant funding for the post-Katrina resilience study was obtained from Louisiana Board of Regents. We began recruiting former coastal parish residents who had relocated permanently to the greater Baton Rouge area. St. Bernard Parish (5 miles southeast of New Orleans) was selected as the catchment area for current coastal residents who had returned to rebuild their homes after Katrina. This new project was proceeding smoothly with concurrent testing sites established in Baton Rouge and St. Bernard when 4 months into data collection something terrible happened.

On April 20, 2010, the Deepwater Horizon Oil Rig exploded in the Gulf of Mexico, killing 11 workers and setting into motion an oil spill, which would continue unabated for a 3-month period. This technological disaster was the worst marine oil spill in US history, with threats to the seafood industry and local culture for commercial fishers for whom water was a way of life (Cherry et al., 2015a; Lyon, Nezat, Cherry, & Marks, 2015). As time went on, it became clear that the unexpected oil spill would alter the course of post-Katrina recovery for coastal residents, especially those whose livelihoods were tied to the commercial fishing industry. Once again, a new sudden science opportunity presented itself, this time, in the form of a technological disaster.

As is the case with exercises in sudden science, a quick response was required. Once again, Cherry sought a modification of the existing IRB protocol and additional funding to extend the post-Katrina resilience study. Briefly, three design changes were made. We added the following: (1) two comparison groups – a non-coastal control group (Baton Rouge area residents) and a group of commercial fishers (current coastal residents put out of work by the oil spill); (2) a coping measure to assess how fishers and their families were coping with stress from the oil spill and an open-ended question to assess the impact of the oil spill on the community; and (3) Plaquemines Parish to enlarge the catchment area for a more diverse sampling of commercial fishers in the shrimp and oyster industries. By adding fishers to the design and enlarging the catchment area, we were in a new position to measure post-Katrina resilience while accounting for the impact of the oil spill. A challenging situation, where sudden science gave way to a new realization – commercial fishers represented a double disaster exposure group (Cherry et al., 2017c; Cherry, Marks, et al., 2015b).

To summarize, the post-Katrina resilience study, which was intended to be a small-scale project, was enlarged greatly in response to an embedded technological disaster. Instead of completing the post-Katrina resilience study within a 6-month time frame as proposed, we finished strong over a 30-month interval with a final sample size that was nearly double what had been proposed. With more time, resources, and participants, the expanded scope of work allowed us to generate a wider swath of knowledge concerning resilience characteristics (e.g., optimism and hope, see Cherry, Sampson, Galea, et al., 2017b); spirituality and humor (see Cherry et al., 2018); perceived silver linings (Hatch, Cherry, Lu, & Marks, 2015b; Stanko et al., 2015) and long-term health impacts for older Katrina survivors (Cherry et al., 2015c; Cherry, Sampson, Galea, et al., 2017a, 2017b; Cherry, Sampson, Nezat, et al., 2017c). Some serendipitous findings were also brought to light concerning religious coping (see Marks, Hatch, Lu, & Cherry, 2015a; but see Marks, Lu, Cherry, & Hatch, 2015b; Stanko et al., 2018). With two sudden science research exercises now completed with peer-reviewed articles, book chapters, and two edited volumes published, the time seemed right to give disaster assessments a rest - until it started raining in south Louisiana, as dicussed next.

Louisiana's Historic Flooding in 2016

On Monday, August 8, 2016, a storm system lingered over south Louisiana. What began as a typical rainy day accelerated to a relentless downpour that lasted for several days. Local schools were cancelled to protect teachers, staff, and students from the perils of torrential rain and flash flooding. Rivers and creeks overtopped their banks flooding private homes, schools, places of worship, and businesses alike.

Directly impacted persons were boated out of their flooded homes and neighborhoods to staging areas and then airlifted to higher ground. By Monday, August 15, 2016, an estimated 7.1 trillion gallons of water had fallen over a 22-parish (county) area (Sergent, Padilla, Loerkhe, Petras, & Thorson, 2016). By comparison, Katrina dumped an estimated 2.3 trillion gallons of water on south Louisiana in 2005. An astonishing event, the 2016 flooding brought over three times as much rain as did Katrina in 2005. Once again, Louisiana residents faced the excruciating reality of catastrophic flooding, lives lost, and property damage estimated in the billions of dollars.

It may be difficult to grasp precisely how much water fell from the sky across those fateful days. Technically speaking, 1 trillion is equivalent to 1 thousand billion, or 1,000,000,000,000. To understand the flooding, one must take 7.1 multiplied by that large number, to determine the amount of rain that fell, culminating in the 2016 flood. For many individuals, 12 zeros is difficult to conceptualize in real terms, but perhaps it helps to clarify the vast amount of water that fell in such a short time. Additionally, the heavy rain was not expected, as is often the case with a named storm like a hurricane. We make these points to emphasize the psychosocial consequences of this historic environmental event for people who lost homes and property in the 2016 flood. Furthermore, the Baton Rouge metropolitan region was presumably a safer and higher ground for former coastal residents who had relocated permanently to this area after Katrina. Flooded in 2005, now flooded again in 2016 without advance warning, the psychosocial consequences of double disaster exposure warrant humanitarian concern as well as scientific attention.

The Louisiana State University (LSU) Flood Study

Thinking about the 2016 flooding, seeing the devastation day in and day out, driving past miles of mountainous debris piles by the side of the road, while trying to carry on as residents of a disaster zone was challenging and painful. From an academic point of view, however, this flood brought a new sudden science opportunity right to our doorsteps. In this spirit, we modified the existing IRB protocol from the prior study on post-Katrina resilience and quickly put together a new research proposal focusing on disaster stressors and cognition across the adult lifespan. The National Science Foundation (NSF) has a program to support disaster-related research that must be carried out quickly, and our proposal received positive feedback. With the new funding secured, we embarked on the latest sudden science research endeavor, the LSU Flood Study.

In a nutshell, the LSU Flood Study addresses adult cognition within a framework relating disaster exposure to performance. The three-parish (county) catchment area included East Baton Rouge, Livingston, and Ascension, which were the hardest hit and most severely damaged areas in the 2016 flooding. The three principle comparison groups included the following: (1) a non-flooded control group, (2) once-flooded group of adults who had structural damage due to water in their homes and loss of property in 2016, and (3) a twice-flooded group who lost homes and property in 2005 (Katrina and Rita) and flooded again in 2016. The principle dependent measures include paper and pencil measures of cognition and crystalized intelligence, along with computerized measures of working memory span and episodic memory. Two waves of testing are included in the design to examine potential changes in responses over time. At the time of this writing, data collection for the LSU Flood Study is nearing completion as we eagerly anticipate the data analyses and manuscript preparation phases of this project.

Conclusions and Implications

Our primary objective in this chapter has been to introduce the concept of "sudden science," referring to the practice of quickly pulling together and executing a research project in the immediate aftermath of a disaster. Three examples of sudden science were presented for illustration. Amid the many challenges faced and workarounds developed over the last 15 years of doing sudden science, there have been risks taken, punishments received, and rewards realized that warrant brief mention. In the sections that follow, we present three broad categories of lessons learned based on the sudden science examples presented earlier. The chapter closes with a reminder of the vulnerabilities as well as resilient outcomes that can be seen in survivors after a disaster. Such observations are important, in that they may be pivotal for future research, as well as serving to remind that life goes on after a disaster (Cherry, 2020).

Lessons Learned

Table 8.1 presents a broad-brush of three categories of concern that may arise during the conduct of sudden science.

The discussion that follows is confined to several examples per category (ethical, methodological, pragmatic) owing to space limitations. Interested readers are referred elsewhere for more extensive coverage of ethical and methodological issues associated with disaster research (Benight & McFarlane, 2007; Ferreira, Buttell, & Ferreira, 2015; Galea, Maxwell, & Norris, 2008a; Knack et al., 2006).

Ethical lessons learned Ethics in research warrant careful consideration under ordinary circumstances. In the case of sudden science, ethical concerns may be exponential. Inspection of Table 8.1 (first panel) lists several issues that may come into play. Participants in disaster research, by definition, have been recently exposed to a catastrophic event, where losses can be substantial. Interestingly, disaster status does not currently fall under the IRB designation of vulnerable (i.e., less than 18 years of age, pregnant, or incarcerated), yet participants who have lived through

Table 8.1 Sudden science categories of concern

Ethical concerns	Methodological concerns	Pragmatic concerns
1. Stressed participants While they may not meet an IRB definition of vulnerable, disaster-exposed participants may be frustrated and possibly short-tempered in connection with disaster-related stressors. They may be exhibiting symptoms of depression, anxiety, and post-traumatic stress	*1. Prospective designs are ideal in principle but atypical in practice* Unless one is in the middle of an ongoing project when a disaster strikes, researchers are unlikely to have reliable indicators of pre-disaster functionality. Lack of pre-disaster baseline is more likely scenario	*1. Establishing community-based testing sites* Disaster assessments may mean fieldwork for research teams quite literally. For scientists who are accustomed to the comfort and privacy of their lab, conducting a sudden science project in a field setting, outside of the laboratory, can be daunting and fraught with logistical challenges
2. Research isn't therapy There are many reasons why people sign up to participate in research. Disaster-exposed participants may wrongly assume that research sessions are therapy or have therapeutic value. They may inadvertently treat the research assistant as a therapist or seek a shoulder to cry on, which is a primary ethical concern	*2. Mixed method approaches yield rich sources of data but tend to be labor-intensive* A mixed method approach allows researchers to capture participants' responses quantitatively (with surveys and questionnaires) and collect their unique experiences and insights (qualitatively). Depending on the final sample size, it may take a long time to transcribe and content analyze narrative responses Qualitative findings (i.e., emergent themes that arise from open and axial coding of narrative text) may be useful to guide future research efforts in areas where extant literature may be lacking	*2. Locating, scheduling, and testing participants in a disaster context* Disaster-exposed participants may not be able to finish a research session, because they must meet with a contractor or they have other demands related to reconstruction of damaged homes and properties. They may cancel and reschedule many research visits. They may be willing to complete some (but not all) surveys and computer tasks, or discontinue completely
3. Psychological First Aid (PFA) PFA referrals may be needed for distressed research participants who want to talk about their experiences with trained counselors who will listen and can provide assistance with resource linkage	*3. Longitudinal assessments are desirable but time-consuming and expensive* Ideally, participants should be followed overtime. A longitudinal design, with at least three points of reference, is desirable to show changes in responses over time and to determine trajectories of recovery	*3. Training and supervising competent research assistants* Disaster assessments are labor-intensive. Depending on the sample size and testing format, it may be necessary to have many research assistants to carry out a project. These assistants should be properly vetted, carefully chosen, and monitored often to avoid experimenter drift and to minimize error variance due to irregularities or idiosyncrasies in administering the protocol
4. Participants have developmental histories Answering questions about the current disaster may prompt memories of prior disasters or other traumatic experiences. Measuring prior lifetime trauma is a key methodological consideration, yet participants may balk at being asked to disclose prior trauma exposure(s)	*4. Identifying an appropriate non-disaster control group isn't easy* Assigning participants to directly versus indirectly affected comparison groups status isn't as easy as one might think, depending on the scope of the disaster	

a disaster may be understandably stressed or traumatized. Among the many stressors that disaster exposed individuals and families may experience are the loss of homes and property, vehicles, jobs, neighbors, and lifestyles. Some survivors may have witnessed death. Behaviorally, disaster survivors may be frustrated, ill mannered, and possibly short-tempered. Clinically, they may show elevations on symptom counts for depression, anxiety, and post-traumatic stress, a common occurrence in the first 2 years after a disaster.

Another grim reality is that survivors may be necessarily cautious due to the perceived risk of fraud or scams in connection with unscrupulous contractors or other charlatans who take advantage of the chaotic post-disaster circumstances to cheat survivors for personal gain. These realities complicate fieldwork. Researchers who live in or work near a disaster site may already have an established presence in the community. Prior relationships that predate a disaster can be leveraged to counter the inherent distrust, which may stymie the research effort. Building on preexisting community partnerships facilitates recruitment and may be key to the success of a sudden science project. Even when researchers do not have prior relationships to build on, they can still foster trust and communicate legitimacy in support of a sudden science project in other ways. For instance, the LSU disaster research team wears name tags that bear the LSU logo and department affiliation. We also routinely wear the traditional LSU purple (or gold) polo shirt for recruitment events and during field testing. Our experience has been that a research team clad in LSU's school colors brings both a legitimacy and a certain sense of familiarity with the university brand that engenders trust and encourages participation. Correspondence on departmental letter head stationary, as well as providing information about the research project on websites hosted by the university, also frames sudden science within the larger context of LSU as a familiar and trusted research institution.

Participants in a sudden science project may want to talk about their disaster experiences. Indeed, some questions in a typical disaster assessment will solicit this type of information to assess disaster impacts and losses. Nonetheless, a prudent social scientist will remember the maxim that "research isn't therapy." While participants may want to transform a research meeting into a therapeutic session, doing so would be unethical. Sometimes all a participant may want is reassurance that things will get better, or possibly a shoulder to cry on. A sensitive researcher may pick up on participants' need to talk about what they have experienced and may also provide validation support in the form of active listening. While not providing therapy in a strict clinical sense, listening to participants in a supportive manner and assisting with resource linkage and referrals when appropriate are in line with best practices for survivors after a disaster or other crisis situation (WHO, 2011). Even still, researchers should tread lightly on this sacred ground of survivors' personal narratives and feelings in connection with reporting their disaster experiences. Clarifying the purpose of sudden science research from the outset and throughout the duration of the project may help in this regard. It is also up to the researchers to dispel myths or misconceptions that may creep in along the way – research sessions are not therapy, even if participants perceived some therapeutic value by virtue of answering research-based questions.

Lastly, it bears mentioning that participants come to a research session with unique, varied, and sometimes tragic developmental histories. Answering questions about the current disaster may prompt memories of prior disasters or other traumatic experiences. Measuring prior lifetime trauma is a methodological necessity, yet participants may balk at being asked to disclose prior lifetime trauma exposure(s). We turn to a discussion of methodological matters in sudden science next.

Methodological lessons learned As noted in Table 8.1 (second column), the use of a prospective design is ideal but not always possible. Prospective designs, where salient participant characteristics are known beforehand, permit clearer and less ambiguous inferences about post-disaster outcomes. The LHAS hurricane assessment was unusual in that we had pre-Katrina data and were fortunate to demonstrate many nonsignificant differences in the pre- to post-Katrina comparisons with LHAS data. With many sudden science endeavors (like the Deepwater Horizon Oil Spill assessment and LSU Flood Study), researchers may collect data immediately after an environmental event has happened without the benefit of access to pre-disaster knowledge about participants. Thus, a researcher cannot unequivocally attribute outcomes to the disaster, as these differences may have existed before the event.

Another methodological concern is the choice of a design. Both the LHAS hurricane assessment and the post-Katrina Resilience Study used mixed method designs that captured participants' responses quantitatively (with surveys and questionnaires) and their unique experiences and insights qualitatively (with open-ended questions). Survey measures with well-established psychometric properties are ideal, given the ease of administration, data scoring, and analyses. Open-ended questions are beneficial, too, in that participants may answer with as much or as little detail as they want, which is important when dealing with sensitive topics. Further, qualitative assessments may be especially valuable to explore areas where prior research may be minimal or nonexistent. Responses to open-ended questions are digitally recorded and transcribed verbatim, and then content is analyzed for meaning and recurrent themes (Marks, 2015). Depending on the sample size, transcribing and content analyzing narrative responses can be labor intensive and may take years to complete.

Lastly, other design choices include whether to follow participants over time in a longitudinal design. Although time consuming and expensive, longitudinal assessments are desirable in that they permit inferences on causality. Furthermore, a longitudinal design, with at least three points of reference, is necessary to determine trajectories of stress responses over time (see Norris, Tracy, & Galea, 2009). A related design issue concerns the question of whether a control group should be included in the design and, if so, who the control (non-disaster exposed) participants might be. When a disaster happens, the entire communities are impacted. Those who have not lost homes in a disaster may seem to be an appropriate control group, yet they often have family and friends who were directly impacted. They may also be providing shelter to displaced family members and friends and are thus indirectly affected by disaster.

Pragmatic lessons learned There are a host of logistical concerns and practicalities associated with sudden science (see Table 8.1, last column). Depending on the type of the disaster, establishing community-based testing sites may be desirable to allow a wide sampling of potential participants. Disaster assessments as fieldwork can literally take the form of research assistants on foot with clipboard and pens fanning out through a disaster impacted community (in pairs or triads as a safety precaution, see Knack et al., 2006 for a related discussion). Our experiences have included testing participants in places of worship and public libraries. We have found that it is vitally important to the success of sudden science recruitment efforts to reach people in their communities. Our outreach efforts have included establishing new connections through local churches and civic organizations, giving talks and workshops, disseminating flyers, sending letters to former research participants to generate word of mouth referrals, and creating an online screening tool posted to social media. For scientists who are accustomed to the familiarity and privacy of their lab, conducting a sudden science project in a field setting can be daunting and fraught with challenges ranging from scheduling space to lack of control while testing sessions are in progress.

Locating, scheduling, and testing participants after a disaster brings a new set of logistical challenges for sudden science researchers to overcome. Disaster-exposed participants may not be able to finish a planned research session, because they must meet with a contractor or other demands on their time related to reconstruction of damaged homes and properties. They may cancel and reschedule many research visits. They may discontinue or only be willing to complete some (but not all) surveys and computer tasks.

Lastly, selecting, training, and supervising competent research assistants who will be responsible for carrying out a sudden science exercise is vitally important to the overall success of the project. The realities of experimenter drift, where overfamiliarity with the protocol leads to sloppy data collection, add to the response burden for the participant as well as introducing unnecessary experimental error, a no-win situation.

Sidestepping Puddles of Vulnerability Discovering Pockets of Resilience

People who have lost homes in severe weather events, like hurricanes or floods, understand the pain and heart-wrenching grief that comes with seeing a lifetime of personal possessions lying in debris piles amid crumbling sheet rock, ruined furniture, appliances, and clothing at the end of the driveway. Researchers who conduct sudden science have heard many stories of loss and grief, too. With ethical safeguards in place and a heart to learn from disaster-exposed adults, researchers in the field collecting data from people who have suddenly lost everything must be mindful to step around puddles of vulnerability with distressed participants. At the same time, researchers should also be aware of pockets of resilience that may also be discovered after a disaster. For instance, neighbors helping neighbors is a common sight as communities come together after a disaster has happened (i.e., the Cajun Navy). From the early hours and days of sprinting to save lives in the immediate aftermath to the marathon of long-term disaster recovery, which may involve wrestling with insurance claims, unscrupulous contractors, accepting painful losses, and discovering a *new normal*, in whatever form, it may take after a disaster.

In closing, natural disasters and severe weather events are happening with increasing frequency today, which brings both tragic happenings and new opportunities for sudden science endeavors to advance the literature. Thoughtful, carefully designed, and well-executed research on individual and community-level risk and resilience characteristics will ultimately lead to evidence-based best practices, an important challenge that awaits future sudden science research.

Acknowledgments Correspondence concerning this article should be addressed to Katie E. Cherry, Department of Psychology, Louisiana State University, Baton Rouge, LA 70803-5501 (telephone: 225-578-8745, fax: 225-578-4125; e-mail: pskatie@lsu.edu).

We express our sincere thanks to Dr. S. Michal Jazwinski, Program Project Director of the Louisiana Healthy Aging Study, for his vision and guidance in developing the LHAS Hurricane Assessment. We are also grateful to Dr. Sandro Galea, Dean of the School of Public Health at Boston University, for his continued support of the disaster research program at LSU. We thank the scores of undergraduate and graduate research assistants at LSU for their many and varied contributions to the research effort.

This research was supported by grants from the National Institute on Aging, the National Science Foundation, the Louisiana Board of Regents, and the Louisiana State University Office of Research and Economic Development. This support is gratefully acknowledged.

References

Bacher, R., Devlin, T., Calongne, K., Duplechain, J., & Pertuit, S. (2005). *LSU in the eye of the storm: A university model for disaster response*. Baton Rouge, LA: LSU Press.

Benight, C. C., & McFarlane, A. C. (2007). Challenges for disaster research: Recommendations for planning and implementing disaster mental health studies. *Journal of Loss and Trauma, 12*(5), 419–434.

Bonanno, G. (2004). Loss, trauma, and human resilience: Have we underestimated the human capacity to thrive after extremely aversive events? *American Psychologist, 59*, 20–28.

Bonnano, G. A., Brewin, C. R., Kaniasty, K., & La Greca, A. M. (2010). Weighing the costs of disaster: Consequences, risks, and resilience in individuals, families, and communities. *Psychological Science in the Public Interest, 11*, 1–49.

Bonanno, G. A., Galea, S., Bucciarelli, A., & Vlahov, D. (2007). What predicts psychological resilience after disaster? The role of demographics, resources, and life stress. *Journal of Consulting and Clinical Psychology, 75*, 671–682.

Bonnano, G. A., & Gupta, S. (2009). Resilience after disaster. In Y. Neria, S. Galea, & F. H. Norris (Eds.), *Mental health and disasters* (pp. 145–160). New York: Cambridge University Press.

Cherry, K. E. (Ed.). (2009). *Lifespan perspectives on natural disasters: Coping with Katrina, Rita and other storms*. New York: Springer.

Cherry, K. E. (Ed.). (2015). *Traumatic stress and long-term recovery: Coping with disasters and other negative life events*. New York: Springer.

Cherry, K. E. (2020). *The other side of suffering: Finding a path to peace after tragedy*. New York: Oxford University Press.

Cherry, K. E., Allen, P. D., & Galea, S. (2010a). Older adults and natural disasters: Lessons learned from Hurricanes Katrina and Rita. In P. Dass-Brailsford (Ed.), *Crisis and disaster counseling: Lessons learned from hurricane Katrina and other disasters* (pp. 115–130). Thousand Oaks, CA: Sage.

Cherry, K. E., & Galea, S. (2015). Resilience after trauma. In D. Ajdukovic, S. Kimhi, & M. Lahad (Eds.), *Resiliency: Enhancing coping with crisis and terrorism: NATO science for peace and security series – E human and societal dynamics* (pp. 35–40). Amsterdam: IOS Press.

Cherry, K. E., Galea, S., & Silva, J. L. (2008). Successful aging and natural disasters: Role of adaptation and resiliency in late life. In M. Hersen & A. M. Gross (Eds.), *Handbook of clinical psychology: Volume 1* (pp. 810–833). Hoboken, NJ: Wiley.

Cherry, K. E., Galea, S., Su, L. J., Welsh, D. A., Jazwinski, S. M., Silva, J. L., et al. (2010b). Cognitive and psychosocial consequences of Hurricanes Katrina and Rita on middle aged, older, and oldest-old adults in the Louisiana Healthy Aging Study (LHAS). *Journal of Applied Social Psychology, 40*, 2463–2487.

Cherry, K. E., Lyon, B. A., Marks, L. D., Nezat, P. F., Adamek, R., Fitzgerald, K. B., et al. (2015a). After the BP Deepwater Horizon Oil Spill: Financial and health concerns among coastal residents and commercial fishers. *Current Psychology, 34*, 576–586.

Cherry, K. E., Marks, L. D., Adamek, R., & Lyon, B. A. (2015b). Younger and older coastal fishers face catastrophic loss after Hurricane Katrina. In K. E. Cherry (Ed.), *Traumatic stress and long-term recovery: Coping with disasters and other negative life events* (pp. 327–348). New York: Springer.

Cherry, K. E., Sampson, L., Galea, S., Marks, L. D., Baudoin, K. H., & Nezat, P. F. (2017a). Health-related quality of life in older coastal residents after multiple disasters. *Disaster Medicine and Public Health Policy, 11*(1), 90–96. https://doi.org/10.1017/dmp.2016.177

Cherry, K. E., Sampson, L., Galea, S., Marks, L. D., Nezat, P. F., Baudoin, K. H., et al. (2017b). Optimism and hope after multiple disasters: Relationships to health-related quality of life. *Journal of Loss and Trauma, 22*(1), 61–76.

Cherry, K. E., Sampson, L., Nezat, P. F., Galea, S., Marks, L. D., & Lyon, B. A. (2017c). Prior hurricane and other lifetime trauma predict coping style in older commercial fishers after the BP Deepwater Horizon oil spill. *Journal of Applied Biobehavioral Research, 22*, e12058. https://doi.org/10.1111/jabr.12076

Cherry, K. E., Sampson, L., Galea, S., Marks, L. D., Stanko, K. E., Nezat, P. F., et al. (2018). Spirituality, humor, and resilience after natural and technological disasters. *Journal of Nursing Scholarship, 50*(5), 492–501.

Cherry, K. E., Sampson, L., Nezat, P. F., Cacamo, A., Marks, L. D., & Galea, S. (2015c). Long-term psychological outcomes in older adults after disaster: Relationships to religiosity and social support. *Aging & Mental Health, 19*, 430–443.

Cherry, K. E., Silva, J., & Galea, S. (2009). Natural disasters and the oldest-old: A psychological perspective on coping and health in late life. In K. E. Cherry (Ed.), *Lifespan perspectives on natural disasters: Coping with Katrina, Rita and other storms* (pp. 171–193). New York: Springer.

Cherry, K. E., Silva Brown, J. L., Marks, L. D., Galea, S., Volaufova, J., Lefante, C., et al. (2011). Longitudinal assessment of cognitive and psychosocial functioning after Hurricanes Katrina and Rita: Exploring disaster impact on middle-aged, older, and oldest-old adults. *Journal of Applied Biobehavioral Research, 16*, 187–211.

Cutter, S. L., Emrich, C. T., Mitchell, J. T., Boruff, B. J., Gall, M., Schmidtlein, M. C., et al. (2006). The long road home: Race, class, and recovery from Hurricane Katrina. *Environment, 48*, 10–20.

Ferreira, R., Buttell, F., & Ferreira, S. (2015). Ethical considerations for conducting disaster research with vulnerable populations. *Journal of Social Work Values and Ethics, 12*(1), 29–40.
Galea, S., Maxwell, A. R., & Norris, F. (2008a). Sampling and design challenges in studying the mental health consequences of disasters. *International Journal of Methods in Psychiatric Research, 17*(S2), S21–S28.
Galea, S., Tracy, M., Norris, F., & Coffey, S. (2008b). Financial and social circumstances, the incidence and course of PTSD in Mississippi during the first two years after Hurricane Katrina. *Journal of Traumatic Stress, 21*, 357–368.
Hatch, T. G., Cherry, K. E., Kytola, K. L., Lu, Y., & Marks, L. (2015a). Loss, challenge, chaos, and failure of systems: The storm after the storm. In K. E. Cherry (Ed.), *Traumatic stress and long-term recovery: Coping with disasters and other negative life events* (pp. 231–245). New York: Springer.
Hatch, T. G., Cherry, K. E., Lu, Y., & Marks, L. D. (2015b). On seeing silver linings after the 2005 Hurricanes Katrina and Rita: Meaning making, positive adaption, and psychological growth. In K. E. Cherry (Ed.), *Traumatic stress and long-term recovery: Coping with disasters and other negative life events* (pp. 389–402). New York: Springer.
Kessler, R. C., Galea, S., Gruber, M. J., Sampson, N. A., Petukhova, M., & Wang, P. S. (2009). Hurricane Katrina. In Y. Neria, S. Galea, & F. H. Norris (Eds.), *Mental health and disasters* (pp. 419–440). New York: Cambridge University Press.
Knack, J. M., Chen, Z., Williams, K. D., & Jensen-Campbell, L. A. (2006). Opportunities and challenges for studying disaster survivors. *Analyses of Social Issues and Public Policy, 6*(1), 175–189.
Kytola, K. L., Cherry, K. E., Marks, L. D., & Hatch, T. G. (2015). When neighborhoods are destroyed by disaster: Relocate or return and rebuild? In K. E. Cherry (Ed.), *Traumatic stress and long-term recovery: Coping with disasters and other negative life events* (pp. 211–229). New York: Springer.
Lyon, B. A., Nezat, P. F., Cherry, K. E., & Marks, L. D. (2015). When multiple disasters strike: Louisiana fishers in the aftermath of hurricanes and the British Petroleum Deepwater Horizon oil spill. In K. E. Cherry (Ed.), *Traumatic stress and long-term recovery: Coping with disasters and other negative life events* (pp. 57–70). New York: Springer.
Marks, L. (2015). A pragmatic, step-by-step guide for qualitative methods: Capturing the disaster and long-term recovery stories of Katrina and Rita. *Current Psychology, 34*, 494–505.
Marks, L. D., Cherry, K. E., & Silva, J. L. (2009). Faith, crisis, coping and meaning making after Katrina: A qualitative, cross-cohort examination. In K. E. Cherry (Ed.), *Lifespan perspectives on natural disasters: Coping with Katrina, Rita and other storms* (pp. 195–215). New York: Springer.
Marks, L. D., Hatch, T. G., Lu, Y., & Cherry, K. E. (2015a). Families and faith-based communities after a disaster: Successes and failures in the wakes of Hurricanes Katrina and Rita. In K. E. Cherry (Ed.), *Traumatic stress and long-term recovery: Coping with disasters and other negative life events* (pp. 247–270). New York: Springer.
Marks, L. D., Lu, Y., Cherry, K. E., & Hatch, T. G. (2015b). Faith and coping: Spiritual beliefs and religious practices after Hurricane Katrina. In K. E. Cherry (Ed.), *Traumatic stress and long-term recovery: Coping with disasters and other negative life events* (pp. 369–387). New York: Springer.
Newman, E., & Kaloupek, D. G. (2004). The risks and benefits of participating in trauma-focused research studies. *Journal of Traumatic Stress, 17*, 383–394.
Norris, F. H., Tracy, M., & Galea, S. (2009). Looking for resilience: Understanding the longitudinal trajectories of responses to stress. *Social Science & Medicine, 68*, 2190–2198.
Richardson, R. C., Plummer, C. A., Barthelemy, J. J., & Cain, D. S. (2009). Research after natural disasters: Recommendations and lessons learned. *Journal of Community Engagement and Scholarship, 2*(1), 3–11.
Sergent, J. Padilla, R., Loerkhe, J. Petras, G., & Thorson, M. (2016, August 23). *Louisiana's historic floods*. Retrieved from http://www.usatoday.com/pages/interactives/la-floods-august-2016/

Silva, J. L., Marks, L. D., & Cherry, K. E. (2009). The psychology behind helping and prosocial behaviors: An examination from intention to action. In K. E. Cherry (Ed.), *Lifespan perspectives on natural disasters: Coping with Katrina, Rita and other storms* (pp. 219–240). New York: Springer.

Silva Brown, J., Cherry, K. E., Marks, L. D., Jackson, E. M., Volaufova, J., Lefante, C., et al. (2010). After hurricanes Katrina and Rita: Gender differences in health and religiosity in middle-aged and older adults. *Health Care for Women International, 31*, 997–1012.

Stanko, K. E., Cherry, K. E., Ryker, K. S., Mughal, F., Marks, L. D., Silva Brown, J., et al. (2015). Looking for the silver lining: Benefit finding after Hurricanes Katrina and Rita in middle-aged, older, and oldest-old adults. *Current Psychology, 34*, 564–575.

Stanko, K. E., Cherry, K. E., Marks, L. D., Sampson, L., Ryker, K. S., Barrios, B., et al. (2018). When reliance on religion falters: Religious coping and post-traumatic stress symptoms in older adults after multiple disasters. *Journal of Religion, Spirituality, and Aging, 30*, 292–313. https://doi.org/10.1080/15528030.2018.1434853

Tausch, C., Marks, L. D., Silva Brown, J. L., Cherry, K. E., Frias, T., MacWilliams, Z., et al. (2011). Religion and coping with trauma: Qualitative examples from Hurricanes Katrina and Rita. *Journal of Religion, Spirituality and Aging, 23*, 236–253.

Weisler, R. H., Barbee, J. G., & Townsend, M. H. (2006). Mental health and recovery in the Gulf Coast after Hurricanes Katrina and Rita. *The Journal of the American Medical Association, 296*, 585–588.

World Health Organization, War Trauma Foundation and World Vision International. (2011). *Psychological first aid: Guide for field workers*. Geneva, Switzerland: WHO. Retrieved from https://www.who.int/mental_health/publications/guide_field_workers/en/

Chapter 9
Role of Trauma-Informed Care in Disasters

Nancy Kusmaul

Introduction

While all geographic areas may experience all types of disasters, certain areas are more prone to specific types of disasters. The Gulf Coast of the United States, for example, is a region known to be at risk for hurricanes. But the chances of an individual hurricane hitting a particular area are less clear. In October 2018, Hurricane Michael became the strongest hurricane in history to strike the Florida Panhandle and the fourth strongest hurricane to make landfall in the US history. Tiny Mexico Beach, Florida, was all but completely flattened. According to the National Weather Service (2018), "Hurricane Michael was truly a unique event." Two months after the hurricane, news outlets reported that only 500 of 2700 dwellings in the city were habitable (Landeck, 2018).

California has a long history of fast moving, destructive wildfires (see Phillips, Chap. 10, this volume). The combination of steep terrain, cyclical droughts, and seasonal winds makes it susceptible to these conditions. In more recent years, population growth and development in these susceptible areas have increased the destructiveness of these blazes as well as their impacts on large populations (Nauslar, Abatzoglou, & Marsh, 2018). Nauslar et al. (2018) cite the interconnection of climate, weather, available naturally occurring fuels, and terrain in determining the occurrence and intensity of large wildfires. In November 2018, years of extreme drought and a pattern of warmer, drier falls combined to create one of the most intense fire seasons in California's history (Cornell, 2018). One of two major fires, the so-called Camp Fire burned through the Northern California foothills, leveling everything in its wake, including nearly the entire city of Paradise, California. In the days and weeks that followed, residents described harrowing experiences, fleeing with the flames licking at the sides of their cars, running down mountains on foot

N. Kusmaul (✉)
School of Social Work, University of Maryland, Baltimore County, Baltimore, MD, USA
e-mail: nkusmaul@umbc.edu

© Springer Nature Switzerland AG 2021

K. E. Cherry, A. Gibson (eds.), *The Intersection of Trauma and Disaster Behavioral Health*, https://doi.org/10.1007/978-3-030-51525-6_9

when roads were blocked, leaving behind possessions and pets, and about friends and neighbors who did not make it out (Winsor, 2019).

Hurricane Michael and the Camp Fire are just two examples of the many events that threaten the life and safety of individuals who live and work in affected areas that occur each year. Areas that experience these events also experience catastrophic changes in housing, employment, and other life areas which then impact many of these same individuals (see Mosby, Chap. 13, this volume). This destruction impacts physical and psychological community, both previous potential sources of support. In this sense, community is an aggregation of people with something in common. The physical community includes the places in common (McKnight & Block, 2012)—those places that are sometimes lost during a disaster such as schools, churches, and other gathering places. The psychological community is the experience of connectedness (McKnight & Block, 2012) that may be felt in engaged communities.

Globally, climate change is leading to rising ocean levels and more severe weather patterns (Jacob, 2015). Development and construction are permanently altering the landscape, sometimes removing naturally occurring barriers and protections (see Birch, Henry, and Nelson, Chap. 14, this volume). Disastrous events are occurring with greater intensity and frequency every year. Areas previously not considered disaster prone are now experiencing natural disasters. In some places, events once considered "100-year" floods, meaning that a flood of that magnitude is likely to occur only once in 100 years, have experienced these floods multiple times in the last several years (i.e., Ellicott City, Maryland; Columbia, South Carolina).

People in areas not directly affected by these major disasters may see disturbing footage of floods, firestorms, tsunamis, earthquakes, hurricanes, tornadoes, and countless other disasters and close calls or poor outcomes that occur during those events. While seeing something on TV is explicitly excluded from the Diagnostic and Statistical Manual of Mental Disorders, 5th Edition (DSM 5) definition of trauma (APA, 2013, pp. 271), it can certainly cause distress in others. For example, people who live in areas that are currently considered safe have seen areas that were considered safe affected by disasters. That may cause people to worry about potential disasters in their own areas. Secondary traumatic stress is the experiences of trauma-related symptoms in those individuals who do not directly experience the trauma but experience it secondhand through stories and long-term exposure to emotionally demanding situations (Institute of Medicine, 2015). Previously applied to health and social services workers (Institute of Medicine, 2015), there have been some reports of it appearing in the general public as well as a result of widespread news coverage (Rogers, 2013).

One does not need to be a mental health clinician to realize that these are potentially traumatic events (PTEs). While the terms trauma, disaster, PTEs, and others appear frequently in the disaster science literature, social workers, psychologists, psychiatrists, and other mental health professionals who diagnose mental disorders rely on the DSM 5, or the Diagnostic and Statistical Manual of Mental Disorders, 5th Edition (APA, 2013). This manual is assembled and updated through a review of current research and assembled through task forces and work groups. The most recent update occurred with the release of the 5th edition in 2013, after several years of task forces and committee work. The DSM 5 defines trauma as "exposure to actual or threatened death, serious injury, or sexual violence" (American Psychiatric Association, 2013, pp. 271). The DSM 5 further states that an individual has to have directly experienced the event, witnessed it, learned that a close friend or relative experienced it, or indirectly learned about the details, particularly through the course of one's work (VA, 2018). The DSM 5 defines trauma in the context of diagnosing trauma and stress-related disorders, so it makes sense that the general public also conflates post-traumatic stress disorder (PTSD) with trauma. While exposure to trauma is a required component of the diagnosis of PTSD, trauma by itself is not PTSD. PTSD refers to a specific set of negative consequences that occur as the result of being exposed to trauma, including reexperiencing the trauma, avoiding trauma-related stimuli, negative thoughts or feelings that began or worsened after the trauma, and trauma-related arousal that causes distress and occurs over a specified time frame (VA, 2018). Just how many people exposed to trauma go on to develop PTSD varies significantly depending on the population, the nature of the trauma, the co-occurrences of other traumas, and individual-level factors (Kessler et al., 2017). Further, when trauma is portrayed as only being about PTSD, those portrayals often focus on specific at-risk populations such as veterans or holocaust survivors. In reality, trauma refers to the experience of the event itself, not the outcome, and includes a much wider set of events and outcomes than the common perceptions would suggest. As Kor (2015) describes, the physical, psychological, and emotional outcomes of survivors of PTEs can vary widely.

Disasters, in this chapter, refer to large-scale events that occur with no, or minimal, warning, cannot be prevented, and result in widespread destruction and/or loss of life. These are often naturally occurring disasters such as hurricanes, tornadoes, floods, and fires but can also be man-made, as in the case of the Oklahoma City Bombing in 1995 or the attacks on the World Trade Center and the Pentagon in 2001. Natural and man-made disasters may meet the definition of trauma, particularly in cases where they threaten death or serious injury (APA, 2013, pp. 274).

While the DSM 5 (APA, 2013) defines trauma as events that threaten life, other researchers have included additional types of traumatic events that have lasting impacts. Felitti et al. (1998) called them "adverse childhood experiences" (ACEs) and included situations such as living with a parent or parents with mental health issues, parental substance use, witnessing one's mother being abused by her partner, parental separation/divorce, and a variety of abuses. Morey and Segerstrom (2015) call this nexus of events early childhood adversity and include both abuse and adversity. Whatever we call it, we know that early childhood adversity is linked to a range of physical health issues (Morey & Segerstrom, 2015), beyond the long-term psychological effects.

Gray, Litz, Hsu, and Lombardo (2004) assessed for exposure to all types of potentially traumatic events including natural disasters, fire/explosion, transportation accidents, exposure to toxic substances, assault (of various types), military combat, captivity, and suffering. It is also important for providers to consider the frequency of such events. *Was the trauma a one-time event, or did it occur frequently or over time, as in the case of ongoing abuse or living in a neighborhood prone to violence* (SAMHSA, 2014)? Felitti et al. (1998) also considered the co-occurrence of events. People in the ACEs study who endorsed one type of experience were more likely to have experienced multiple types of trauma. So our consideration of traumatic events needs to incorporate all of these things: What happened, how often it happened, and how many other types of potentially traumatic experiences did the person encounter.

SAMHSA (2014) describes these considerations as the three E's: event, experience, and effect. *What event was experienced?* While the experience of traumatic events alone does not guarantee that someone has suffered adverse psychological outcomes, it is important to consider everyone might have. SAMHSA (2014) connects this to the DSM criteria, which discusses psychological distress that follows exposure to a "traumatic or stressful event (APA, 2013, pp. 265)" as quite variable. The diagnosis of many types of "Trauma- and Stressor-Related Disorders" (APA, 2013, pp. 265) requires having been exposed to a traumatic event. Reactive attachment disorder (APA, 2013, pp. 266) and disinhibited social engagement disorder (APA, 2013, pp. 269) are included in this category but do not require exposure to a traumatic event. But for our purposes, the first question is to understand what it was that happened.

The second question SAMHSA (2014) asks is: *What was the individual's experience of that event*? This concept takes into consideration individual level factors that allow two people to experience the same event and have different outcomes. This question is critical in disaster response when multiple family members, for example, might have had the same experience and yet be coping in very different ways. When working with individuals, it is important to understand each person's ability to understand and cope with the current challenging situation in the context of challenging circumstances they may have experienced in the past. Individual level factors include how the person labels and assigns meaning to the event, as well as the level of physical and psychological disruption they experience (SAMHSA, 2014). It is important to determine their understanding of, "Why me?"

Finally, SAMHSA's third question is: *What was the effect*? This question examines the short- and long-term effects of a past traumatic experience on an individual. Again, while adverse effects are often exclusively associated with the development of "Trauma- and Stressor-Related Disorders" (APA, 2013, pp. 265), considering adverse effects can and should be broader than this. SAMHSA (2014) describes that the effects of traumatic events can occur in many ways, from long-term to short-term effects, including effects that occur immediately, and those that occur after a long period has elapsed from the exposure to the traumatic event. As I will discuss in the remainder of the chapter, when looking for the effects of traumatic events in disaster response, responders must consider both effects of the disasters and the effects of previous traumas. They may manifest as people to have the inability to cope with normal daily stressors, trouble establishing trust in relationships, difficulty paying attention, or behaviors that seem out of proportion to the situation at hand (SAMHSA, 2014), which could be difficult to evaluate in disaster response, when disaster-related events are not normal daily stressors.

By necessity, disaster response tends to focus on meeting immediate needs for shelter, food, clothing, and medical attention. Disaster services are less likely to address the psychological trauma of disasters or the fact that disasters may cause not only individual losses but the losses of community and social support. Trauma is ubiquitous to the human experience. Felitti et al. (1998) found that more than half of his sample (52%) had experienced at least one of a variety of adverse childhood experiences. In this chapter, I frame trauma-informed care from the perspective that disaster survivors may have experienced traumatic events during their disaster survival, that these disaster events may trigger reactions related to previous traumas, and that disaster response services need to be trauma-informed. I also discuss the effects of disaster response on responders and the benefit of trauma-informed services for them.

What Is Trauma-Informed Care?

Trauma-informed care is an approach that was first introduced in community social service agencies and inpatient psychiatric units nearly 20 years ago (Bloom, 2013; Harris & Fallot, 2001). This approach assumes that everyone who encounters the agency, including both staff and clients, might have had a past traumatic experience (Bloom, 2013). A trauma-informed environment treats all individuals in ways that protect trauma survivors from re-traumatization much like universal precautions presume that all blood or body fluid might contain a blood-borne disease (Kusmaul & Anderson, 2018). Given Felitti et al.'s (1998) finding of more than half of their participants having had some kind of past traumatic experience, trauma is widespread (Kessler et al., 2017), and anyone who works with people might never know who has experienced it or where it might be found.

The Substance Abuse and Mental Health Services Administration (SAMHSA, 2014) operationalizes trauma-informed care into six organizing principles. They are safety; trustworthiness and transparency; peer support; collaboration and mutuality; empowerment, voice, and choice; and cultural, historical, and gender issues. These principles were chosen because traumatic experiences are often about being in life- or health-threatening situations beyond an individual's control. Traumatic events often include a power differential—something, someone, some force, and even the force of nature with more power than the individual exerted power over that individual (SAMHSA, 2014). These principles are about restoring a sense of control and not repeating it through coercive or rigid policies and practices. While traditional approaches may promote trauma-informed principles, it is not in a systemized, thoughtful way that specifically considers trauma.

Safety

The safety component of trauma-informed care encompasses both physical and emotional safety (Kusmaul, Wilson, & Nochajski, 2015). At an organizational level, all staff and clients should be able to feel physically and psychologically safe (SAMHSA, 2014). There are many factors that influence feelings of safety, but for service recipients, both staff actions when negative events take place and staff attitudes are an important part of the experience of safety (Kusmaul, Wolf, Sahoo, Green, & Nochajski, 2019).

Safety has long been acknowledged as a fundamental part of human motivation. Maslow (1943) posited that after physiological needs, such as food, water, shelter, etc., safety needs are the next to be addressed, as "practically everything looks less important than safety" (Maslow, 1943, pp. 376). In the early days after a disaster, both resources to address physiological needs, such as food, water, and shelter, and safety, may be in short supply. Once physiological needs have been met, safety needs to be addressed. In the trauma-informed framework, no principle is more important than the others, but for disaster survivors, safety and assurances of safety are often the area of greatest risk. Depending on the circumstances of the disaster, they may have narrowly escaped life-threatening circumstances or have witnessed those who were unable to escape, threatening their sense of safety.

Trustworthiness and Transparency

Trustworthiness and transparency in service delivery is about delivery of services as promised. Many times in the days after a disaster, survivors are made promises about services available to them or benefits they will receive only to learn later that they do not qualify or such benefits are not available. It is critical that disaster responders (the individual workers) are able to provide clear, accurate information about services and that providers (larger agencies) create transparent criteria for eligibility.

This is especially important in cases where individuals have lost property or whose homes were damaged or destroyed as a result of the disaster. These two conditions have been found to be associated with depression in the aftermath of a disaster (Tang, Liu, Liu, Xue, & Zhang, 2014). Restoration of property and housing can ease distress but only if services are provided in a trustworthy way. Communication during the application process enhances the transparency of the process and increases trust in the service rather than being told after a long and secretive evaluation that eligibility criteria are not met.

Peer Support

Peer support in the SAMHSA (2014) model refers to opportunities to connect with others who are having or have had similar experiences. This can take the form of support groups, but that is not the only kind of peer support that is effective. In addition, peer support is an area that may need specific attention in the wake of a disaster when survivors may be dispersed from neighborhoods of origin and traditional sources of social support. In Hurricane Katrina, workers at a senior center that reopened after the disaster described driving through the city to determine whether members had survived, relocated, or died (Croom & Jenkins, 2007). For seniors who had relied on this peer support for daily functioning prior to the disaster, their community had been dispersed, removing established peer support networks at a time when their need for support was even higher (Roberto, Kamo, & Henderson, 2009).

Preexisting peer support, as in the case of community cohesion, has been shown to reduce the occurrence of post-disaster PTSD even in the face of significant losses during the disaster (Hikichi, Aida, Tsuboya, Kondo, & Kawachi, 2016). Further, social cohesion enhances community resilience post disaster (Hikichi et al., 2016). The opposite effect is observed when individuals must relocate because of a disaster, resulting in not just the loss of a familiar living space but the loss of "daily life patterns, social networks, and supports" (Johnson, 1996 as cited by Sanders, Bowie, & Bowie, 2003, pp. 26).

Collaboration and Mutuality

Collaboration and mutuality is about "…leveling of power differences between staff and clients…" (SAMHSA, 2014, pp. 11). In disaster response, there are complicated layers of responsibility with many different agencies having control over the same geographic area. Response may be provided by local, state, and federal government agencies, as well as national and local nonprofits (Institute of Medicine, 2015) all with varying degrees of control over response policies, timing of services, access to affected areas, and power over the provision of recovery dollars. People working directly with disaster survivors also have varying degrees of power. Volunteers such as at shelters have the lowest levels of power—limited training and limited access to information. Disaster response services should be provided in ways that level the power differentials as much as possible and allow responders and survivors to work together toward common goals.

Empowerment, Voice, and Choice

Empowerment, voice, and choice are about the ability to direct one's own future. Empowerment involves giving service recipients, in this case, disaster survivors, the power to make decisions about their own disaster recovery. Voice is the ability to tell one's own story. Choice in service provision implies that clients have the option to choose between different providers or services to address unmet needs. In disaster response, this is not always the case as when resources are stretched and a limited number of services options are available. The question of choice in disaster is also larger than that and more complicated. As the time from the disaster increases, clients have the choice of returning to devastated areas to rebuild or moving to other areas to start again. Whether or not someone believes they have a choice in that depends on whether they have resources. People with no savings, or no insurance, or inadequate funding from FEMA or similar response agencies may not have the funds to rebuild or relocate, depending on the relative costs of either.

The other consideration related to choice is related to where/how someone was during the disaster event. How much choice did individuals have in whether to evacuate and what choices did they make? Municipalities face the conflict between individual choice and their responsibility to keep the public safe. There is much debate about whether the decision to evacuate or leave an unsafe area should be a choice (Burson, Kalin, McAdam, & Weerasinghe, 2018). If it was an expected event such as a hurricane or flood, and someone chose not to evacuate which led to dire circumstances or dramatic rescue, there may be greater disaster-related trauma than if someone had followed evacuation orders. There are many factors that are associated with the choice to evacuate including gender, perceived threat, education level, and household size, especially the number and ages of children in the household (Kyne, Lomeli, Donner, & Zuloaga, 2018). In the aftermath of the disaster, survivors and responders alike may be questioning choices that were made by themselves or others that exacerbated negative situations.

Cultural, Historical, and Gender Issues

Gender may influence disaster decision-making as one study found that females are more likely to evacuate or not know what they would do in particular circumstances than males (Kyne et al., 2018). However, when it comes to trauma-informed service delivery, this category is broader and encompasses the idea that a person's experience of the event may vary due to cultural factors such as past experiences with power and power differentials (SAMHSA, 2014). This also encompasses both the individual and the community level. On the individual or family level, one example is that decisions may have been made about evacuation or response based on preexisting power differentials in the household, which may be related to culture or gender norms.

On the community level, areas most impacted by disasters are often those that have been neglected prior to disasters. While disasters can be indiscriminate, the reality is that the ill effects of disasters often impact most significantly those who have experienced previous victimization. Those who are most vulnerable in disasters have often been victimized by environmental injustice. That is, communities experience different "vulnerability, impacts, response, recovery, and resiliency" (Adeola & Picou, 2017, pp. 233), rooted in preexisting social inequality. These environmental risks include living in communities with more exposure to chemical plants, landfills, and other types of pollution, and in particular in low-lying areas (Adeola & Picou, 2017). Buffalo Creek, West Virginia, in the 1970s was a low-lying area at the base of an unstable dam created by waste from coal mining operations. The conditions of financial exploitation in Appalachia that made Buffalo Creek more vulnerable to flood were appalling. Townspeople with economic interests (bankers, politicians, real estate, and law enforcement officials), together with local and national coal companies, focused on financial gain without regard for the safety of mountain communities for many years (Erikson, 1976). This exploitation led to conditions that led to a catastrophic dam break and subsequent flood that killed 118 people and wiped entire communities off the map.

A Trauma-Informed Disaster Approach

These trauma-informed principles were developed to be implemented in a variety of settings, including behavioral health, youth services, child welfare systems, and juvenile justice systems (SAMHSA, 2014). Disaster response can be the same and different than standard types of social services. Disaster response can be provided in a variety of settings. In the immediate aftermath, disaster response may be provided by local service agencies operating in emergency modes, until larger (regional/national) agencies are able to come in to provide backup support. After some time, local agencies may revert to normal operations but continue to address disaster-related needs. In the next section, I share a fictionalized case study about a woman, “Maria,” and her family. Throughout the remainder of the chapter, this story, which illustrates a typical experience with a disaster (hurricane), will show the different ways trauma can be experienced and reexperienced during disasters and how behavior health can address trauma through trauma-informed interventions.

Case Study

The Scene: Maria has lived in her own home in Morehead City, North Carolina, for nearly 50 years, alone there since her husband died about 5 years ago. Her children moved to Raleigh (3 hours away) for their jobs, and even though they visit her about twice a month, she is mostly on her own in Morehead City. She has a car, but most of her driving is limited to her neighborhood—the grocery store and church. Hurricane Amelia is working its way up the east coast of the United States, and the forecast has varied. Maria’s children would like her to come to them, but she is not comfortable driving that far, and so far the predictions have not been that bad.

The Storm: Hurricane Amelia made an unexpected turn for the coast. Despite mandatory evacuation orders, Maria was unable to get out before the storm hit. As a result of the evacuation orders, her children were unable to get to her, and many of her neighbors with children left hours earlier and didn’t realize she was still there. She was trapped in her house for nearly 24 hours—8 hours as the winds howled and trees and debris fell around her and another 16 as the waters rose and she moved up to the second floor to stay dry.

The Rescue: She was rescued by the local fire department in a boat and brought to a shelter. Most of the homes in her neighborhood have been damaged, the power is out indefinitely, and there is no sign of when she will return home. The area is restricted to rescuers so her kids can't get to her for now. Red Cross staff are doing their best to care for her and comfort her.

Trauma and Trauma History in Disasters

Past Experiences Triggered by Disaster

Felitti et al. (1998) brought our awareness about the prevalence of trauma in the general population to light. Their groundbreaking study of medical health seekers at Kaiser Permanente in California revealed widespread experience of seven different categories of adverse childhood experiences (Felitti et al., 1998). They also learned that people who have experienced one type of trauma are much more likely to experience other types of trauma. When working with individuals in disasters, those with preexisting mental health problems are most likely to experience problems post disaster (Dirkzwager, Grievink, van der Velden, & Yzermans, 2006). Even in children, those with prior experiences of trauma are more likely to experience PTSD post disaster (Gordon-Hollingsworth, Yao, Chen, Qian, & Chen, 2018). These negative effects are not just limited to PTSD. Prior traumatic experiences coupled with new experiences of natural disasters are associated with greater risks of major depressive episodes and substance use disorders as well (Adams et al., 2015). The new experiences may trigger memories of past experiences, including both experiences that are similar to the new experiences, and events that are completely different. Surprisingly, a history of dissimilar traumatic events has been linked to more disaster-related distress than similar events (Dougall, Herberman, Delahanty, Inslicht, & Baum, 2000).

Responders need to be aware of both the symptoms of traumatic reactions such as difficulty coping with seemingly small stressors. They also need to consider the complexity of these reactions because they may not only reflect the present circumstances but may also be reflective of prior lifetime trauma.

New Traumatic Experiences in Disasters

Erikson (1976) wrote about the dual traumas that emerged from a large scale community-wide flood. He wrote about individual trauma and collective trauma. The specifics of the traumas may vary with the specifics of the disasters, but the common themes are around fear of conditions that remind them of the events and controlling the environment to prevent the loss of control that occurred during the disaster. According to Erikson (1976), individual trauma can be so powerful that it may destroy one's resolve in a moment, impeding a person's ability to protect himself or herself from a future sudden threat. Traumatic reactions occur in individuals when events overwhelm an individual's internal and external resources (Bloom & Farragher, 2010).

Tang et al. (2014) studied mental health problems following natural disasters and found that in children who had survived a disaster, experiencing trauma during the disaster was a significant risk factor for developing depression. Specifically, children who were injured or scared experienced depression more than twice as often as those who had not. Those who had been trapped or had witnessed injury or death of others were also at great risk (Tang et al., 2014). Interestingly, Erikson (1976) noted that many survivors attributed their post-flood difficulties directly to the Buffalo Creek disaster. For example, he described survivors as being afraid when it rained, as the flood occurred when a dam burst after a period of wet weather, or children who slept fully clothed in case they needed to escape quickly. Maria of our case study is certainly at risk for this kind of reaction in subsequent rains or winds. What if she had experienced this before?

Recall that Erikson (1976) described collective trauma in terms of the disruption of social relations across an entire community: the shattering of bonds of friendship and affection that underlie community well-being. There are many factors that contribute to social connectedness and social structures in a community, from the size of the community, the similarities and differences of the people who live there, and the socioeconomic status of the neighborhood (Lenzi, Vieno, Santinello, & Perkins, 2013). Neighborhood connectedness includes trust between residents, the ability to help and be helped by neighbors, neighborhood friends, and relationships with neighbors (Lenzi et al., 2013). In certain types of disasters, neighbors are essential responders, particularly close neighbors who have relationships with each other, because they can respond when others are unable to access the area, such as in fires or floods. Maria's neighbors weren't there when she needed them, but she misses the daily close contact she had with them. Disasters create collective trauma, particularly in close-knit communities. This collective trauma may be exacerbated by disaster response efforts, as detailed in the next section.

Re-traumatization by Disaster Response and Recovery

The prevalence of post-disaster depression in adults ranges from 5.8% to 54% and from 7.5% to 44.8% in children (Tang et al., 2014). A large meta-analysis of multiple studies across multiple disasters, including hurricanes, earthquakes, tsunamis, tornadoes, and floods, showed that the most significant risk factor for developing this depression was a history of prior trauma. People with a history of prior trauma were more than twice as likely to develop depression as a result of the new disaster experience (Tang et al., 2014), highlighting the importance of providing trauma-sensitive and trauma-informed services to this population.

In talking about the loss of community as a result of disaster response procedures, Buffalo Creek survivors spoke of their displacement experience and disruption of familiar social bonds. They were forced to relocate quickly without regard to personal preferences. Survivors had no choice as to where or with whom they had to share living spaces. Because federal officials at that time were in such a hurry to resolve the homelessness situation faced by so many, provisions after the Buffalo Creek disaster had strangers living virtually on top of strangers (Erikson, 1976). An important takeaway from this disaster concerns the importance of preserving familiar social bonds, whenever possible, after a disaster. Allowing residents to stay in close proximity with friends and former neighbors may promote a sense of community that facilitates post-disaster recovery. As the Buffalo Creek disaster so clearly illustrates, post-disaster relocation can erode social capital and social support. When individuals are able to relocate together, group bonds are strengthened, resulting in greater socializing and social participation than before the disaster (Hikichi et al., 2017). Those that may relocate initially but are able to remain connected or return to their communities fare better, particularly those who are able to assist and support others (Hrostowski & Rehner, 2012).

Disaster survivors often want to return to familiar neighborhoods despite destruction and loss because of access to informal support systems and previous community resources and providers (Kytola, Cherry, Marks, & Hatch, 2015; Sanders et al., 2003). If disaster response prevents this return due to lack of funds for rebuilding, condemnation of destroyed areas, or other types of forced relocation, disaster survivors risk reexperiencing previous traumas due to the powerlessness from lack of choice and lack of supports they may have established to cope with previous trauma. In Maria's case, different families received different levels of reimbursement from insurance and FEMA, and some families were unable to return. Maria's children are reluctant to have her return to Morehead City. She wants to continue living on her own but is not sure she has the strength to fight with her children about this.

Trauma-Informed Care as an Individual Responder: Universal Precautions

When thinking about trauma-informed disaster response services, we need to consider services provided in the immediate aftermath, such as shelters and temporary housing. Longer-term services are also a critical consideration, such as social services in the home community that deal with aftereffects in the weeks and months that follow a disaster. Maslow (1943) will tell us the most important needs are physiological needs and safety needs. Trauma-informed care tells us that while safety is important to secure, all services need to be provided in ways that address all of the trauma-informed principles. Services may include shelters, transportation, and distribution of supplies. Even evacuation orders need to be thought about in a trauma-informed way: Are officials frightening or bullying people in an attempt to promote evacuation? Are they addressing the barriers to true choices? Many people might want to evacuate but don't feel they have a choice because of transportation, or concern about pets, or lack of trust in shelter options. While recent efforts have been made to improve evacuation support in regard to transportation, pets, lack of trust in shelter, individuals may lack awareness or understanding about these resources which ultimately still influences one's decisions to evacuate.

While trauma-informed disaster preparedness is important, disaster response services need to be trauma-informed too. In some types of disasters, particularly those that occur without warning, all services are disaster response services. These services should encompass all of the trauma-informed care principles. Here are some examples of how. Safety is essential for disaster survivors, particularly those who experienced threats to health and safety during the disaster. When helping survivors to plan their future, survivors should be empowered to use their voice to be clear about hopes and wants for recovery or relocation. People should be provided with more than one option about rebuilding or relocating. Programs should be trustworthy, including a concerted effort by communities and disaster responders to address shady or untrustworthy contractors who take advantage of disaster survivors.

Trauma-Informed Care as an Individual Responder: Specific Interventions

Disaster responders need to be alert for overt symptoms of trauma and to have resources available to address them. For instance, symptoms of adverse effects of trauma include an inability to concentrate and an inability to handle every day stressors, problems with memory, thinking, and emotional reactions (SAMHSA, 2014). Other symptoms include intrusive memories, vigilance, and a reexperiencing of past experiences (Erikson, 1976). This second list is particularly important to be aware of in disaster survivors as these may occur as a result of past traumas or of the experiences of survival that took place during the disaster.

In general, disaster services such as shelters and temporary, immediate responders are not equipped to handle acute trauma. In these cases, workers and volunteers should have resources available to which to refer survivors. As the time from the disaster increases and community agencies take over more of the long-term response, staff should be alert for trauma-related symptoms and either address them through their trauma treatment services, if they have them, or refer clients in need to other agencies. Emergency management coordinators and local officials should be aware of the limited capacity of agencies which may be taxed with the increased volume of clients in need as the community returns to normal operations (see Davis, Chap. 3, this volume).

Trauma-Informed Care for Disaster Workers and Disaster Organizations

Trauma-informed care has its origins as an organizational culture model that assumes that everyone in the organization, including staff, may have had a past traumatic experience that we (the collective we) may or may not know about (Bloom, 2013). This is a critical perspective, but disaster response makes it more challenging. Disaster response is populated by a combination of volunteers and staff and by locals and visiting responders, all with different levels of trauma exposure. Local responders may have experienced ill effects of the trauma along with community members. These local volunteers may be well meaning but may lack the training that professional responders have to deal with the things they see or hear about in the course of their work. Out-of-town responders often come prepared with previous disaster experience and more extensive training but may be unaware of past community level traumas or understanding this community's unique needs. This combination makes disaster responders potentially unprepared to respond to the varied needs of community residents.

Disaster responders are also vulnerable to the effects of trauma themselves (Gill, 2007; Hyman, 2005; Institute of Medicine, 2015). They may have experienced the disaster along with community members. They may develop secondary traumatic stress as a result of repeatedly hearing harrowing tales of survival, rescue, and death that survivors may have witnessed. While secondary traumatic stress is not the same as a direct experience of trauma, the APA includes the exposure to repeated stories through work as a criteria for diagnosing PTSD (APA, 2013), and studies have linked it indirectly to adverse health outcomes (Lee, Gottfried, & Bride, 2018).

In this context, disaster response agencies such as the American Red Cross, the Salvation Army, and local, state, and federal government entities need to be trauma-informed for their paid and volunteer workforce. They need to examine ways to ensure the physical and emotional safety of workers such as limits on the number of hours they can work, physical security and protections, and appropriate support and resources for workers. Agencies need to be transparent and trustworthy for workers. Make sure they know about what resources they will be able to access, and be clear about the support they will receive and that they will have available to provide to survivors. Provide opportunities for disaster responders to have peer support to cope with the stresses and strains of disaster response. Ensure that disaster responders are able to have their voices heard. They are often the eyes and ears, reporting on the reality of conditions on the ground, and should be taken seriously. They should be provided with choice regarding assignment and when they need rest or space away from the stress of the disaster response situation. Culture, history, and gender may dictate how responders cope with the stress of disaster response. For example, male first responders, particularly those from certain ethnic groups, may have constrained opportunities in which to express emotions (McAllister, Callaghan, & Fellin, 2019) because of prescribed gender roles in their culture. Factors such as having children are also associated with risk of burnout and PTSD (Boland et al., 2018). Agencies and coworkers should be aware of these challenges and collaborate with all of their workers to ensure they have the support they need to cope with difficult work to continue to support survivors.

Conclusion and Implications

Trauma-informed care is strengths based and promotes resilience. This chapter has laid out the need for trauma-informed care in disaster response and recovery as a means to avoid adverse outcomes for survivors of disasters, particularly those with past experiences of trauma. It is important to emphasize that using trauma-informed care does not mean that trauma survivors are weaker or in need of special care. On the contrary, trauma-informed care operates from the strengths perspective (see Wilks, Choate, Brown, Du, and Curry, Chap. 2, this volume). The strengths perspective counters the traditional problem-based approach and instead focuses on a client's abilities, talents, and resources. Further, interventions look to use those abilities, talents, and resources to move toward client-identified goals (Kim, 2013).

Identifying client strengths and goals promotes resilience. For resilience to occur, an individual has to have experienced a trauma and emerged stronger as a result of the challenge. One theory is that for each person who experiences a traumatic event, they have a pivotal moment when they move from survivor or victim to resilient person (Assimakopoulos, 2001). Further, positive support following the traumatic event fosters the reestablishment of psychological well-being (Assimakopoulos, 2001). Trauma-informed disaster response services can be this positive support which can, in turn, enhance the ability of disaster survivors to become resilient in the face of their challenges.

References

Adams, Z., Danielson, C., Sumner, J., McCauley, J., Cohen, J., & Ruggiero, K. (2015). Comorbidity of PTSD, major depression, and substance use disorder among adolescent victims of the spring 2011 tornadoes in Alabama and Joplin, Missouri. *Psychiatry, 78*, 170–185. https://doi.org/10.1080/00332747.2015.1051448

Adeola, F., & Picou, J. S. (2017). Hurricane Katrina-linked environmental injustice: Race, class, and place differentials in attitudes. *Disasters, 41*(2), 228–257. https://doi.org/10.1111/disa.12204

American Psychiatric Association (APA). (2013). *Diagnostic and statistical manual of mental disorders* (5th ed.). Arlington, VA: American Psychiatric Publishing.

Assimakopoulos, P. (2001). *The pivotal moment: A qualitative investigation into resilience.* (Doctoral dissertation). Retrieved from ProQuest Dissertations & Theses Global. (Accession No. 3012741).

Bloom, S. (2013). *Creating sanctuary: Toward the evolution of Sane Societies (Revised Edition).* New York: Routledge.

Bloom, S. L., & Farragher, B. (2010). *Destroying sanctuary: The crisis in human service delivery systems*. New York: Oxford University Press.

Boland, L., Kinzy, T., Myers, R., Fernstrom, K., Kamrud, J., Mink, P., et al. (2018). Burnout and exposure to critical incidents in a cohort of emergency medical services workers from Minnesota. *Western Journal of Emergency Medicine, 19*(6), 987–995. https://doi.org/10.5811/westjem.8.39034

Burson, B., Kalin, W., McAdam, J., & Weerasinghe, S. (2018). The duty to move people out of harm's way in the context of climate change and disasters. *Refugee Survey Quarterly, 37*, 379–407. https://doi.org/10.1093/rsq/hdy015

Cornell, M. (2018, November 13). How catastrophic fires have raged through California. *National Geographic*. Retrieved from https://www.nationalgeographic.com/environment/2018/11/how-california-fire-catastrophe-unfolded/

Croom, C., & Jenkins, P. J. (2007). The story of the senior center: Meeting immediate needs. *Generations, 31*(4), 47–48.

Dirkzwager, A. J., Grievink, L., van der Velden, P. G., & Yzermans, C. J. (2006). Risk factors for psychological and physical health problems after a man-made disaster. Prospective study. *British Journal of Psychiatry, 189*, 144–149. https://doi.org/10.1192/bjp.bp.105.017855

Dougall, A., Herberman, H., Delahanty, D., Inslicht, S., & Baum, A. (2000). Similarity of prior trauma exposure as a determinant of chronic stress responding to an airline disaster. *Journal of Consulting and Clinical Psychology, 68*(2), 290–295. https://doi.org/10.1037//0022-006x.68.2.290

Erikson, K. (1976). *Everything in its path: Destruction of community in the Buffalo Creek Flood*. New York: Simon and Schuster.

Felitti, V., Anda, R., Nordenberg, D., Williamson, D., Spitz, A., Edwards, V., et al. (1998). Relationship of childhood abuse and household dysfunction to many of the leading causes of death in adults the adverse childhood experiences (ace) study. *American Journal of Preventive Medicine, 14*, 245–258. https://doi.org/10.1016/S0749-3797(98)00017-8

Gill, D. A. (2007). Secondary trauma or secondary disaster? Insights from Hurricane Katrina. *Sociological Spectrum, 27*, 613–632.

Gordon-Hollingsworth, A., Yao, N., Chen, H., Qian, M., & Chen, S. (2018). Understanding the impact of natural disasters on psychological outcomes in youth from Mainland China: A meta-analysis of risk and protective factors for post-traumatic stress disorder symptoms. *Journal of Child and Adolescent Trauma, 11*, 205–226. https://doi.org/10.1007/s40653-015-0051-2

Gray, M., Litz, B., Hsu, J., & Lombardo, T. (2004). Psychometric properties of the life events checklist. *Assessment, 11*(4), 330–341. https://doi.org/10.1177/1073191104269954

Harris, M., & Fallot, R. (2001). Using trauma theory to design service systems. In *New directions for mental health services*. San Francisco: Jossey-Bass.

Hikichi, H., Aida, J., Tsuboya, T., Kondo, K., & Kawachi, I. (2016). Can community social cohesion prevent posttraumatic stress disorder in the aftermath of a disaster? A natural experiment from the 2011 Tohoku earthquake and tsunami. *American Journal of Epidemiology, 183*(10), 902–910. https://doi.org/10.1093/aje/kwv335

Hikichi, H., Sawada, Y., Tsuboya, T., Aida, J., Kondo, K., Koyama, S., et al. (2017). Residential relocation and change in social capital: A natural experiment from the 2011 Great East Japan Earthquake and Tsunami. *Science Advances, 3*(7), e1700426. https://doi.org/10.1126/sciadv.1700426

Hrostowski, S., & Rehner, T. (2012). Five years later: Resiliency among older adult survivors of Hurricane Katrina. *Journal of Gerontological Social Work, 55*, 337–351. https://doi.org/10.1080/01634372.2011.639055

Hyman, O. (2005). Perceived social support and secondary traumatic stress symptoms in emergency responders. *Journal of Trauma Stress, 17*(2), 149–156. https://doi.org/10.1023/B:JOTS.0000022621.27014.0e

Institute of Medicine. (2015). *Healthy, resilient, and sustainable communities after disasters: Strategies, opportunities, and planning for recovery*. Washington, DC: The National Academies Press. https://doi.org/10.17226/18996

Jacob, K. (2015). Sea level rise, storm risk, denial, and the future of coastal cities. *Bulletin of the Atomic Scientists, 71*(5), 40–50. https://doi.org/10.1177/0096340215599777

Johnson, R. A. (1996). The meaning of relocation among elderly religious sisters. *Western Journal of Nursing Research, 18*, 172–181.

Kessler, R., Aguilar-Gaxiola, S., Alonso, J., Benjet, C., Bromet, E. J., Cardoso, G., et al. (2017). Trauma and PTSD in the WHO World Mental Health Surveys. *European Journal of Psychotraumatology, 8*(sup5), 1353383. https://doi.org/10.1080/20008198.2017.1353383

Kim, J. (2013). Strengths perspective. In *Encyclopedia of social work*. Retrieved from http://oxfordre.com/socialwork/view/10.1093/acrefore/9780199975839.001.0001/acrefore-9780199975839-e-382. https://doi.org/10.1093/acrefore/9780199975839.013.382

Kor, E. M. (2015). Triumph over tragedy: The healing power of forgiveness. In K. E. Cherry (Ed.), *Traumatic stress and long-term recovery: Coping with disasters and other negative life events* (pp. 419–430). New York: Springer.

Kusmaul, N. & Anderson, K. (2018). Applying a Trauma-Informed Perspective to Loss and Change in the Lives of Older Adults. *Social Work in Health Care, 57*(5), 355–375. https://doi.org/10.1080/00981389.2018.1447531

Kusmaul, N., Wilson, B., & Nochajski, T. (2015). The infusion of trauma informed care in organizations: Experience of agency staff. *Human Service Organizations: Management, Leadership, & Governance, 39*, 25–37. https://doi.org/10.1080/23303131.2014.968749

Kusmaul, N., Wolf, M., Sahoo, S., Green, S., & Nochajski, T. (2019). Client experiences of trauma-informed care in social service agencies. *Journal of Social Service Research, 45*(4), 589–599. https://doi.org/10.1080/01488376.2018.1481178

Kyne, D., Lomeli, A., Donner, W., & Zuloaga, E. (2018). Who will stay, who will leave: Decision-making of residents living in potential hurricane impact areas during a hypothetical hurricane event in the Rio Grande Valley. *Journal of Homeland Security and Emergency Management, 15*(2), 33. https://doi.org/10.1515/jhsem-2017-0010

Kytola, K. L., Cherry, K. E., Marks, L. D., & Hatch, T. G. (2015). When neighborhoods are destroyed by disaster: Relocate or return and rebuild? In K. E. Cherry (Ed.), *Traumatic stress and long-term recovery: Coping with disasters and other negative life events* (pp. 211–229). New York: Springer.

Landeck, K. (2018, December 10). Two months after Hurricane Michael, Mexico Beach 'beating strong.' *Panama City News Herald*. Retrieved from https://www.newsherald.com/news/20181210/two-months-after-hurricane-michael-mexico-beach-beating-strong

Lee, J., Gottfried, R., & Bride, B. (2018). Exposure to client trauma, secondary traumatic stress, and the health of clinical social workers: A mediation analysis. *Clinical Social Work Journal, 46*(3), 228–235. https://doi.org/10.1007/s10615-017-0638-1

Lenzi, M., Vieno, A., Santinello, M., & Perkins, D. (2013). How neighborhood structural and institutional features can shape neighborhood social connectedness: A multilevel study of adolescent perceptions. *American Journal of Community Psychology, 51*, 451–467. https://doi.org/10.1007/s10464-012-9563-1

Maslow, A. H. (1943). A theory of human motivation. *Psychological Review, 50*(4), 370–396. https://doi.org/10.1037/h0054346

McAllister, L., Callaghan, J., & Fellin, L. (2019). Masculinities and emotional expression in UK servicemen: 'Big boys don't cry'? *Journal of Gender Studies, 28*(3), 257–270. https://doi.org/10.1080/09589236.2018.1429898

McKnight, J., & Block, P. (2012). *The abundant community: Awakening the power of families and neighborhoods*. San Francisco: Berrett-Koehler Publishers, Inc.

Morey, J., & Segerstrom, S. (2015). Physiological consequences: Early hardship and health across the life span. In K. Cherry (Ed.), *Traumatic stress and long-term recovery* (pp. 151–176). New York: Springer Publishing Company.

National Weather Service. (2018). *Hurricane Michael 2018*. Retrieved from https://www.weather.gov/tae/HurricaneMichael2018

Nauslar, N., Abatzoglou, J., & Marsh, P. (2018). The 2017 North Bay and Southern California fires: A case study. *Fire, 1*(1), 18. https://doi.org/10.3390/fire1010018

Roberto, K., Kamo, Y., & Henderson, T. (2009). Encounters with Katrina: Dynamics of older adults' social support networks. In K. E. Cherry (Ed.), *Lifespan perspectives on natural disasters* (pp. 133–152). New York: Springer Publishing Company.

Rogers, D. (2013, January 23). *Secondary traumatic stress in the general public following disasters: A personal experience.* Retrieved from https://www.prainc.com/secondary-traumatic-stress-in-the-general-public-following-disasters-a-personal-experience/

Sanders, S., Bowie, S., & Bowie, Y. (2003). Lessons learned on forced relocation of older adults: The impact of Hurricane Andrew on health, mental health, and social support of public housing residents. *Journal of Gerontological Social Work, 40*(4), 23–35. https://doi.org/10.1300/J083v40n04_03

Substance Abuse and Mental Health Services Administration (SAMHSA). (2014). *SAMHSA's concept of trauma and guidance for a trauma-informed approach. HHS Publication No. (SMA) 14-4884.* Rockville, MD: Substance Abuse and Mental Health Services Administration.

Tang, B., Liu, X., Liu, Y., Xue, C., & Zhang, L. (2014). A meta-analysis of risk factors for depression in adults and children after natural disasters. *BMC Public Health, 14*, 623. http://www.biomedcentral.com/1471-2458/14/623

Veterans Administration. (VA). (2018). *PTSD and DSM 5.* [Web Page] Retrieved from https://www.ptsd.va.gov/professional/treat/essentials/dsm5_ptsd.asp

Winsor, M. (2019, February 8). *Paradise residents recall horrifying escape from the Camp Fire: 'This should never happen again'.* Retrieved from https://abcnews.go.com/US/paradise-residents-recall-horrifying-escape-camp-fire-happen/story?id=59933153

Chapter 10
The Impact of Social Support After a Disaster

Judith R. Phillips

Introduction

Although disasters emerge in many different forms such as wildfires, hurricanes, tornadoes, floods, and earthquakes, disasters often share common outcomes for survivors; disasters threaten the physical health, psychological health, and psychosocial functioning of community-residing adults exposed to a disaster. Well-documented post-disaster responses include increased anxiety and stress reactions, decreased quality of everyday life and well-being, and development of psychopathology. There is property loss, economic loss, and sometimes loss of human and companion pet lives. There may have been a threat of death or injury to an individual or others during the actual disaster. In addition, communities are impacted and often changed which further impacts survivors' well-being. Disaster survivors usually face all of these outcomes in the days, weeks, and months after a disaster (Acierno, Ruggiero, Kilpatrick, Resnick, & Galea, 2006; Bryant et al., 2014; Camilleri et al., 2010; Felix & Afifi, 2015; Gibbs et al., 2016; Marshall, Schell, Elliott, Rayburn, & Jaycox, 2007; McFarlane, Clayer, & Bookless, 1997; Sanders, Bowie, & Bowie, 2003; Tyler, 2006).

Because of these threats, losses, and outcomes, poor short-term psychological health consequences are usually seen in community-residing adults in the immediate days and weeks post disaster. For example, Marshall et al. (2007) found poor psychological health and evidence of psychopathology such as post-traumatic stress disorder and major depression in disaster survivors 3 months after a wildfire. Poor psychological health for disaster survivors may also persist several years after a disaster. For example, Bryant et al. (2014), Camilleri et al. (2010), McFarlane et al. (1997), and Gibbs et al. (2016) found various levels of psychological distress (post-traumatic stress disorder, depression) and some dissatisfaction about quality of life in some disaster survivors at 20 months, 3 years, 4 years, and 5 years after different

J. R. Phillips (✉)
Department of Psychology, California State University San Marcos, San Marcos, CA, USA
e-mail: jphillip@csusm.edu

© Springer Nature Switzerland AG 2021

K. E. Cherry, A. Gibson (eds.), *The Intersection of Trauma and Disaster Behavioral Health*, https://doi.org/10.1007/978-3-030-51525-6_10

disaster events although many of the respondents reported good mental health at the time of assessment.

With these short-term and long-term consequences evident after a disaster, psychological health professionals and researchers began to look for protective factors that would aid in the reduction of these outcomes. The research literature documents that receiving social support after a disaster could serve as a protective factor for disaster survivors' psychological health and lessen their vulnerability post disaster (Acierno et al., 2006; Camilleri et al., 2010; Felix & Afifi, 2015; Haines, Hurlbert, & Beggs, 1996; Miller et al., 2012; Norris & Kaniasty, 1996; Sanders et al., 2003; Weems et al., 2007). This confirmation includes investigations after various disaster events such as wildfires (Camilleri et al., 2010; Felix & Afifi, 2015), after hurricanes (Acierno et al., 2006; Haines et al., 1996; Norris & Kaniasty, 1996; Sanders et al., 2003; Weems et al., 2007), and after tornadoes (Miller et al., 2012).

The purpose of this chapter is to examine the impact of post-disaster social support on the psychological health of community-residing adults. The first objective is to introduce the literature on received social support and psychological health of those recovering from disasters. Explanations of the types and sources of social support and the different statues of disaster victims/survivors will be reviewed. This chapter also discusses the concept of a disaster survivor's "comfortableness" in seeking and providing social support and how this variable may impact willingness to seek and provide social support. The second objective is to present select data from studies conducted on the 2007 and 2014 northern San Diego County, California, wildfires. Specifically, this chapter reports the outcomes for social support, comfortableness in seeking and providing social support, and psychological health of a total of 313 adults aged 18–94 years who participated in these studies. The final objective is to reflect on how organizations and individuals who are responsible for the welfare of citizens after a disaster can implement these findings into their disaster planning, preparations, and programs to assist with post-disaster coping.

Literature Review: The Protective Factor of Received Social Support

Social support refers to help that is provided to an individual or a group of individuals who need assistance. Received social support has been found to be beneficial for survivors trying to recover from the impact of disasters, and previous studies have found the adults of all ages cope better and have better psychological health with the support of others. Therefore, the existence of social support post disaster is important in facilitating recovery of all who are affected by disasters (Acierno et al., 2006; Camilleri et al., 2010; Felix & Afifi, 2015; Kaniasty & Norris, 1995; Sanders et al., 2003).

Empirical research has also shown that whether in the short term or in the long term, received social support post disaster is beneficial for psychological health. For example, when 1543 adults aged 18–60+ years were interviewed in 6–8 months after the 2004 Florida hurricane season, low levels of social support predicted higher levels of post-traumatic stress disorder, major depression, and generalized anxiety disorder. In this study, received social support referred to emotional social support ("someone available to love you"), instrumental social support ("someone available to help you"), and appraisal social support ("someone available to give good advice in crisis") (Acierno et al., 2006). In another study assessing 402 community-residing adults aged 18–94 years at 5 and 11 months (short-term period) after wildfires occurred in California from 2008 to 2009, having high social support at the time of the fire predicted better mental health for adults, especially for those adults who had to be evacuated because of the closeness of the wildfires. Social support was defined as structural social support ("How many people are so close that you can count on them for serious problems?") and functional social support ("How easy was it to get help from neighbors if you should need it?") (Felix & Afifi, 2015).

Using qualitative methods, Camilleri et al. (2010) investigated received social support and psychological health 3 years (long-term period) after the 2003 Canberra, Australia, bushfires. Five hundred community-residing adult participants indicated that received social support such as kindness, understanding, and offers of housing and financial aid facilitated their recovery. Family, friends, neighbors, and governmental and community agencies provided such social support. At the time of assessment (3-years post disaster), over half of participants (52%) were classified as having low levels of psychological distress, and almost half (47%) reported that their lives were the same as before the bushfires. Camilleri et al. (2010) explains that the Recovery Center model, a "one-stop model" where bushfire survivors have access to multiple services and resources all at one location, and community involvement were key features of recovering from this disaster, suggesting a "relationship" between psychological health and received social support.

Types and Sources of Social Support

Social support is generally seen as a multidimensional concept, one that encompasses emotional, informational, and tangible social support (Kaniasty & Norris, 1995, 2000). Emotional social support describes actions which show feelings toward another person such as one person expressing concern and interest about another's well-being and/or one person being comforted with a hug by another person (Kaniasty & Norris, 1995, 2000). Tangible social support refers to "objective" support such as someone receiving money, appliances, furniture, tools, food, and equipment or someone being offered a place to stay (Kaniasty & Norris, 1995, 2000). Informational social support defines actions where information is dispersed. Examples include having one person suggest to another person an action to take and/or one person being given information on how to do something by another (Kaniasty & Norris, 1995, 2000).

Empirical and anecdotal literature divide the sources/providers of social support into three categories: (1) family members or those linked by blood or marriage; (2) friends, neighbors, and coworkers; and (3) those outside an individual's immediate social circle such as community members, voluntary organizations, and professional service providers (Kaniasty & Norris, 1995, 2000).

Disaster Victim/Survivor Status Groups

Disasters trigger different challenges for different groups of disaster survivors, dependent on whether they live in the direct path of the disaster or live in a surrounding area. Community-residing adults who live in the direct path of the disaster and who experience property damage and/or structural damage (total or partial) to their residences as a direct result of the disaster are referred to as primary victims/disaster survivors (Bolin, 1985; Kaniasty & Norris, 2000; Kaniasty, Norris, & Murrell, 1990; Norris & Murrell, 1988).

Adults who live in the assumed path of the wildfire and must evacuate as the disaster approaches but whose residences are not damaged or destroyed are identified as secondary victims/disaster survivors (Bolin, 1985; Kaniasty et al., 1990; Kaniasty & Norris, 2000; Norris & Murrell, 1988).

Non-victim status is assigned to those who live around the disaster area but not in the direct path of the disaster; they are not asked to evacuate as the disaster approaches and do not have any structural or property damage to their residences (Bolin, 1985; Kaniasty et al., 1990; Kaniasty & Norris, 2000; Norris & Murrell, 1988).

Literature Review for Disaster Victim/Survivor Status Variables

Few empirical studies have investigated all four variables: the three types of social support received, the three types of disaster survivors, the three types of social support providers, and their impact on psychological health. Watanbe, Okumura, Chiu, and Wakai (2004) assessed psychological heath and sources of social support. They conducted interviews at 6 and 12 months with 54 older adults, who were displaced by the 1999 Chi-Chi earthquake in Taiwan. Social support from a child living with the parent or from a child living apart and social support from extended family were associated with improved psychological health at 6 months. At 12 months post disaster, social support from neighbors and extended family was linked to improved psychological health. Watanabe and colleagues suggested that social support from the immediate family was essential in the early months after the disaster as gaining the social support of new neighbors in their displaced housing took time. Yet when it was established, this social support from new neighbors was effective in improving psychological health (Watanbe et al., 2004).

When reviewing disaster survivor status and types and sources of received social support, primary disaster survivors receive the most support, whether it is emotional, tangible, or informational support (Bolin, 1985; Kaniasty & Norris, 1995; Norris & Murrell, 1988; Phillips, Cruz, Cuadra, Martinez, & Perdomo, 2013). The most common types of received social support are emotional and tangible support, and the most common sources are from family as well as friends, neighbors, and coworkers (Kaniasty et al., 1990; Kaniasty & Norris, 1995; Phillips et al., 2013).

When exploring the literature for types and sources of received social support post disaster for all impacted individuals, the most common type of received social support is emotional support (Kaniasty & Norris, 1995). Kaniasty and Norris (1995) asked community residents-survivors of Hurricane Hugo about social support after the hurricane. They found that while these community residents received and provided all three types of social support, emotional support such as assurance and interest was received and provided most often. Family, friends, and neighbors provided such support.

Bolin and Klenow (1983) interviewed residents who survived two tornadoes in Texas. These survivors reported that they received assistance in the form of emotional support such as concern and tangible support such as physical assistance with cleanup, temporary shelter, and offers of food. Providers were family, friends, neighbors, and governmental organizations such as FEMA and the Small Business Administration. Sanders et al. (2003) found that after Hurricane Andrew, informal support was identified as emotional and tangible support. These types of social support were provided by family, friends, and formal assistance from social workers and governmental and nonprofit organizations.

Characteristics of social support providers include those who had minimal disaster-related needs or did not have any disaster-related needs. These individuals are identified as secondary disaster survivors (those who reported that they were evacuated from their homes because they lived in close proximity to the path of the wildfire, but they did not experience any structural damage to their residence) or non-victims (those community residents who did not have to evacuate and did not have any property damage, yet they lived in the area of the disaster) (Bolin, 1985; Haines et al., 1996; Kaniasty & Norris, 1995; Norris & Murrell, 1988).

Comfortableness in Seeking Social Support

From past studies, we see how receiving various types of social support from various sources reduces the impact of an aversive event, but it is not well studied why some disaster survivors will seek social support while other disaster survivors will not (Kaniasty & Norris, 2000). According to Kaniasty and Norris, how comfortable an individual is in seeking social support may predict received social support, if comfortableness in asking for help is an individual trait (Kaniasty & Norris, 2000). When disaster survivors do not receive social support, is it because no one responded to their requests, or did they not seek out help and social support at all? Kaniasty and Norris (2000) measured how comfortable and ready 404 participants were to seek and accept social support after a disaster (Hurricane Andrew). They found that comfortableness in seeking social support predicted received social support for all ethnic groups within the sample. Kaniasty and Norris then developed a model in which having a need for support and being comfortable in seeking help predicted received social support. Therefore, whether one receives social support could be based on whether that disaster survivor is comfortable in seeking out help or is not comfortable requesting help.

Comfortableness in Providing Social Support

As comfortableness in seeking or requesting social support is related to received social support, could comfortableness in providing support to disaster survivors also be related to whether one is comfortable in providing such support? While no studies appear to be published on comfortableness in providing social support, Lebowitz (2017) investigated psychological health (depression) and satisfaction which came from providing support to others after the Fukushima power plant disaster which was damaged by the 2011 Japan earthquake. The data was collected within 1 year of the damage to the power plant. His findings revealed an inverse relationship between providing social support to others and mental health levels such that as one provides support to others, the provider's own mental health issues (depression) decreases. He suggested that providing social support to others might provide a buffering effect for a disaster survivor's own mental health issues (Lebowitz, 2017). Thus, it is possible that being comfortable in providing social support and providing support to others post disaster may help to lessen the provider's own post-disaster psychological health issues. This will be discussed more fully subsequently in the context of two wildfire events that occurred in San Diego County, California, in October of 2007 and May 2014. These wildfire events prompted our research on the social support variables of type and source of social support and comfortableness in seeking and providing social support and their impact on the psychological health of community-residing adults who experienced specific wildfires within northern San Diego County, California, in 2007 and 2014.

The Disasters: Northern San Diego County Wildfires in 2007 and 2014

The Witch Creek wildfire which began on October 21, 2007, was the wildfire of focus in our first study. This wildfire began with a small fire near Ramona, a city in the foothills northeast of the city of San Diego, merged within hours with another fire, and moved west toward the coast. It was uncontrolled for over a week. It burned nearly 200,000 acres of the county with over 1100 residences destroyed, many of them in the Rancho Bernardo community. Two community residents died. It is still the ninth largest and the sixth most destructive wildfire in California history (CAL FIRE, 2019).

Our second study concentrated on 14 separate wildfires which burned simultaneously within northern San Diego County between May 13 and May 18, 2014. The Cocos, Poinsettia, and Bernardo wildfires were the largest. The Cocos wildfire erupted within yards of California State University San Marcos during exam week and forced cancellation of exams and postponement of the university's graduation. This wildfire and the 13 others burned approximately 26,000 acres and destroyed 65 homes. Two deaths were attributed to these wildfires (May 2014 San Diego Wildfires After Action Report, 2014).

Method

The primary aim of Study 1 (the 2007 wildfire) was to investigate the types and sources of social support received by the three victim/survivor groups identified in this sample. We also wanted to explore if received social support impacted psychological well-being. The aims of Study 2 (the 2014 wildfire) were to explore the relationships between comfortableness in seeking social support and receiving support as well as comfortableness in providing social support and providing support to others, the psychological well-being of the disaster victim/survivors, and the types and sources of the received and provided social support.

Participants

Study 1: Data from this study came from 191 community-residing adults who lived in northern San Diego County at the time of the 2007 wildfire and in 2012 at the time of data collection, approximately 4.5 year after the wildfire. Adults needed to be at least 25 years at the time of the wildfire to participate in the study. Forty-six participants provided narrative data to a prompt at the end of the questionnaire which asked for additional comments related to the participants' experiences after the wildfires.

Study 2: Participants were 122 community-residing adults who lived in and around the areas affected by the May 2014 northern San Diego County wildfires. At the time of data collection in 2017 which was approximately 3 years after the wildfires, they also lived in the area. Participants needed to be at least 18 years at the time of the data collection (2017) to complete a questionnaire.

Procedure

The snowball sampling technique was used to disperse the 2007 questionnaire to community-residing adults in the areas near and around the wildfire. Individuals known to me, my student assistants, my family, and my friends gave out questionnaires to area residents they knew who lived in or around the areas impacted for this wildfire and who met the age requirements. Questionnaires were distributed from

May to July 2012. When completed, participants mailed their questionnaire to me in the provided stamped and addressed envelope. The response rate was 45.7%.

Qualitative data from the 2007 wildfire (Study 1) was independently analyzed for content by two researchers who followed a codebook divided into two themes: received social support (informational, emotional, and tangible) and helping/providing social support behavior toward others (informational, emotional, and tangible). Each of these themes was examined as to who were the sources of the received social support (family, friends, neighbors, and coworkers and those outside the immediate social circle) and who was the provider of the helping social support behavior (family, friends, neighbors, and coworkers and those outside the immediate social circle). The victim/survivor status of each receiver was also identified. The codebook definitions were taken from Kaniasty and Norris (1995, 2000). Content analysis was found to be 83.0%. Differences between the two researchers were discussed and an agreement reached.

Study 2 had two types of distribution. The same snowball technique was also used to distribute the questionnaire for those who lived in and around the May 2014 wildfires. All questionnaires were distributed between March and October 2017. Those who completed the questionnaire mailed it to me in the provided stamped and addressed envelope. In addition, students at California State University San Marcos were invited to participate in the study from March to early April 2017 to fulfill human participant pool class requirements. Overall, the response rate was 39.61%.

An information sheet, indicating California State University San Marcos IRB approval of each study, was attached to each questionnaire.

Measures

The 2007 and the 2014 questionnaires contained demographic questions that assessed age, gender, marital status at the time of the disasters, racial background, total household income for 2011/2016, and self-reported physical health status at two points (the time of the disaster and currently). Questions also measured participants' property damage and evacuation experience during each wildfire; these two items identified disaster survivor status.

The *Inventory of Post-Disaster Social Support (*IPSS*)* scale measures received social support (Kaniasty & Norris, 2000; Norris, Murphy, Kaniasty, Perilla, & Ortis, 2001). Participants were asked to estimate how often they received each type of social support and from what source within a clearly established timeframe anchored between the October 2007 and around New Year's Day 2008 for Study 1 and between May 2014 and July 4, 2014, for Study 2. This support was related to the wildfires. They were instructed to indicate a reasonable estimate and not to try to count the number of times things happened.

Three types of social support were assessed: emotional, informational, and tangible. Emotional social support assessed actions such as being shown signs of affection and concern about your well-being being shown. Informational social support measured being given information on how to do something. Tangible social support referred to being given/loaned/offered money, furniture tools, food, and a place to stay. Three distinct sources of support were measured: (1) family; (2) friends, neighbors, and coworker; and (3) outsiders, including people outside respondent's immediate social circle such as community leaders, voluntary organizations, and professional service providers. The responses were scored on a four-option response set [(0) = never, (1) = once or twice, (2) = a few times, (3) = many times]. An example from this set of questions is: "Between the 2014 May wildfires and around the 4th of July, as a direct result of your experience with the wildfires, how often did *family members* help you understand the situation you were in or give you information about how to do something such as, but not limited to, who to contact for insurance forms?"

Study 1 contained three questions assessing received emotional support, three questions assessing informational support, and four questions assessing tangible support. Within each of these 10 questions, there was a question related to each of the three possible sources that a particular support could be received from. Study 2 had three questions assessing each type of received support, and within each question, there was a question related to each of the three possible sources of that received support.

The questionnaire for Study 2 (the May 2014 wildfires) amended the IPSS scale to included questions pertaining to what type of social support did participants provide to (1) family; (2) friends, neighbors, and coworkers; and (3) people outside your immediate social circle such as community leaders, voluntary organizations, and professional service providers. Questions also asked about what type of help was provided: tangible, emotional, and informational. The same response set of (0) = never, (1) = once or twice, (2) = a few times, and (3) = many times was used (Kaniasty & Norris, 1995, 2000). This help was related to support after the wildfires. An example of these questions is: "Between the 2014 May wildfires and around the 4th of July, as a direct result of your experience with the wildfires, how often did *you* help family members understand the situation they were in and/or give them information about how to do something such as, but not limited to, who to contact for insurance forms?" Study 2 had three questions assessing each type of support that might be provided, and within each question, there was a question related to each of the three possible sources that the support was provided to.

In Study 2 additional questions assessed participants' level of comfortableness in seeking social support; the prompt was "The next set of questions is about how comfortable you were in requesting or seeking help from others as a direct result of the wildfires." Participants were again told to indicate a reasonable estimate and not to try to count the number of times things happened. Nine questions asked about comfortableness in seeking tangible, emotional, and informational support from the three sources: (1) family; (2) friends, neighbors, and coworker; and (3) outsiders, including people outside respondent's immediate social circle such as community leaders, voluntary organizations, and professional service providers. The responses were scored on a five-option response set [(0) = not applicable to me, (1) = very uncomfortable, (2) = somewhat uncomfortable, (3) = somewhat comfortable, (4) = very comfortable] (Kaniasty & Norris, 2000). For example, "Between the May 2014 wildfires and July 4, 2014, as a direct result of your experience with the wildfires, how comfortable were you in requesting items such as, but not limited to, money, tools, appliances, furniture, and a place to stay from family members?"

Study 2 also assessed how comfortable participants were in providing help to others as a direct result of the wildfires. Nine questions assessing tangible, emotional, and informational help provided to (1) family; (2) friends, including neighbors and coworkers; and (3) people outside your immediate social circle such as nonprofit organizations and/or community members were used (e.g., "Between the May 2014 wildfires and July 4, 2014, as a direct result of your experience with the wildfires, how comfortable were you in providing items such as, but not limited to, money, tools, appliances, furniture, and a place to stay to family members?") Responses were scored on a five-option response set [(0) = not applicable to me, (1) = very uncomfortable, (2) = somewhat uncomfortable, (3) = somewhat comfortable, (4) = very comfortable] (Kaniasty & Norris, 2000).

The *Center for Epidemiologic Studies Depression Scale* (CES-D; Radloff, 1977) assessed nonclinical depressive mood and symptoms. It is a frequently used and well-documented scale and intended for the general community-based population. A 20-item scale was used for both the 2007 and 2014 questionnaires and scored on a four-option response set [(0) = rarely or none of the time, (1) some of the time [1–2 days], (2) = occasionally or moderate amount of time [3–4 days], (3) = most or all of the time]. A score of 16 or higher suggests a high level of depressive symptoms, while a score of 16 or lower reflects a low level of depressive symptoms (Myers & Weissman, 1980).

We will now turn our attention to our findings from Studies 1 and 2 where we will report who our participants are, the levels of their psychological well-being, and the types of received and provided social support as well as the sources of this received and provided social support. In addition, we will describe the levels of comfortableness in seeking social support and the relationship to receiving social support as well as the relationship between comfortableness in providing social support and providing such support.

Results

Demographic Statistics

In Study 1 (2007 wildfire), 191 community-residing adult participants ranged in age from 30 to 94 years in 2012 ($M = 63.25$ years, $SD = 14.59$). One hundred and ten were female (57.60%). One hundred and sixty-eight identified themselves as White (non-Hispanic) (88.00%). Eight identified themselves as Hispanic/Latino, five as Asian, five as American Indian, three as Black/African American, and two as Native Hawaiian. At the time of the wildfire, 147 participants were married (77.00%), 16 were widowed, and 10 were divorced. The remaining sample identified themselves as either separated, never married, or cohabiting. Within this sample, 36 were primary disaster survivors (18.85%). Within the primary disaster survivor group, 27 reported total loss of their residence and 9 reported partial loss of their residence. One hundred participants (52.36%) were classified as secondary disaster survivors as they indicated that they had to evacuate but their residences were not damaged. Fifty-four were non-victims (28.27%) as they did not have to evacuate and did not have any structural damage to their residences. Eighty-four percent (44.00%) of the sample has an annual income of over $100,000 in 2011. The remaining sample had incomes from $15,000 per year to $100,000 per year. Table 10.1 presents details of respondents' severity of exposure (disaster survivor status) and age in decades for 2007 wildfire participants.

The 46 participants who provided narrative data in Study 1 ranged in age from 28 to 83 in 2007 ($M = 58.04$). Thirty-two were female (69.57%). Most were married at the time of the disaster (78.28%). Twelve respondents were primary disaster survivors (26.09%), 22 were secondary disaster survivors (47.83%), and 12 were non-victims (26.09%).

In Study 2 (2014 wildfire), 122 participants ranged in age from 18 to 80 ($M = 38.80$, $SD = 17.68$) and 80 were female (65.60%). Thirty-seven of the participants (30.33%) were California State University San Marcos students who participated by choice for human participant pool class requirement credit. The remaining sample (85 participants; 69.67%) were community-residing adults. Over half were White (non-Hispanic) (53.30%) and over a quarter were Hispanic (26.20%). At the

Table 10.1 Age and disaster survivor status from 2007 wildfire (study 1)

	Disaster survivor status		
Age in decades	Primary	Secondary	Non-victim
30–39	2	4	5
40–49	3	10	8
50–59	16	19	16
60–69	7	22	13
70–79	3	26	8
80–89	5	13	4
90–94	0	6	0
Total	36	100	54

Table 10.2 Age and disaster survivor status from 2014 wildfire (study 2)

	Disaster survivor status	
Age in decades	Secondary	Non-victim
Teens	1	14
20–29	12	25
30–39	5	8
40–49	7	13
50–59	5	15
60–69	2	8
70–79	1	5
80–89	0	1
Total	33	89

time of these wildfires, 54 were married (44.3%) and 49 were single and had never been married (40.2%). The remaining participants identified themselves as either widowed, divorced, separated, or cohabiting. Thirty-three respondents were identified as secondary disaster survivors (27.05%) and 89 were classified as non-victims (72.95%). No primary disaster survivors participated in this study. Forty-seven respondents (38.5%) reported an annual income in 2016 of over $100,000 while 21 reported an annual income of $25,001–$50,000. Table 10.2 presents details of respondents' severity of exposure (disaster survivor status) and age in decades for 2014 wildfire participants.

Psychological Health

The mean for respondents in both studies for CES-D scores was well below the cutoff for depressive symptomology: Study 1 (2007 wildfire): $M = 8.88$, $SD = 7.86$, range = 0–46; Study 2 (2014 wildfire): $M = 11.00$, $SD = 8.86$, range = 0–45. These results suggest that most participants' mental health was good at the time of participation in each study.

Study 1 Primary disaster survivors from the 2007 wildfire did report more depressive symptomology during a week in 2012 ($M = 10.11$, $SD = 10.42$, range = 0–46) than secondary disaster survivors ($M = 8.01$, $SD = 6.79$, range = 0–30) and non-victims ($M = 9.65$, $SD = 7.66$, range = 0–37), but there was no significant effect between the groups and CES-D scores, $F(2, 163) = 1.51$, $p = 0.22$.

Age comparisons disclosed a significant effect for age in decades, $F(6, 163) = 2.91$, $p = 0.01$. Post hoc comparisons revealed that the mean CES-D score for participants in their 60s ($M = 10.97$, $SD = 10.37$, range = 0–46) was significantly different from participants in their 70s ($M = 5.27$, $SD = 4.47$, range = 0–15), $p = 0.03$. There were no other significant differences found between the other decades. Table 10.3 presents details of respondents' psychological health and age in decades for 2007 wildfire participants.

Table 10.3 Mean CES-D, standard deviation scores, and age in decades for 2007 participants (study 1)

Age in decades	Mean CES-D scores (SD)
30	10.64 (7.32)
40	7.40 (6.79)
50	9.37 (7.46)
60	10.97 (10.37)
70	5.27 (4.47)
80	8.32 (6.20)
90	15.33 (10.29)

Table 10.4 Mean CES-D, standard deviation scores, disaster survivor status, and gender for 2007 participants (study 1)

Disaster survivor status/gender	CES-D mean (SD)
Primary	
Male	5.55 (5.05)
Female	12.83 (11.81)
Secondary	
Male	6.95 (5.22)
Female	8.79 (7.77)
Non-victim	
Male	11.09 (7.58)
Female	8.48 (7.57)

Table 10.5 Mean CES-D, standard deviation scores, and age in decades for 2014 participants

Age in decades	Mean CES-D scores (SD)
Teens	17.53 (11.57)
20	13.56 (9.46)
30	10.77 (6.99)
40	7.53 (5.64)
50	8.42 (7.28)
60	7.7 (7.53)
70	6.33 (2.73)
80	0.00

Participants' gender, disaster survivor status, and psychological health *(*CES-D*)* scores indicated no main effect, but an interaction was found, $F(2, 176) = 3.93$, $p = 0.02$. Women respondents' psychological health scores decreased as their exposure to the wildfire lessened; men's psychological health scores increased as their exposure to the wildfires lessened. Table 10.4 presents CES-D mean scores, standard deviation, gender, and disaster survivor status data for 2007 wildfire participants.

Study 2 Secondary disaster survivors from the 2014 wildfire did report more depressive symptomology during a week in 2017 ($M = 11.33$, $SD = 8.15$, range = 0–30) than non-victims ($M = 10.87$, $SD = 9.16$, range = 0–45), but there was no significant effect between the groups and CES-D scores ($t(117) = 0.253$, $p = 0.80$, two-tailed independent samples t-test). Table 10.5 presents details of respondents' psychological health and age in decades for 2014 wildfire participants.

Psychological Health and Received Social Support from the 2007 Wildfire Study

Weak correlations were found between psychological health (as measured by CES-D in 2012) and post-disaster total social support (as measured when the scores of the three types of social support, emotional, informational, and tangible, were combined). Similarly, there were no significant correlations between psychological health (as assessed in 2012) and each of the three different types of post-disaster social support (emotional, tangible, and informational).

There were only weak correlations between the psychological health of primary disaster survivors (as assessed in 2012) and the different types and different sources of social support they received post disaster. This held true also for secondary disaster survivors and non-victims.

Received Social Support from the 2007 Wildfire Study

The total social support received by primary disaster survivors (M = 49.9, SD = 19.22) was significantly different from that received by secondary disaster survivors (M = 19.68, SD = 15.45) and non-victims (M = 15.87, SD = 12.94), $F(2,168) = 54.17$, $p = 0.00$. Secondary disaster survivors' total received social support did not differ significantly from that of non-victims.

Types of received social support There were three different types of received social support—emotional, tangible, and informational—explored in Study 1 (2007 wildfire). There were significant effects of disaster survivor status on all three types of received social support with primary disaster survivors receiving significantly more emotional, tangible, and informational social support than secondary disaster survivors and non-victims across all age groups. Scores ranged from 0 to 27 for received emotional social support and informational social support and 0 to 36 for received tangible social support (Kaniasty & Norris, 1995, 2000).

Emotional social support Primary disaster survivors (M = 20.28, SD = 5.31) received more emotional social support than secondary disaster survivors (M = 11.04, SD = 8.12) and non-victims (M = 8.85, SD = 7.50). A primary disaster survivor mentioned in the narrative data that the "biggest encouragement was the outpouring of concern…from friends, even those outside my immediate circle," while another primary disaster survivor stated that "creating a circle of support" with other primary disaster survivors who understood her needs and emotions "who are walking the same mile" was the type of received emotional social support that was very important to her recovery. Another primary disaster survivor wrote of the "very kind" hotel manager who greeted them every day during the months they lived there.

Even secondary disaster survivors and non-victims reported in the narrative data that they received emotional social support from others. A secondary disaster survivor wrote of supportive family and friends being worried about her, and another secondary disaster survivor mentioned how their community came together immediately after the wildfire and how "it reinforced our belief that we live in a community of caring people." A non-victim wrote that out-of-state family called and emailed to make sure they were safe, even though they were not directly impacted.

Tangible social support Primary disaster survivors ($M = 16.81$, $SD = 8.6$) received more tangible social support than secondary disaster survivors ($M = 4.10$, $SD = 4.47$) and non-victims ($M = 1.98$, $SD = 2.52$). In the narrative data, one non-victim respondent stated that he took a truckload of replacement items such as clothing and a TV to a friend who lost their home; "I felt like sharing what I had was the best approach." Primary disaster survivors wrote of the outpouring of tangible help such as furniture, clothing, and toys by individuals, churches, and other organizations. Primary disaster survivors were also matched with a volunteer who functioned as a "personal shopper" and assessed their needs and helped gather furniture, clothing, and other necessaries.

In this analysis, secondary disaster survivor received significantly more tangible social support than non-victims did. One secondary disaster survivor, for example, wrote in the narrative data that as she evacuated her home, she fell and hurt her back. A neighbor took her to the hospital. When that hospital was evacuated and she was eventually sent by cab to the County Evacuation Center at QUALCOMM stadium, the taxi driver asked for a wheelchair for her. "In two minutes I had three wheelchairs," she stated. She stated that she needed a bed and not a cot to sleep on; although a bed wasn't available, the volunteers found her a mattress. She had to stay there 3 days before she could return home. She said that the volunteer help was "tremendous."

Informational social support Primary disaster survivors ($M = 14.87$, $SD = 7.25$) received more informational social support than secondary disaster survivors ($M = 4.93$, $SD = 5.69$) and non-victims ($M = 4.18$, $SD = 4.85$). One primary disaster survivor mentioned in the narrative data that a nonprofit organization gave them invaluable information on filing insurance claims. Primary disaster survivors had available to them a *Fire Survivor's Handbook*, "a how-to and what-to-do guide," that was the product of those who lost their homes to the San Diego Cedar Creek wildfire in 2003.

Sources of social support Three types of sources of social support—family members; friends, neighbors, and coworkers; and those outside an individual's immediate social circle such as community members, voluntary organizations, and professional service providers—were examined. Each was analyzed with the three disaster survivor statuses: primary, secondary, and non-victim. There were significant effects of disaster survivor status on all three sources of received social support with primary disaster survivors receiving significantly more social support from family, friends, and neighbors and from outsiders than secondary disaster survivors and non-victims across all age groups. Scores ranged from 0 to 30 (Kaniasty & Norris, 1995; Kaniasty & Norris, 2000).

Social support from family Primary disaster survivors (M = 16.75, SD = 8.56) received social support from family members more often than secondary disaster survivors (M = 8.00, SD = 6.44) and non-victims (M = 7.26, SD = 5.66) did. One primary disaster survivor mentioned in the narrative data that "they were blessed to have had…great family and friends. So many people went out of their way to help us."

In the narrative data, there were also examples of secondary disaster survivors and non-victims receiving support from family. For example, one secondary disaster survivor mentioned "supportive family" in the days after the wildfire, and a non-victim wrote of out-of-state family who called and emailed to check about their safety.

Social support from friends Primary disaster survivors (M = 18.75, SD = 6.90) received more social support from friends, neighbors, and coworkers than did secondary victims (M = 7.69, SD = 5.95) and non-victims (M = 5.25, SD = 4.97). There were several comments in the narrative data from primary disaster survivors related to received social support from friends, neighbors, and coworkers. For example, one primary disaster survivor wrote that her husband and she were blessed to have had great friends to help them. She went on to say that this "taught my husband and me what matters." Another primary disaster survivor wrote of "great friends" who went out of their way to help them.

Social support from others outside one's social circle Primary disaster survivors (M = 16.53, SD = 7.92) received more social support from those outside their immediate social circle such as community leaders, voluntary organizations, and professional service providers than secondary disaster survivors (M = 4.36, SD = 4.8) and non-victims (M = 2.94, SD = 3.73). A primary disaster survivor stated that there was a group called Fired Up Sisters (FUS) that was created to support women whose homes were destroyed or damaged and that this support system helped her tremendously. She stated, "We cried, we solved problems, we moved forward. After a couple of months, friends and families didn't want to hear about our constant problems, so with our FUS group, we found a 'new' normal." Another primary disaster survivor mentioned that the outpouring of concerns and help from those outside her immediate social circle encouraged her each day.

In the narrative data, there was also mention by secondary disaster survivors of support from those outside ones' social circle. For example, one secondary disaster survivor wrote that "the kindness of strangers was completely overwhelming" after the wildfire.

Received and Provided Social Support from the 2014 Wildfire Study

Received social support The total social support received by secondary disaster survivors (M = 11.10, SD = 6.63) was significantly more than the social support received by non-victims (M = 5.10, SD = 5.24), $t(47.79) = 4.68$, $p = 0.001$. Scores ranged from 0 to 27 (Kaniasty & Norris, 1995, 2000). Emotional support was most frequently received; the next most frequently received support was tangible support, and the third was informational support with secondary disaster survivors receiving significantly more of each type than non-victims, $p = <0.005$. The most frequent source of received social support was from family; the next most frequent source of received social support was from neighbors, friends, and coworkers and then from others outside one's immediate social circle with secondary disaster survivors receiving from each source significantly more than non-victims, $p = <0.001$.

Provided social support The total social support that secondary disaster survivors provided to others (M = 10.67, SD = 7.77) was significantly more than the total social support that non-victims provided to others (M = 7.31, SD = 6.62), $t(119) = 2.37$, $p = 0.019$. Scores ranged from 0 to 27 (Kaniasty & Norris, 1995, 2000). The most frequent types of social support provided by secondary disaster survivors and non-victims were emotional support and tangible support; there were no significant differences between the amount of these two social support types and the two disaster status groups. Secondary disaster survivors provided significantly more informational social support than non-victims, $p = 0.02$. The most frequent source that secondary disaster survivors provided social support to was neighbors, friends, and coworkers; secondary disaster survivors (M = 4.36, SD = 3.24) significantly provided more support to neighbors, friends, and coworkers than non-victims (M = 2.88, SD = 2.52), $p = 0.02$. Family was the next source that 2014 secondary disaster survivors provided social support to and then to those outside one's immediate social circle with no significant difference in providing to these groups between secondary disaster survivors and non-victims. Non-victims provided more social support to family, then to friends, neighbors, and coworkers, and lastly to others outside one's immediate social circle.

Comfortableness in Seeking Social Support from the 2014 Wildfire Study

The relationships between feeling comfortable in seeking a type of social support from a certain source and receiving a type of social support from a certain source revealed strong positive correlations for both secondary disaster survivors and non-victims. For secondary disaster survivors, feeling comfortable seeking informational support was associated with receiving informational social support, $r = 0.73$,

$p = 0.00$. For secondary disaster survivors, feeling comfortable seeking support from others outside their immediate social circle was associated with receiving support from others outside their immediate social circle, $r = 0.56$, $p = 0.00$. An independent samples *t*-test revealed a significant difference in comfortableness in seeking social support between secondary disaster survivors ($M = 13.73$, $SD = 8.30$) and non-victims ($M = 7.74$, $SD = 9.03$), $t(118) = 3.31$, $p = 0.001$. Scores ranged from 0 to 36 (Kaniasty & Norris, 2000). Secondary disaster survivors were significantly more comfortable in seeking emotional and tangible social support from family, neighbors, friends, and coworkers and others outside their immediate social group than non-victims, $p < 0.05$. Secondary disaster survivors and non-victims were not significantly different in comfortableness in seeking informational support.

Comfortableness in Providing Social Support from the 2014 Wildfire Study

The relationships between being comfortable in providing a type of social support to a certain source and actually providing a type of social support to a certain source revealed strong positive correlations for both secondary disaster survivors and non-victims. For example, for secondary disaster survivors, feeling comfortable providing tangible support was associated with actually providing tangible support, $r = 0.57$, $p = 0.00$. For non-victims, feeling comfortable providing support to family was associated with actually providing support to family, $r = 0.52$, $p = 0.00$. There was no difference between secondary disaster survivors ($M = 19.79$, $SD = 11.08$) and non-victims ($M = 16.42$, $SD = 12.62$) in comfortableness in providing social support, $t(119) = 1.35$, $p = 1.80$. Scores ranged from 0 to 36 (Kaniasty & Norris, 2000). Secondary disaster survivors and non-victims were not significantly different in comfortableness in providing emotional, tangible, or informational social support to family; neighbors, friends, and coworkers; and others outside their immediate social group. Mean scores did indicate that both secondary disaster survivors and non-victims were most comfortable providing emotional support to family and neighbors, friends, and coworkers.

Conclusion and Implications

Although disasters have an immediate impact on the psychological health of the community-residing adults exposed to them, the findings from the 2007 and 2014 northern San Diego wildfire studies showed a potential for long-term recovery and good psychological health. Each disaster status group from these two wildfire studies revealed good psychological health 4 years and 3 years, respectively, after the disaster events. These results were similar to many of the respondents who reported good mental health levels in other studies (Bryant et al., 2014; Camilleri et al., 2010;

Gibbs et al., 2016). Anecdotal confirmation comes from two primary disaster survivor participants from the 2007 wildfire study who wrote that they were proof that with time, individuals do recover from the tragedy of losing a home to a wildfire and life can return to normal.

The results from these two studies showed little evidence of an association between total social support received and psychological well-being (depressive symptomatology). Perhaps the 4 years and 3 years that existed between the two wildfires and the collection of data were too long to establish a direct relationship. Acierno et al. (2006) and Felix and Afifi (2015) took their assessments within 5–11 months after the disaster.

When participants though were asked to think only about the time of the wildfire and to think about social support that they received, there was significance in many variables, for example, the social support received most frequently was emotional social support, and the disaster status group who received the most social support was the disaster group with the most severe exposure to the wildfire. For the 2007 wildfire, that was the primary disaster survivor group, and for the 2014 study, the secondary disaster survivor group. Our 2007 results supported Camilleri et al.'s (2010) study. Primary disaster survivors were aided significantly more than the secondary disaster survivors and non-victims, and they received the three types of social support (informational, emotional, and tangible) more often than secondary disaster survivors and non-victims. Primary disaster survivors were also helped significantly more by family; friends, neighbors, and coworkers; and others outside their immediate social circle than secondary disaster survivors and non-victims. In the 2014 study, it was the secondary disaster survivors, the group who had the most severe exposure to the wildfire, who received significantly more social support and from all sources than the non-victim group.

After the wildfire in Canberra, Australia, community residents formed new groups to help with recovery (Camilleri et al., 2010). The 2007 primary disaster survivors also formed a new organization, the Fired Up Sisters. Organized in November 2007, this support group was created by women who were primary disaster survivors of the October 2007 wildfire. This organization provided an emotional support group for primary disaster survivors (Rossi, 2010). In the narrative data, a primary disaster survivor wrote that she had amazing support and friendship from all her Fired Up Sisters. Because we were not able to reach any primary disaster survivors in the 2014 wildfire, we don't know if a group was established to aid primary disaster survivors, but in a later wildfire, the 2017 Lilac wildfire, the Fired Up Sisters organization did reach out to and help form a support group for primary disaster survivors of the Lilac wildfire (Fired Up Sisters, n.d.).

In the 2014 study, community-residing adult participants were asked whom they provided social support to. It was the secondary disaster survivors who provided significantly more social support to others, and they significantly provided more informational social support than the non-victim group. Both disaster status groups did provide emotional and tangible social support though. Interestingly, secondary disaster survivors provided more social support to friends, neighbors, and coworkers than to family.

The 2014 study explored comfortableness in seeking and providing social support and how this related to actually receiving and providing social support. Our results for comfortableness in seeking social support supported Kaniasty and Norris (2000) as there were strong positive correlations between seeking and receiving social support. This applied to both disaster status groups and for the three types of social support and the three sources of social support. For example, if a community-residing adult was comfortable seeking emotional social support, they tended to receive emotional support. Our 2014 results also found the same for comfortableness in providing social support and actually providing social support. If an individual felt comfortable in providing social support, they tended to provide it. We did not find an association with psychological health as did Lebowitz (2017). The timing of assessments might have played a role in this.

How do these findings translate into federal, state, and local planning, preparation, and programs to aid with post-disaster coping of community-residing adults? Since received social support appears to be a crucial point in adults' psychological recovery from a disaster such as a wildfire (Acierno et al., 2006; Camilleri et al., 2010; Felix & Afifi, 2015; Gibbs et al., 2016) and Kaniasty and Norris (2000) and this study found that those who feel comfortable seeking support are more likely to receive more support, increasing an impacted community adult's feelings of comfortableness to seek social support should be a priority. Possibly this could be accomplished if federal, state, and local programs made use of already established social networks within affected communities. Local churches, neighborhood schools, and residential homeowner association offices and those individuals associated with these sites such as ministers, teachers, and administrative staff would be already known entities to those impacted by a disaster. A familiar place and a familiar face could increase comfortableness in those who need support such that they will ask for support.

These churches, schools, and/or offices also could serve as a "one-stop service/donation center," allowing neighbors, friends, and organizations a place to bring tangible items and informational items to one central location to help those who have lost and/or are displaced from their homes. This would ease the immediate confusion of "Where do I go to find…?" for those most impacted by the disaster, and this might increase comfortableness in seeking support as it would decrease stress in a familiar setting. Several respondents from the 2007 wildfire mentioned that they volunteered to be "personal shoppers." They were matched with families or individuals to assess their needs and gather what the family/individual needed. This could be a service offered at the "one-stop service/donation center" and would add a more personal touch between community individuals in the taxing times directly after the disaster.

The 2007 respondents also mentioned another service needed at a "one-stop service/donation center"—that of being matched with someone who had in a previous wildfire lost their home. Previous primary disaster survivors could answer questions and provide reassurance about the future that only those who have been there can provide. Also "creating a circle of support," an organization similar to Fired Up Sisters, could be drawn from the primary disaster survivors visiting the center.

Overall, these ideas would provide the third type of social support: emotional support. Therefore, all three types of social support could be found at the "one-stop service/donation center."

But where do we find volunteers, those individuals who feel comfortable providing help to others and do volunteer at sites such as "one-stop service/donation centers"? This is a second key point as seen from the results of this study and Lebowitz's study (2017); those who feel comfortable providing help are more likely to provide help. Volunteers could definitely come from local churches and neighborhood schools where a large number of roles are filled by volunteers, but more individuals might be needed. Marketing on websites such as NextDoor.com internet community and the neighborhood school's, local church's, and local organizations' websites would provide opportunities to reach more community adults comfortable with providing help who in turn would be willing to step in to provide help. Even past disaster experience could increase comfortableness in helping behavior. A respondent from the 2007 wildfire mentioned that after that wildfire, her group of neighbors "realized that they needed to become neighbors helping neighbors. Now they have very active neighborhood watch program," looking out for and providing help to each other. Comfortableness in helping could come for common disaster experiences.

In closing, our findings offer a contribution to the growing body of literature on the impact of social support after a disaster and psychological health. Future research should continue to explore these relationships to offer to community leaders evidence-based suggestions to aid in the development of interventions and further the recovery of community-residing adults impacted by a disaster.

Acknowledgments I would like to thank former students Edith Jimenez, Moe Perdomo, David Martinez, Sean Cuadra, Araceli Cruz, and Margaret Deitering for their help in the development of these studies, their help in entering and checking the data, and their help in the analyses of the data. I am also grateful to these students, Joe Phillips, and all the individuals who distributed the questionnaires. Two California State University San Marcos CHABSS Lecturer Professional Development Grants supported these research studies. This support is gratefully acknowledged.

References

Acierno, R., Ruggiero, K. J., Kilpatrick, D. G., Resnick, H. S., & Galea, S. (2006). Risk and protective factors for psychopathology among older versus younger adults after the 2004 Florida hurricanes. *The American Journal of Geriatric Psychiatry, 14*, 1051–1059. https://doi.org/10.1097/01.JGP.0000221327.97904.b0

Bolin, R. (1985). Disaster characteristics and psychosocial impacts. In B. J. Soder (Ed.), *Disasters and mental health: Selected contemporary perspectives* (pp. 3–28). Rockville, MD: National Institute of Mental Health.

Bolin, R., & Klenow, D. (1983). Response of the elderly to disaster: An age-stratified analysis. *The International Journal of Aging and Human Development, 16*, 283–296.

Bryant, R. A., Waters, E., Gibbs, L., Gallagher, H. C., Pattison, P., Lusher, D., et al. (2014). Psychological outcomes following the Victorian Black Saturday bushfires. *Australian and New Zealand Journal of Psychiatry, 48*, 634–643. https://doi.org/10.1177/0004867414534476

CAL FIRE Stats and Events. (2019). *California Department of Forestry and Fire Prevention*. Retrieved from https://www.fire.ca.gov/stats-events/

Camilleri, P., Healy, C., Macdonald, E. M., Nicholls, S., Sykes, J., Winkworth, G., et al. (2010). Recovery from bushfires: The experience of the 2003 Canberra bushfires three years after. *Journal of Emergency Primary Health Care, 8*, 1–15.

Felix, E. D., & Afifi, W. (2015). The role of social support on mental health after multiple wildfire disasters. *Journal of Community Psychology, 43*(2), 156–170. https://doi.org/10.1002/jcop.21671

Fired Up Sisters. (n.d.). *Lilac wildfire survivor support*. Retrieved from http://firedupsisters.com/lilac-wildfire-survivor-support/

Gibbs, L., Bryant, R. A., Harms, L., Forbes, D. Block, K., Gallagher, H. C., Ireton, G, et al. (2016). *Beyond bushfires: Community, resilience and recovery-Final report*. Retrieved from https://mspgh.unimelb.edu.au/__data/assets/pdf_file/0008/3043187/Beyond-Bushfires-Final-Report-2016.pdf

Haines, V. A., Hurlbert, J. S., & Beggs, J. J. (1996). Exploring the determinants of support provision: Provider characteristics, personal networks, community contexts, and support following life events. *Journal of Health and Social Behavior, 37*, 252–264.

Kaniasty, K., & Norris, F. H. (1995). In search of altruistic community: Patterns of social support mobilizations following Hurricane Hugo. *American Journal of Community Psychology, 23*, 447–477.

Kaniasty, K., & Norris, F. H. (2000). Help-seeking comfort and receiving social support: The role of ethnicity and context of need. *American Journal of Community Psychology, 28*, 545–581. https://doi.org/10.1023/A:1005192616058

Kaniasty, K. Z., Norris, F. H., & Murrell, S. A. (1990). Received and perceived social support following natural disaster (1990). *hmayJournal of Applied Psychology, 20*, 85–114.

Lebowitz, A. J. (2017). Relational satisfaction from providing and receiving support is associated with reduced post-disaster depression: Data from within one year of the 2011 Japan Triple Disaster. *Journal of Community Mental Health, 53*, 202–2014. https://doi.org/10.1007/s10597-016-9995-4

Marshall, G. N., Schell, T. E., Elliott, M. N., Rayburn, N. R., & Jaycox, L. H. (2007). Psychiatric disorders among adults seeking emergency disaster assistance after a wildland-urban interface fire. *Psychiatric Services, 58*, 509–514. https://doi.org/10.1176/appi.ps.58.4.509

May 2014 San Diego Wildfires After Action Report. (2014). Retrieved from http://www.sdfire-foundation.org/commandpost/San-Diego-County-Wildfires-May-2014.pdf

McFarlane, A. C., Clayer, J. R., & Bookless, C. L. (1997). Psychiatric morbidity following a natural disaster: An Australian bushfire. *Social Psychiatry and Psychiatric Epidemiology, 32*, 261–268. https://doi.org/10.1007/BF00789038

Miller, P. A., Roberts, N. A., Zamora, A. D., Weber, D. J., Burleson, M. H., Robles, E., et al. (2012). Families coping with natural disasters: Lessons from wildfires and tornadoes. *Qualitative Research in Psychology, 9*, 314–336. https://doi.org/10.1080/14780887.2010.500358

Myers, J. K., & Weissman, M. M. (1980). Use of a self-report symptom scale to detect depression in a community sample. *The American Journal of Psychiatry, 137*, 1081–1084.

Norris, F. H., & Kaniasty, K. (1996). Received and perceived social support in time of stress: A test of the social support deterioration deterrence model. *Journal of Personality and Social Psychology, 71*, 498–511. https://doi.org/10.1037/00223514.71.3498

Norris, F. H., Murphy, A. D., Kaniasty, K., Perilla, J. L., & Ortis, D. C. (2001). Postdisaster social support in the United States and Mexico: Conceptual and contextual considerations. *Hispanic Journal of Behavior Sciences, 23*, 469–496. https://doi.org/10.1177/0739986301234008

Norris, F. H., & Murrell, S. A. (1988). Prior experience as a moderator of disaster impact on anxiety symptoms in older adults. *American Journal of Community Psychology, 16*, 665–683. https://doi.org/10.1007/BF00930020

Phillips, J. R., Cruz, A., Cuadra, S., Martinez, D., & Perdomo, M. (2013). Predictors of mental health after a disaster: A study of older adults four years after the 2007 San Diego County wildfires. *The Gerontologist, 53*(S1), 109.

Radloff, L. S. (1977). The CES-D scale: A self-report depression scale for research in the general population. *Applied Psychological Measurement, 1*(3), 385–401. https://doi.org/10.1177/014662167700100306

Rossi, N. (2010, May 15). Fired Up Sisters was born out of tragedy: Wildfire support becomes relief group. *San Diego Union Tribune*. Retrieved from http://www.utsandiego.com/

Sanders, S., Bowie, S., & Bowie, Y. (2003). Lessons learned on forced relocation of older adults: The impact of Hurricane Andrew on health, mental health, and social support of public housing residents. *Journal of Gerontological Social Work, 40*, 23–35.

Tyler, K. A. (2006). The impact of support received and support provision on changes in perceived social support among older adults. *The International Journal of Aging and Human Development, 62*, 21–38.

Watanbe, C., Okumura, J., Chiu, T., & Wakai, S. (2004). Social support and depressive symptoms among displaced older adults following the 1999 Taiwan earthquake. *Journal of Traumatic Stress, 17*, 63–67. https://doi.org/10.1023/B:JOTS.0000014678.79875.30

Weems, C. F., Watts, S. E., Marsee, M. A., Taylor, L. K., Costa, N. M., Cannon, M. F., et al. (2007). The psychosocial impact of Hurricane Katrina: Contextual difference in psychological symptoms, social support, and discrimination. *Behaviour Research and Therapy, 45*, 2295–2306. https://doi.org/10.1016/j.brat.2007.04.013

Chapter 11
Evidence-Based Treatment for Mental Health Care Post-Disaster

Emma L. Lucas and Lisa M. Brown

Introduction

In November 2018, California experienced the deadliest and most devastating wildfire season since the early 1900s. The Camp Fire took the lives of 88 people and resulted in approximately 16.5 billion USD in damages. To date, the Camp Fire is the costliest wildfire in recorded history (Centre for Research on the Epidemiology of Disasters [CRED], 2019). Three months prior, California experienced the sixth most destructive wildfire in the state's history. One year prior, the state experienced the second most destructive wildfire in its history. From 2000 to 2009, natural disasters alone have increased threefold in comparison to 1980 to 1989 (Leaning & Guha-Sapir, 2013). Given the increasing frequency and growing magnitude, the importance of disaster mental health services in the short and long term is evident. Although there are a number of laypeople and mental health professionals trained in evidence-informed and evidence-based practices, improved infrastructure is needed to address the immediate mental health needs of disaster-affected populations and to support communities in their long-term recovery efforts.

According to CRED, in 2018, more than 95.6 million people worldwide were affected by natural disasters (CRED, 2018). Disasters are traumatic events that result in adverse consequences to large groups of people (Neria, Nandi, & Galea, 2008), disrupt the infrastructure of communities (Norris, Friedman, & Watson, 2002), and affect physical and mental health outcomes to those impacted (Neria et al., 2008). Mental health providers often play an integral role in helping communities prepare for, respond to, and recover from disasters. In this chapter, we review an array of interventions and treatments that disaster mental health programs typically use when caring for affected populations.

E. L. Lucas · L. M. Brown (✉)
Palo Alto University, Palo Alto, CA, USA
e-mail: lbrown@paloaltou.edu

© Springer Nature Switzerland AG 2021

K. E. Cherry, A. Gibson (eds.), *The Intersection of Trauma and Disaster Behavioral Health*, https://doi.org/10.1007/978-3-030-51525-6_11

To date, there are no concrete estimations of the prevalence of mental health morbidity in disaster-affected populations. Prevalence is difficult to measure because it is often unknown if mental health symptoms were present before the disaster, if preexisting conditions were exacerbated, or if the symptoms were a new onset of illness as a direct result of disaster exposure. However, systematic reviews have attempted to assess the magnitude of mental health sequelae from disasters (North & Pfefferbaum, 2013; Neria et al., 2008). For human-made (e.g., terrorist attacks), technological (e.g., factory explosions), and natural (e.g., tornadoes, earthquakes) disasters, studies suggest that people who were directly affected by the disaster have the highest prevalence of post-traumatic stress disorder (PTSD) followed closely by rescue workers and then members of the general population (Neria et al., 2008).

The Continuum of Care Model of Behavioral Health and the Research Base

Relief organizations and behavioral health providers use appropriate evidence-based, evidence-informed, or evidence-supported practices. In 2015, the Substance Abuse and Mental Health Services Administration (SAMHSA 2015) developed an inventory of disaster behavioral health interventions that included the research evidence available for each intervention and treatment and then organized them into different levels (i.e., Tier 1, universal; Tier 2, selective; and Tier 3, indicated). Tier 1 includes universal interventions such as psychoeducation and outreach. Tier 2 or targeted interventions includes short-term interventions. Tier 3 includes more intensive treatments such as psychiatric services and long-term psychological care (Pynoos, Goenjian, & Steinberg, 1998). Within these defined levels, the mental health approaches reviewed in this chapter will be described by their research base as evidence-based practices or evidence-informed practices. For most disaster survivors, early and immediate interventions will alleviate mental health symptoms resulting from the event. However, there are a small number of individuals who will need more intensive or long-term interventions. According to the American Psychological Association (APA), "evidence-based practice in psychology is the integration of the best research available with clinical expertise in the context of patient characteristics, culture, and preferences" (APA, 2006). Evidence-informed or evidence-supported practices are informed by empirical research and clinical understanding (Nevo & Slonim-Nevo, 2011). For example, Psychological First Aid is an evidence-informed practice that has not been evaluated in its entirety after a disaster. Individual core components comprising Psychological First Aid have been independently assessed (e.g., breathing exercises, psychoeducation, and cognitive restructuring).

Universal (or Early) Interventions (Tier 1)

Immediately after a disaster, individuals, families, and communities begin to mobilize to assess the damage incurred, reconnect with loved ones, support others in need, and initiate recovery. Although several interventions have been developed to address individual, family, and community mental health needs at this early stage, Psychological First Aid (PFA) has accrued the most evidence to date in support of its usefulness post-disaster. PFA is an evidence-informed intervention developed to assist people immediately after a disaster (see Cherry, Calamia, & Elliott, this volume, Chap. 8). The intervention was designed to reduce initial distress, promote short- and long-term adaptive functioning, and primarily focus on individual strengths. PFA assumes that most survivors of disaster will not develop long-term mental health problems. It uses a modular approach where specific core components can be used with individuals across the life span and from diverse cultures. Importantly, it can be delivered by trained laypeople in various settings, thereby making it both scalable and economical (SAMHSA, 2015). One of the primary considerations when providing PFA is entering the setting. The intervention was developed to be delivered in a public shelter where the people to be approached are not known to the responders. PFA providers will need to assess the environment and the individuals affected by the disaster, in particular paying close attention to potentially vulnerable subgroups of the population (Ruzek et al., 2007).

In 2006, the PFA Field Operations Guide (PFA Guide) was developed to provide updated guidelines for disaster relief practitioners responding to the immediate needs of individuals, groups, and communities impacted by disaster (Brymer et al., 2006; Vernberg et al., 2008). Trained volunteers are instructed to select one or more of the PFA eight core actions to use with disaster-exposed people: (1) contact and engagement, (2) safety and comfort, (3) stabilization, (4) information gathering, (5) practical assistance, (6) connection with social supports, (7) information on coping supports, and (8) linkage with collaborative services (Ruzek et al., 2007). In addition to supporting communities after disaster, the PFA Guide also acknowledges the challenges volunteers encounter when providing first aid services in post-disaster settings and encourages appropriate professional development and self-care (Vernberg et al., 2008). First responders have reported that use of the PFA model is beneficial when assisting disaster survivors and that they are satisfied with their experiences implementing the model (Allen et al., 2010). Despite these positive impressions, more evaluation research needs to be conducted on PFA.

Given the existing feedback on PFA post-disaster, several variations of the model have been developed, tested, and disseminated. *Group Psychological First Aid (Group-PFA)* involves several stages that are implemented after conducting an initial assessment following a disaster (Everly, Phillips, Kane, & Feldman, 2006). The stages comprise an introduction that includes group leaders, group purpose, expected duration, and "group ground rules," incident review or briefing, clarification of the facts of the event, psychoeducation, activities to support group cohesion and resiliency, and facilitation of connections with support systems. After the implementation of *Group-PFA*, post-group activities may include continuity of care for group members and, as appropriate, leader closure and self-care and group evaluation (Everly et al., 2006). A group format supports broad application of PFA when the availability of trained volunteers is limited, making it unfeasible to deliver one-on-one PFA intervention. Also, group interventions provide an alternative to individual treatment for disaster survivors. When clusters of individuals are impacted by the same traumatic event, group interventions may be clinically indicated, while disseminating treatment individually may hinder treatment progress (Everly et al., 2006). However, group intervention may not be best suited for some individuals (e.g., severe mental illness, families, etc.).

After Hurricane Katrina, mental health programming for children was implemented in traditional and nontraditional settings, for example, schools (Plummer, Cain, Fisher, & Bankston, 2008; Speier, Osofsky, & Osofsky, 2009) including a pilot study testing a six-week group intervention for children (Plummer et al., 2008). Although the intervention did not strictly adhere to the *Group-PFA* model proposed by Everly et al. (2006), the intervention did ascribe to the PFA guidelines but was specifically tailored for children who were exposed to disaster. Children participating in the group perceived the intervention to be helpful, and areas for improvement were highlighted for future group implementation with children exposed to disasters. Both traditional and nontraditional methods of mental health programming for children present challenges including obtaining parental consent, delays in service delivery, considerations of developmental level, severity of need among the children, continued presence of ongoing stressors from the disaster, and insufficient infrastructure and support for the community (Plummer et al., 2008; Speier et al., 2009).

Other adaptations of PFA have also been developed for use post-disaster. Although the evidence base is lacking, they are worth noting as they pull from the PFA model and may be useful in unique situations. For example, the *PFA Rapid Model* focuses on providing face-to-face training in PFA to public health professionals with no prior mental health education or experience. The training uses concepts and skills related to reflective listening, assessment, prioritization, intervention, and disposition for public health professionals participating in disaster response (SAMHSA, 2015). The *Nebraska PFA* program is an adaptation of the Community-Based Psychological Support developed by the International Red Cross. Similar to other PFA programs, the Nebraska model can also be delivered by non-mental health professionals. The seven skill areas in this program are (1) psychological support, (2) stress and coping, (3) supportive communication, (4) promoting community self-help, (5) functional needs, (6) helping the helper, and (7) de-escalation (SAMHSA, 2015; Nebraska Disaster Behavioral Health, n.d.).

The PFA intervention has also been adapted for school administrators, educators, and staff to support children, adults, and families after disasters. For the workplace, the PFA model has been modified to train management and employees to use PFA with a brief online course entitled *Building Workforce Resilience Through the Practice of Psychological First Aid: A Course for Supervisors and Leaders (PFA-L)*. Lastly, the *Listen, Protect, Connect* Model of PFA focuses on supporting trauma-exposed children within the school setting. There is promising evidence for the *Listen, Protect, Connect* PFA intervention in decreasing psychological distress and increasing school connectedness and social support (Ramirez et al., 2013). Having a significant number of people trained and ready to use PFA will improve timeliness in delivery of appropriate mental health care as well as increase access to needed services by disaster-affected individuals and groups.

Targeted Interventions or Intermediate Interventions (Tier 2)

After the initial crisis intervention phase, communities move into the beginning phases of recovery. This stage may begin days or weeks after a disaster depending on the existing infrastructure of the community post-disaster. Similar to early interventions post-disaster, an array of targeted or intermediate interventions have been developed for use post-disaster. Below, five interventional models are reviewed along with their research base. Each of these models focus on the adjustment of individuals to their new post-disaster reality.

Skills for Psychological Recovery (SPR) is a modular intervention that was specifically designed to be used after PFA with those who need additional support during the recovery phase. SPR includes a variety of skill-building components from existing mental health treatments, such as problem-solving, positive activities scheduling, managing reactions, helpful thinking, and building healthy social connections (SAMHSA, 2015). Program evaluations of SPR trainings in New Zealand and Australia found that the training program was largely successful in building practitioner competence and confidence to implement the intervention (Forbes et al., 2010; Wade et al., 2014). To date, there is limited evaluative and outcome research of SPR post-disaster.

A second post-disaster intervention, Cognitive Behavioral Therapy for Post-Disaster Distress (CBT-PD), is a manualized treatment consisting of ten sessions focusing on identifying and challenging maladaptive disaster-related beliefs. There are four components: psychoeducation, breathing retraining, behavioral activation, and cognitive restructuring (SAMHSA, 2015). Bryant proposed research to explore potential modifications to CBT to improve the delivery of the treatment to larger disaster-affected populations (2007). At present, identified challenges with bringing the treatment to scale and implementing with large numbers of disaster-affected individuals include the amount of resources required, especially after large-scale events (Bryant, 2007). These resources include time and staffing to provide training and supervision to providers and then maintaining adequate provider levels to meet the ongoing needs of the community. Significant reductions in PTSD symptoms have been demonstrated in both national (Hamblen et al., 2009) and international subgroups of the population who have received CBT-PD treatment (Leiva-Bianchi, Cornejo, Fresno, Rojas, & Serrano, 2018).

A third approach to treatment post-disaster are mind/body therapies. Mind/body therapies are complementary health approaches that are offered to individuals or groups to minimize their distress and increase resilience. Mindfulness-Based Cognitive Therapy (MBCT) is a treatment that includes mindfulness. Mindfulness-Based Stress Reduction (MBSR) is a group intervention that includes mindfulness practices such as meditation and yoga. MBSR can effectively reduce symptoms related to PTSD including depression and anxiety (Peterson & Pbert, 1992; Hofmann, Sawyer, Witt, & Oh, 2010). Alternative therapies offer a variety of options for individuals who reject or are not ready for more traditional therapeutic approaches.

Two final interventions for the intermediate phase post-disaster are Specialized Crisis Counseling Services (SCCS) and Writing for Recovery. SCCS is a five-step model that includes 1) assessing current needs of survivors; 2) reviewing cases through team meetings to determine appropriate services; 3) providing services and resources to survivors; 4) monitoring and supporting staff through ongoing training, supervision, and consultation; and 5) reviewing successes and further resource needs with survivors during the final visit. The model was developed to engage survivors through outreach and then empower them to take an active role in the disaster response services they receive, such as solution-focused, skill-based interventions and other interventions that facilitate recovery (SAMHSA, 2015). The model was implemented in the aftermath of Hurricane Katrina with promising outcomes (Jones, Allen, Norris, & Miller, 2009; Hansel et al., 2011).

The Writing for Recovery model uses narrative therapy techniques to help normalize distressing reactions to traumatic events. Individuals complete writing assignments to process their thoughts and feelings (SAMHSA, 2015). In a 2004 study, participants who engaged in narrative exposure therapy, a short-term therapeutic approach founded on cognitive behavioral and testimony therapy, had lower endorsement of PTSD symptoms compared to participants who received supportive counseling or psychoeducation. This study suggests that narrative therapy is a possible treatment for individuals who have been exposed to trauma (Neuner, Schauer, Klaschik, Karunakara, & Elbert, 2004).

Intensive Interventions or Long-Term Interventions (Tier 3)

In the general population, it is estimated that approximately 6% to 20% of trauma-exposed individuals will eventually develop PTSD (Kessler, Sonnega, Bromet, Hughes, & Nelson, 1995). Individual and group treatments focused on processing the traumatic event are available to treat trauma-exposed individuals.

Cognitive behavioral therapy (CBT) has been found to be an effective treatment for many mental health disorders (Butler, Chapman, Forman, & Beck, 2006; Tolin, McKay, Forman, Klonsky, & Thombs, 2015). Since the development of cognitive behavioral therapy, a number of variations of this model have been adapted to enhance potential usefulness with survivors of disasters.

The two gold standard treatments for PTSD are Cognitive Processing Therapy (CPT) and Prolonged Exposure Therapy (PE). CPT is a combination of cognitive therapy and exposure therapy delivered over 12 sessions (SAMHSA, 2015). CPT has been found to be effective for treating PTSD and depression among sexual assault survivors (Resick & Schnicke, 1992; Resick, Nishith, Weaver, Astin, & Feuer, 2002; Resick, Williams, Suvak, Monson, & Gradus, 2012; Lenz, Bruijn, Serman, & Bailey, 2014). PE aims to help reduce the level of fear and anxiety connected with trauma reminders, thereby also reducing avoidance, anxiety, and fear when encountering a reminder (SAMHSA, 2015). Among the veteran population, PE has been found effective in reducing symptom of PTSD and depression (Eftekhari et al., 2013) and may translate well to disaster survivors. These treatments provide options for disaster survivors requiring longer-term support after the initial and intermediate post-disaster phases.

PTSD often co-occurs with other disorders. Skill Training in Affect and Interpersonal Regulation (STAIR) is an evidence-based cognitive behavioral therapy for individuals with PTSD and co-occurring disorders. STAIR was developed to teach clients skills in emotion regulation and interpersonal functioning. In a 2007 study of survivors of mass trauma, STAIR treatment that was augmented with a modified version of prolonged imaginal exposure (MPE) reduced maladaptive coping strategies while encouraging the use of social support. STAIR's flexible application of its manualized treatment for survivors of mass violence makes it a valuable tool (Levitt, Malta, Martin, Davis, & Cloitre, 2007).

There are other forms of therapies with less empirical support but are worth noting. Eye Movement Desensitization and Reprocessing (EMDR) is an integrative information processing approach that combines elements from cognitive behavioral, interpersonal, experiential, and body-centered therapies using an eight-phase approach (SAMHSA, 2015). Studies evaluating EMDR post-disaster demonstrate its efficacy addressing traumatic reactions across the life span post-disaster (Grainger, Levin, Allen-Byrd, Doctor, & Lee, 1997; Jarero, Artigas, Montero, & Lena, 2008).

Psychodynamic approaches to PTSD focus on factors that may influence or cause PTSD symptoms. The intent of psychodynamic therapy is to help people get in touch with and work through painful unconscious feelings (SAMHSA, 2015). Psychodynamic psychotherapies may treat complex PTSD better than other forms of therapy (Schottenbauer, Glass, Arnkoff, & Gray, 2008); however, the research is limited, and the use of specific treatment components may be more effective (Fonagy, 2015). Further research is warranted in expanding the use of these approaches for individual's mental health treatment post-disaster.

Other Direct Practice Interventions

Many psychological interventions are now supported through the use of Internet- and mobile-based applications. Three mobile applications have been developed to provide psychoeducation about PTSD and to support delivery of gold standard treatments for PTSD. The CPT Coach mobile application was developed by the

Veteran Affairs (VA)'s National Center for PTSD, VA's Sierra Pacific (VISN 21), Mental Illness Research Education Clinical Centers of Excellence (MIRECC), and Department of Defense's (DoD) National Center for Telehealth and Technology to enhance the CPT treatment provided by clinicians. Similarly, the PE Coach mobile application was created by the VA's National Center for PTSD and DoD's National Center for Telehealth and Technology. The app is a complementary tool to in-person PE treatment. Lastly, the PTSD Coach app was developed by the VA's National Center for PTSD with the DoD's National Center for Telehealth and Technology. This app focuses on psychoeducation and management of symptoms after trauma and provides a tool to enhance and support therapeutic interventions between sessions.

Consideration for Vulnerable Populations

Vulnerable populations are groups of individuals whose exposure during the pre-disaster phase, disaster event, or post-disaster phase adversely affects them beyond what is expected because of preexisting traits, conditions, or limited resources. Consequently, vulnerable populations need to be considered in disaster preparation, response, and recovery. In addition to designating children, older adults, and people with physical disabilities, illness, mental illness, substance use, and limited English proficiency as vulnerable populations, first responders and journalists are also a higher risk for adverse outcomes. Each of these vulnerable populations will be discussed below along with relevant mental health concerns and intervention and treatment considerations.

First Responders and Recovery Workers

First responders and recovery workers are often among the first witnesses to the scene of a disaster and usually include police officers, firefighters, and emergency medical technicians. First responders receive specialized training to provide assistance at a disaster. Secondary traumatic stress, also known as vicarious trauma, is the negative effects from exposure to traumatic experiences in the context of one's occupation with many of the same symptoms as PTSD (Molnar et al., 2017). A recent systematic review found that the prevalence among first responders for mental health conditions, such as PTSD, depression, suicidality, anxiety, alcohol use, and sleep problems, was higher than the general population (Jones, 2017). First responders to the World Trade Center suffered from PTSD 11 to 13 years post-disaster. PTSD adversely affected their health and psychosocial well-being, specifically their satisfaction with life (Bromet et al., 2016). Predictive factors for first responders incurring adverse mental health consequences include demographic factors (e.g., age, previous trauma exposure), experiences during the event (e.g., sustaining injury), cognitive factors (e.g., tendency to catastrophize), less social support, and various work-related factors (Kleim & Westphal, 2011). Longitudinal research is needed to determine risk and resiliency factors of first responders and their relationship to mental health outcomes (Kleim & Westphal, 2011).

Recovery workers are often volunteers who have received training to support response and recovery efforts such as provision of basic necessities including shelter, food, water, and clothing to disaster survivors. For recovery workers, poor predeployment preparation, lack of support, and role management during and after deployment affect an individual's well-being post-disaster. However, improving management strategies and leadership abilities within disaster response organizations could potentially alleviate many of these concerns (Brooks et al., 2015).

Disaster preparation and ongoing training is a key component to the adequate preparation of first responder and recovery workforces. A 2014 study in Victoria, Australia, found that if a disaster were to occur, only 32% to 42% of mental health providers were ready to provide current best practice interventions (Reifels, Naccarella, Blashki, & Pirkis, 2014). Supportive resources have been developed to aid in the delivery of services. These include application-based resources such as SAMHSA's Behavioral Health Disaster Response App, SAMHSA's Suicide Safe Mobile App, and the Psychological First Aid (PFA) Mobile App. Additional reference and resource materials need to be developed to support not only the implementation of services at the disaster site but to support the physical and mental well-being of the first responder and recovery workforce.

Media and News Reporters

Media and news reporters are often among the first individuals arriving at a disaster scene. In general, journalists lack access to formal and informal support systems to manage the mental health effects of disasters (Osofsky, Holloway, & Pickett, 2005). During Hurricane Maria, journalists were informally called to act as first responders when access to radio was the only communication medium that was operable. Sharing equipment placed a sense of responsibility on journalists to assist disaster-affected communities as first responders (Nieves-Pizarro, Takahashi, & Chavez, 2019). It is recommended journalists receive specialized training and opportunities to take part in disaster planning exercises to prepare for disaster reporting and to enhance their resilience to catastrophic events. Furthermore, research indicates that establishment of peer and professional groups would be beneficial to support journalists who are experiencing mental health issues after disasters (Sreedharan & Thorsen, 2018). Finally, disaster training can be effective in managing many of the mental health consequences experienced by members of the media (Dworznik & Grubb, 2007). For example, the Dart Center for Journalism and Trauma (a project of the Columbia Journalism School) provides education, training, and consultation services that cover a range of trauma-informed practices as they relate to the field of journalism such as developing trauma awareness, monitoring staff trauma exposure, providing support, establishing resiliency, and practicing good self-care ("Dart Training," 2019).

Children and Families

The vulnerability of children (Norris et al., 2002; Kousky, 2016) and families must be considered during disaster planning, response, and recovery. Factors increasing individual vulnerability include low family socioeconomic status, parental high stress levels, poor coping strategies, high exposure to media coverage, and other hardships (Pfefferbaum, Jacobs, Houston, & Griffin, 2015). Parental psychopathology and visible suffering may also exacerbate children's negative and enduring post-disaster experiences (Cobham, McDermott, Haslam, & Sanders, 2016). Not surprisingly, children are susceptible to the development of mental health disorders after a disaster, as well as traumatic grief (see Kozu, this volume, Chap. 6). In a comparison of children exposed to disasters that occurred in two different countries, one-third of the sample met criteria for a mental health condition using a screener that indicated that they would have benefitted from mental health services (Navarro et al., 2016).

Due to the potential adverse effect disasters may have on children, establishing safety and stability post-disaster is imperative as it has been shown to improve long-term outcomes (Gibbs et al., 2015). Children are impacted in primarily three areas post-disaster: 1. physical health, 2. mental health, and 3. educational attainment and achievement. However, these effects will differ depending upon the children's response to the disaster, family characteristics, community and country infrastructure, response to the event, and type of disaster (Kousky, 2016). There are specific treatments for children and families that address each of these areas and support recovery.

Trauma-Focused Cognitive Behavioral Therapy (TF-CBT) is an empirically supported treatment for individuals, families, and direct survivors of trauma. TF-CBT is a phase-based treatment starting with initial stabilization and moving to trauma processing and finally an integration phase. The initial stabilization phase focuses on the development of coping skills so that individuals are able to manage the trauma processing phase focused on understanding the trauma experience. The final integration phase consolidates the individual's experience and builds safety and trust (Cohen, Mannarino, Kliethermes, & Murray, 2012). Project Fleur-de-Lis was based on the TF-CBT model and implemented after Hurricane Katrina. The project was successful in implementing stepped mental health screening and treatment for children affected by the hurricane (Cohen et al., 2009). A program for families, Triple P, is an evidence-based parenting program. The post-disaster version, Disaster Recovery Triple P (DRTP), holds promise as an effective intervention to mitigate parental risk and decrease children's vulnerability post-disaster (Cobham et al., 2016). Trauma-informed services are further discussed in an earlier chapter (see Kusmaul, this volume, Chap. 9).

Children may also obtain support in school settings through intermediate interventions. Cognitive Behavioral Intervention in Schools (CBITS) is a ten-session CBT group intervention that addresses symptoms of PTSD, anxiety, and depression (SAMHSA, 2015). CBITS has been successful in decreasing PTSD symptoms in students who were exposed to violence (Stein et al., 2003). CBITS training and implementation are delivered by a Web-based platform. Initial research findings revealed that the CBITS website garnered more traffic on the implementation support section versus the training section. This finding suggested that after initial training, providers find reference sections for implementation to be useful. Web-based platforms may be very helpful for the continued implementation of evidence-based practices after the initial training phase and should be considered in dissemination plans (Vona et al., 2014).

Older Adults

Some older adults are at higher risk of experiencing PTSD and adjustment disorder symptoms than younger adults when a natural disaster of sudden onset is experienced (Parker et al., 2016); however, other older adults are more resilient than previously thought; in particular, the oldest old may possess stronger coping strategies than previously thought (Cherry, Silva, & Galea, 2009). Risk factors for poor mental health outcomes include severe mental illness, advanced age, cognitive or sensory impairment, complex mental health conditions, physical health conditions or chronic illness, and limited social support (Sakauye et al., 2009). When older adults perceived social cohesion in their community, decreased symptoms of PTSD were reported (Heid, Pruchno, Cartwright, & Wilson-Genderson, 2017). Aside from community dwelling older adults, residents in nursing homes and long-term care settings must also be considered in disaster planning, response, and recovery (Allen & Nelson, 2009; Cherry, Allen, & Galea, 2010). A PFA program that was adapted to meet the needs of nursing home residents demonstrated promise in addressing not only physical safety and health-care needs of this vulnerable population but their mental health concerns as well (Brown et al., 2009). Although older adults are the fastest growing age group in the United States and other countries, there are limited treatment studies examining the efficacy of trauma treatments in this vulnerable population (Dinnen, Simiola, & Cook, 2015), particularly post-disaster.

People with Physical Disabilities or Illness

People with physical disabilities or illness have special considerations to account for in disaster planning. However, very little research addressing the needs of this population has been conducted. People with physical disability or illness are more likely to experience poor health outcomes from natural disasters; however, they are also less likely to implement disaster preparedness measures (Bethel, Foreman, & Burke, 2011; Eisenman et al., 2009). Furthermore, people with physical disabilities or illness have less access to disaster-related information and post-disaster treatment (Stough, 2009). Children with disabilities are particularly vulnerable to poor post-disaster outcomes due to their susceptibility to poverty, exposure to hazards, loss or separation from caregivers, and caregiver strain (Peek & Stough, 2010). To meet the mental health needs of people with physical disabilities post-disaster, future research should be conducted to look at their unique concerns, strengths, and resources.

People with Mental Health and/or Substance Use Problems

In any given year, approximately 20% of the US population is diagnosed with a mental health disorder (SAMHSA, 2016), and about 4% of the US population has a substance use disorder (Grant et al., 2016). For those who use or are in recovery, substances may be turned to as a coping mechanism post-disaster (see Waters & Copeland, this volume, Chap. 17). Post-disaster mental health and substance use screens may be a useful tool in disaster response to detect those at increased risk for adverse outcomes and to address potential concerns among disaster survivors. Screening, Brief Intervention, Referral to Treatment (SBIRT) is a public health approach to the delivery of early intervention and treatment services for people with or at risk for developing substance use disorders (Babor et al., 2007). Along with Motivational Interviewing, SBIRT has potential utility for disaster survivors with preexisting drug and alcohol disorders who experience a worsening or recurrence of symptoms (Palinkas, 2015). Seeking Safety, a present-focused therapy for individuals with comorbid trauma exposure and substance use, may be provided individually or in groups in various settings as an intermediate- and long-term intervention (Najavits, Weiss, Shaw, & Muenz, 1998).

People with Limited English Proficiency

The majority of the population in the United States speak English. Yet according to the 2011 American Community Survey conducted by the US Census Bureau, 22% of the population 5 years old and over reported speaking English "not well" or "not at all" (Ryan, 2013). Disaster literacy is defined as an individual's ability to read, understand, and use information to make informed decisions and follow instructions in the context of mitigating, preparing, responding, and recovering from a disaster (Brown, Haun, & Peterson, 2014). Among those with limited English proficiency are immigrant populations, whose situations during disasters are further complicated by the sociopolitical climate. An immigrant's social and legal status may affect their access to social support and formal services during disaster recovery (Fussell, Delp, Riley, Chavez, & Valenzuela, 2018). Disaster literacy should be incorporated during disaster planning and recovery to maximize engagement and communication with immigrant populations.

Written materials explaining how to prepare for disasters, steps for staying safe during an evolving catastrophic event, and describing how to secure resources post-disaster are not useful if the reader does not have adequate disaster literacy. Older adults, English and non-English speakers alike, are less likely to be literate and able to read and act upon simple print information. It is critical that prior to a disaster, information about mental health services are carefully reviewed to ensure that they are written at or below a sixth grade level, include pictographs that visually depict the written content, and are actionable (Brown et al., 2014). People who are empowered to take appropriate action often fare better after a disaster than those who are passive or dependent upon the verbal opinions and advice of others.

Not surprisingly, people with limited English proficiency in the United States are a vulnerable population during disasters. Studies indicate that the inclusion of minorities, specifically those with limited English proficiency, in disaster preparedness efforts has the potential to improve disaster response by extending improvements beyond the translation of print materials to incorporate cultural and literacy factors (Andrulis, Siddiqui, & Gantner, 2007; Andrulis, Siddiqui, & Purtle, 2011; Brown et al., 2014). Additionally, medical interpreters may have an important role in disaster preparedness and response; however, they need additional training to understand and feel confident providing interpretation services during disasters (Shiu-Thornton, Balabis, Senturia, Tamayo, & Oberle, 2007).

Future Directions for Disaster Mental Health

As the frequency and intensity of disasters continue to increase, research, program, and policy development are imperative to address community needs pre- and post-disaster, particularly in the area of mental health. Vulnerable populations are at a particularly higher risk of experiencing negative mental health consequences after disasters. Improvements to the development and implementation of disaster mental health planning efforts afford the opportunity to educate, establish response plans, and provide resources to communities in advance of potential events. Disaster mental health preparedness policy and planning for vulnerable populations include training, collaboration, and evaluation of existing programs (Roudini, Khankeh, & Witruk, 2017). More research and program evaluation is needed to increase the evidence base of disaster mental health response.

There are additional challenges not highlighted in this chapter that must be incorporated in future disaster mental health improvement endeavors. First, how are the mental health needs of displaced individuals addressed post-disaster? How can displaced individuals be best supported by federal, state, county, and local community agencies? Second, what is the best method to build a network of deployable specialists within an affected community? Research shows that trained mental health responders may not be able to deploy because of unanticipated family or work commitments. How do communities build, prepare, and sustain evidence-based models of response and treatment post-disaster? Finally, how do communities best address the needs of their most vulnerable populations? Can disaster mental health models learn from other public health models implemented to address community health problems?

Aside from the outstanding needs and challenges of disaster mental health, there are promising opportunities and new innovations on the horizon to enhance disaster mental health and response. The use of technology provides opportunities to engage communities and provide immediate assistance if cell towers and electrical systems are operable post-disaster. Aside from the development of disaster mental health applications and websites, the use of social media has also been proposed as an opportunity to address disaster planning and response (Houston et al., 2015) and may have purposes in disaster mental health as well. In the meanwhile, it appears prudent to have planned redundancy when disseminating disaster mental health information. Each approach, paper documents versus electronic apps and Web-based programs, has its own merits and unique weaknesses. Being prepared to deal with the existing infrastructure post-disaster will influence disaster mental health care at all phases of recovery.

References

Allen, B., Brymer, M. J., Steinberg, A. M., Vernberg, E. M., Jacobs, A., Speier, A. H., et al. (2010). Perceptions of psychological first aid among providers responding to Hurricanes Gustav and Ike. *Journal of Traumatic Stress, 23*(4), 509–513.

Allen, P. D., & Nelson, H. W. (2009). Disaster services with frail older persons: From preparation to recovery. In K. E. Cherry (Ed.), *Lifespan perspectives on natural disasters: Coping with Katrina, Rita and other storms* (pp. 153–169). New York: Springer.

American Psychological Association. (2006). Evidence-based practice in psychology. *American Psychologist, 61*(4), 271–285. https://doi.org/10.1037/0003-066X.61.4.271

Andrulis, D. P., Siddiqui, N. J., & Gantner, J. L. (2007). Preparing racially and ethnically diverse communities for public health emergencies. *Health Affairs, 26*(5), 1269–1279.

Andrulis, D. P., Siddiqui, N. J., & Purtle, J. P. (2011). Integrating racially and ethnically diverse communities into planning for disasters: The California experience. *Disaster Medicine and Public Health Preparedness, 5*(3), 227–234.

Babor, T. F., McRee, B. G., Kassebaum, P. A., Grimaldi, P. L., Ahmed, K., & Bray, J. (2007). Screening and Brief Intervention, and Referral to Treatment (SBIRT): Toward a public health approach to the management of substance abuse. *Substance Abuse, 28*(3), 7–30.

Bethel, J. W., Foreman, A. N., & Burke, S. C. (2011). Disaster preparedness among medically vulnerable populations. *American Journal of Preventive Medicine, 40*(2), 139–143.

Bromet, E. J., Hobbs, M. J., Clouston, S. A. P., Gonzalez, A., Kotov, R., & Luft, B. J. (2016). DSM-IV post-traumatic stress disorder among World Trade Center responders 11–13 years after the disaster of 11 September 2001 (9/11). *Psychological Medicine, 46*(4), 771–783. https://doi.org/10.1017/S0033291715002184

Brooks, S. K., Dunn, R., Sage, C. A. M., Amlôt, R., Greenberg, N., & Rubin, G. J. (2015). Risk and resilience factors affecting the psychological wellbeing of individuals deployed in humanitarian relief roles after a disaster. *Journal of Mental Health, 24*(6), 385–413. https://doi.org/10.3109/09638237.2015.1057334

Brown, L. M., Bruce, M. L., Hyer, K., Mills, W. L., Vongxaiburana, E., & Polivka-West, L. (2009). A pilot study evaluating the feasibility of psychological first aid for nursing home residents. *Clinical Gerontologist, 32*(3), 293–308. https://doi.org/10.1080/07317110902895317

Brown, L. M., Haun, J. N., & Peterson, L. (2014). A proposed disaster literacy model. *Disaster Medicine and Public Health Preparedness, 8*(3), 267–275. https://doi.org/10.1017/dmp.2014.43

Bryant, R. A. (2007). Early intervention for post-traumatic stress disorder. *Early Intervention in Psychiatry, 1*(1), 19–26. https://doi.org/10.1111/j.1751-7893.2007.00006.x

Brymer, M., Jacobs, A., Layne, C., Pynoos, R., Ruzek, J., Steinberg, A., Vernberg, E., & Watson, P. (2006). Psychological first aid: Field operations guide, 2nd Edition. National Child Traumatic Stress Network and National Center for PTSD, July, 2006. Available on: www.nctsn.org and www.ncptsd.va.gov

Butler, A., Chapman, J., Forman, E., & Beck, A. (2006). The empirical status of cognitive-behavioral therapy: A review of meta-analyses. *Clinical Psychology Review, 26*(1), 17–31. https://doi.org/10.1016/j.cpr.2005.07.003

Cherry, K. E., Allen, P. D., & Galea, S. (2010). Older adults and natural disasters: Lessons learned from Hurricanes Katrina and Rita. In P. Dass-Brailsford (Ed.), *Crisis and disaster counseling: Lessons learned from Hurricane Katrina and other disasters* (pp. 115–130). Thousand Oaks, CA: Sage.

Cherry, K. E., Silva, J., & Galea, S. (2009). Natural disasters and the oldest-old: A psychological perspective on coping and health in late life. In K. E. Cherry (Ed.), *Lifespan perspectives on natural disasters: Coping with Katrina, Rita and other storms* (pp. 171–193). New York: Springer.

Cobham, V. E., McDermott, B., Haslam, D., & Sanders, M. R. (2016). The role of parents, parenting and the family environment in children's post-disaster mental health. *Current Psychiatry Reports, 18*(6), 53. https://doi.org/10.1007/s11920-016-0691-4

Cohen, J. A., Jaycox, L. H., Walker, D. W., Mannarino, A. P., Langley, A. K., & DuClos, J. L. (2009). Treating traumatized children after Hurricane Katrina: Project Fleur-de Lis™. *Clinical Child and Family Psychology Review, 12*(1), 55–64. https://doi.org/10.1007/s10567-009-0039-2

Cohen, J. A., Mannarino, A. P., Kliethermes, M., & Murray, L. A. (2012). Trauma-focused CBT for youth with complex trauma. *Child Abuse & Neglect, 36*(6), 528–541. https://doi.org/10.1016/j.chiabu.2012.03.007

CRED. (2018). *Natural disasters 2017*. Brussels, Belgium: CRED; 2018 EM-DAT file dated 02/07/2018.

Dart Training. (2019, July 8). Retrieved from https://dartcenter.org/programs/training

Dinnen, S., Simiola, V., & Cook, J. M. (2015). Post-traumatic stress disorder in older adults: A systematic review of the psychotherapy treatment literature. *Aging & Mental Health, 19*(2), 144–150.

Dworznik, G., & Grubb, M. (2007). Preparing for the worst: Making a case for trauma training in the journalism classroom. *Journalism & Mass Communication Educator, 62*(2), 190–210.

Eftekhari, A., Ruzek, J. I., Crowley, J. J., Rosen, C. S., Greenbaum, M. A., & Karlin, B. E. (2013). Effectiveness of national implementation of prolonged exposure therapy in Veterans Affairs care. *JAMA Psychiatry, 70*(9), 949–955. https://doi.org/10.1001/jamapsychiatry.2013.36

Eisenman, D. P., Zhou, Q., Ong, M., Asch, S., Glik, D., & Long, A. (2009). Variations in disaster preparedness by mental health, perceived general health, and disability status. *Disaster Medicine and Public Health Preparedness, 3*(1), 33–41.

Everly Jr., G. S., Phillips, S. B., Kane, D., & Feldman, D. (2006). Introduction to and overview of group psychological first aid. *Brief Treatment & Crisis Intervention, 6*(2). https://doi.org/10.1093/brief-treatment/mhj009

Fonagy, P. (2015). The effectiveness of psychodynamic psychotherapies: An update. *World Psychiatry, 14*(2), 137–150. https://doi.org/10.1002/wps.20235

Forbes, D., Fletcher, S., Wolfgang, B., Varker, T., Creamer, M., Brymer, M. J., et al. (2010). Practitioner perceptions of skills for psychological recovery: A training programme for health practitioners in the aftermath of the Victorian bushfires. *Australian & New Zealand Journal of Psychiatry, 44*(12), 1105–1111. https://doi.org/10.3109/00048674.2010.513674

Fussell, E., Delp, L., Riley, K., Chávez, S., & Valenzuela Jr., A. (2018). Implications of social and legal status on immigrants' health in disaster zones. *American Journal of Public Health, 108*(12), 1617–1620.

Gibbs, L., Block, K., Harms, L., MacDougall, C., Baker, E., Ireton, G., et al. (2015). Children and young people's wellbeing post-disaster: Safety and stability are critical. *International Journal of Disaster Risk Reduction, 14*, 195–201.

Grainger, R. D., Levin, C., Allen-Byrd, L., Doctor, R. M., & Lee, H. (1997). An empirical evaluation of eye movement desensitization and reprocessing (EMDR) with survivors of a natural disaster. *Journal of Traumatic Stress, 10*(4), 665–671.

Grant, B. F., Saha, T. D., Ruan, W. J., Goldstein, R. B., Chou, S. P., Jung, J., ... & Hasin, D. S. (2016). Epidemiology of DSM-5 drug use disorder: results from the National Epidemiologic Survey on Alcohol and Related Conditions–III. JAMA psychiatry, 73(1), 39–47.

Hamblen, J. L., Norris, F. H., Pietruszkiewicz, S., Gibson, L. E., Naturale, A., & Louis, C. (2009). Cognitive behavioral therapy for postdisaster distress: A community based treatment program for survivors of Hurricane Katrina. *Administration and Policy in Mental Health and Mental Health Services Research, 36*(3), 206–214. https://doi.org/10.1007/s10488-009-0213-3

Hansel, T. C., Osofsky, H. J., Steinberg, A. M., Brymer, M. J., Landis, R., Riise, K. S., et al. (2011). Louisiana Spirit Specialized Crisis Counseling: Counselor perceptions of training and services. *Psychological Trauma: Theory, Research, Practice, and Policy, 3*(3), 276–282. https://doi.org/10.1037/a0024644

Heid, A. R., Pruchno, R., Cartwright, F. P., & Wilson-Genderson, M. (2017). Exposure to Hurricane Sandy, neighborhood collective efficacy, and post-traumatic stress symptoms in older adults. *Aging & Mental Health, 21*(7), 742–750. https://doi.org/10.1080/13607863.2016.1154016

Hofmann, S. G., Sawyer, A. T., Witt, A. A., & Oh, D. (2010). The effect of mindfulness-based therapy on anxiety and depression: A meta-analytic review. *Journal of Consulting and Clinical Psychology, 78*(2), 169. https://doi.org/10.1037/a0018555

Houston, J. B., Hawthorne, J., Perreault, M. F., Park, E. H., Goldstein Hode, M., Halliwell, M. R., et al. (2015). Social media and disasters: A functional framework for social media use in disaster planning, response, and research. *Disasters, 39*(1), 1–22.

Jarero, I., Artigas, L., Montero, M., & Lena, L. (2008). The EMDR integrative group treatment protocol: Application with child victims of a mass disaster. *Journal of EMDR Practice and Research, 2*(2), 97. https://doi.org/10.1891/1933-3196.2.2.97

Jones, K., Allen, M., Norris, F. H., & Miller, C. (2009). Piloting a new model of crisis counseling: Specialized crisis counseling services in Mississippi after Hurricane Katrina. *Administration and Policy in Mental Health and Mental Health Services Research, 36*(3), 195.

Jones, S. (2017). Describing the mental health profile of first responders: A systematic review. *Journal of the American Psychiatric Nurses Association, 23*(3), 200–214. https://doi.org/10.1177/1078390317695266

Kessler, R. C., Sonnega, A., Bromet, E., Hughes, M., & Nelson, C. B. (1995). Posttraumatic stress disorder in the National Comorbidity Survey. *Archives of General Psychiatry, 52*(12), 1048–1060.

Kleim, B., & Westphal, M. (2011). Mental health in first responders: A review and recommendation for prevention and intervention strategies. *Traumatology, 17*(4), 17–24. https://doi.org/10.1177/1534765611429079

Kousky, C. (2016). Impacts of natural disasters on children. *The Future of Children*, 73–92. https://doi.org/10.1353/foc.2016.0004

Leaning, J., & Guha-Sapir, D. (2013). Natural disasters, armed conflict, and public health. *New England Journal of Medicine, 369*(19), 1836–1842. https://doi.org/10.1056/NEJMra1109877

Leiva-Bianchi, M., Cornejo, F., Fresno, A., Rojas, C., & Serrano, C. (2018). Effectiveness of cognitive-behavioural therapy for post-disaster distress in post-traumatic stress symptoms after

Chilean earthquake and tsunami. *Gaceta Sanitaria, 32*, 291–296. https://doi.org/10.1016/j.gaceta.2017.07.018
Lenz, S., Bruijn, B., Serman, N., & Bailey, L. (2014). Effectiveness of cognitive processing therapy for treating posttraumatic stress disorder. *Journal of Mental Health Counseling, 36*(4), 360–376.
Levitt, J. T., Malta, L. S., Martin, A., Davis, L., & Cloitre, M. (2007). The flexible application of a manualized treatment for PTSD symptoms and functional impairment related to the 9/11 World Trade Center attack. *Behaviour Research and Therapy, 45*(7), 1419–1433. https://doi.org/10.1016/j.brat.2007.01.004
Molnar, B. E., Sprang, G., Killian, K. D., Gottfried, R., Emery, V., & Bride, B. E. (2017). Advancing science and practice for vicarious traumatization/secondary traumatic stress: A research agenda. *Traumatology, 23*(2), 129.
Najavits, L. M., Weiss, R. D., Shaw, S. R., & Muenz, L. R. (1998). "Seeking safety": Outcome of a new cognitive-behavioral psychotherapy for women with posttraumatic stress disorder and substance dependence. *Journal of Traumatic Stress, 11*(3), 437–456. https://doi.org/10.1023/A:1024496427434
Navarro, J., Pulido, R., Berger, C., Arteaga, M., Osofsky, H. J., Martinez, M., et al. (2016). Children's disaster experiences and psychological symptoms: An international comparison between the Chilean earthquake and tsunami and Hurricane Katrina. *International Social Work, 59*(4), 545–558. https://doi.org/10.1177/0020872814537850
Nebraska Disaster Behavioral Health. (n.d.). Psychological first aid training. Retrieved from http://www.disastermh.nebraska.edu/education/psychological.php
Neria, Y., Nandi, A., & Galea, S. (2008). Post-traumatic stress disorder following disasters: A systematic review. *Psychological Medicine, 38*(4), 467–480.
Neuner, F., Schauer, M., Klaschik, C., Karunakara, U., & Elbert, T. (2004). A comparison of narrative exposure therapy, supportive counseling, and psychoeducation for treating posttraumatic stress disorder in an African refugee settlement. *Journal of Consulting and Clinical Psychology, 72*(4), 579. https://doi.org/10.1037/0022-006X.72.4.579
Nevo, I., & Slonim-Nevo, V. (2011). The myth of evidence-based practice: Towards evidence-informed practice. *British Journal of Social Work, 41*(6), 1176–1197.
Nieves-Pizarro, Y., Takahashi, B., & Chavez, M. (2019). When everything else fails: Radio journalism during Hurricane Maria in Puerto Rico. *Journalism Practice*, 1–18. https://doi.org/10.1080/17512786.2019.1567272
Norris, F. H., Friedman, M. J., & Watson, P. J. (2002). 60,000 disaster victims speak: Part II. Summary and implications of the disaster mental health research. *Psychiatry: Interpersonal and Biological Processes, 65*(3), 240–260.
Norris, F. H., Friedman, M. J., Watson, P. J., Byrne, C. M., Diaz, E., & Kaniasty, K. (2002). 60,000 disaster victims speak: Part I. An empirical review of the empirical literature, 1981–2001. *Psychiatry: Interpersonal and Biological Processes, 65*(3), 207–239.
North, C. S., & Pfefferbaum, B. (2013). Mental health response to community disasters: A systematic review. *JAMA, 310*(5), 507–518. https://doi.org/10.1001/jama.2013.107799
Osofsky, H. J., Holloway, H., & Pickett, A. (2005). War correspondents as responders: Considerations for training and clinical services. *Psychiatry: Interpersonal and Biological Processes, 68*(3), 283–293.
Palinkas, L. A. (2015). Behavioral health and disasters: Looking to the future. *The Journal of Behavioral Health Services & Research, 42*(1), 86–95. https://doi.org/10.1007/s11414-013-9390-7
Parker, G., Lie, D., Siskind, D., Martin-Khan, M., Raphael, B., Crompton, D., et al. (2016). Mental health implications for older adults after natural disasters – A systematic review and meta-analysis. *International Psychogeriatrics, 28*(1), 11–20. https://doi.org/10.1017/S1041610215001210

Peek, L., & Stough, L. M. (2010). Children with disabilities in the context of disaster: A social vulnerability perspective: Children with disabilities in disaster. *Child Development, 81*(4), 1260–1270. https://doi.org/10.1111/j.1467-8624.2010.01466.x

Peterson, L. G., & Pbert, L. (1992). Effectiveness of a meditation-based stress reduction program in the treatment of anxiety disorders. *The American Journal of Psychiatry, 149*(7), 936–943.

Pfefferbaum, B., Jacobs, A. K., Houston, J. B., & Griffin, N. (2015). Children's disaster reactions: The influence of family and social factors. *Current Psychiatry Reports, 17*(7), 57. https://doi.org/10.1007/s11920-015-0597-6

Plummer, C. A., Cain, D. S., Fisher, R. M., & Bankston, T. Q. (2008). Practice challenges in using psychological first aid in a group format with children: A pilot study. *Brief Treatment & Crisis Intervention, 8*(4). https://doi.org/10.1093/brief-treatment/mhn019

Pynoos, R. S., Goenjian, A. K., & Steinberg, A. M. (1998). A public mental health approach to the postdisaster treatment of children and adolescents. *Child and Adolescent Psychiatric Clinics, 7*(1), 195–210.

Ramirez, M., Harland, K., Frederick, M., Shepherd, R., Wong, M., & Cavanaugh, J. E. (2013). Listen protect connect for traumatized schoolchildren: A pilot study of psychological first aid. *BMC Psychology, 1*(1), 26.

Reifels, L., Naccarella, L., Blashki, G., & Pirkis, J. (2014). Examining disaster mental health workforce capacity. *Psychiatry, 77*(2), 199–205.

Resick, P. A., Nishith, P., Weaver, T. L., Astin, M. C., & Feuer, C. A. (2002). A comparison of cognitive-processing therapy with prolonged exposure and a waiting condition for the treatment of chronic posttraumatic stress disorder in female rape victims. *Journal of Consulting and Clinical Psychology, 70*(4), 867.

Resick, P. A., & Schnicke, M. K. (1992). Cognitive processing therapy for sexual assault victims. *Journal of Consulting and Clinical Psychology, 60*(5), 748–756. https://doi.org/10.1037/0022-006X.60.5.748

Resick, P. A., Williams, L. F., Suvak, M. K., Monson, C. M., & Gradus, J. L. (2012). Long-term outcomes of cognitive–behavioral treatments for posttraumatic stress disorder among female rape survivors. *Journal of Consulting and Clinical Psychology, 80*(2), 201.

Roudini, J., Khankeh, H. R., & Witruk, E. (2017). Disaster mental health preparedness in the community: A systematic review study. *Health Psychology Open, 4*(1), 205510291771130. https://doi.org/10.1177/2055102917711307

Ruzek, J. I., Brymer, M. J., Jacobs, A. K., Layne, C. M., Vernberg, E. M., & Watson, P. J. (2007). Psychological first aid. *Journal of Mental Health Counseling, 29*(1), 17–49. https://doi.org/10.17744/mehc.29.1.5racqxjueafabgwp

Ryan, C. (2013). Language use in the United States: 2011 (American Community Survey Reports). U.S. Census Bureau. Retrieved from https://www2.census.gov/library/publications/2013/acs/acs-22/acs-22.pdf

Sakauye, K. M., Streim, J. E., Kennedy, G. J., Kirwin, P. D., Llorente, M. D., Schultz, S. K., et al. (2009). AAGP position statement: Disaster preparedness for older Americans: Critical issues for the preservation of mental health. *The American Journal of Geriatric Psychiatry, 17*(11), 916–924. https://doi.org/10.1097/JGP.0b013e3181b4bf20

Schottenbauer, M. A., Glass, C. R., Arnkoff, D. B., & Gray, S. H. (2008). Contributions of psychodynamic approaches to treatment of PTSD and trauma: A review of the empirical treatment and psychopathology literature. *Psychiatry: Interpersonal and Biological Processes, 71*(1), 13–34. https://doi.org/10.1521/psyc.2008.71.1.13

Shiu-Thornton, S., Balabis, J., Senturia, K., Tamayo, A., & Oberle, M. (2007). Disaster preparedness for limited English proficient communities: Medical interpreters as cultural brokers and gatekeepers. *Public Health Reports, 122*(4), 466–471.

Speier, A. H., Osofsky, J. D., & Osofsky, H. J. (2009). Building a disaster mental health response to a catastrophic event: Louisiana and Hurricane Katrina. In K. E. Cherry (Ed.), *Lifespan perspectives on natural disasters* (pp. 241–260). New York: Springer.

Sreedharan, C., & Thorsen, E. (2018). *Voices from Nepal: Lessons in post-disaster journalism*. Center for the Study of Journalism, Culture and Community, Bournemouth University. Retrieved from http://eprints.bournemouth.ac.uk/31242/1/Voices%20from%20Nepal%20 2018.pdf

Stein, B. D., Jaycox, L. H., Kataoka, S. H., Wong, M., Tu, W., Elliott, M. N., et al. (2003). A mental health intervention for schoolchildren exposed to violence: A randomized controlled trial. *JAMA, 290*(5), 603–611. https://doi.org/10.1001/jama.290.5.603

Stough, L. (2009). *The effects of disaster on the mental health of individuals with disabilities*. New York: Cambridge University Press.

Substance Abuse and Mental Health Services Administration (SAMHSA). (2015, May). Disaster technical assistance center supplemental research bulletin: Disaster behavioral health interventions inventory. Retrieved from https://www.samhsa.gov/sites/default/files/dtac/supplemental-research-bulletin-may-2015-disaster-behavioral-health-interventions.pdf

Substance Abuse and Mental Health Services Administration (SAMHSA). (2016). *Key substance use and mental health indicators in the United States: Results from the 2015 National Survey on Drug Use and Health*. Rockville, MD: Center for Behavioral Health Statistics and Quality.

Tolin, D. F., McKay, D., Forman, E. M., Klonsky, E. D., & Thombs, B. D. (2015). Empirically supported treatment: Recommendations for a new model. *Clinical Psychology: Science and Practice, 22*(4), 317–338. https://doi.org/10.1111/cpsp.12122

Vernberg, E. M., Steinberg, A. M., Jacobs, A. K., Brymer, M. J., Watson, P. J., Osofsky, J. D., et al. (2008). Innovations in disaster mental health: Psychological first aid. *Professional Psychology: Research and Practice, 39*(4), 381–388. https://doi.org/10.1037/a0012663

Vona, P., Wilmoth, P., Jaycox, L. H., McMillen, J. S., Kataoka, S. H., Wong, M., et al. (2014). A web-based platform to support an evidence-based mental health intervention: Lessons from the CBITS web site. *Psychiatric Services, 65*(11), 1381–1384.

Wade, D., Crompton, D., Howard, A., Stevens, N., Metcalf, O., Brymer, M., et al. (2014). Skills for psychological recovery: Evaluation of a post-disaster mental health training program. *Disaster Health, 2*(3–4), 138–145. https://doi.org/10.1080/21665044.2015.1085625

Chapter 12
When Disaster Strikes: The Role of the Family Nurse Practitioner in Providing Primary Health Care

Wanda Raby Spurlock and Sandra C. Brown

Introduction

Disasters affect millions of people around the world every year. The most frequently occurring type of disaster, a natural disaster, is defined as an act of "nature" of such an extent as to produce a catastrophic situation that disrupts day-to-day patterns of life. Including all types of severe weather events, natural disasters can pose a significant threat to human health, well-being, and safety, resulting in destruction of property, displacement, serious injury, and loss of life. Although natural disasters are unpredictable, chances are that they will occur at some point in time and more often in some geographical regions than others. Recognized as the most significant global threat of the twenty-first century, climate change is projected to drive more frequent and intense weather events such as hurricanes and severe floods (Field et al., 2014).

In addition to adverse physical health effects, exposure to a natural disaster can be associated with mental health conditions such as post-traumatic stress, depression, and substance use issues, ranging from mild and transient to severe and chronic (Freedy & Simpson, 2007). When mass evacuations are required, the chaotic and sudden separation or displacement from family and other routine social support networks can trigger psychological and behavioral responses that can lead to ineffective coping. Characteristics or experiences of individuals before, during, and following a disaster may increase vulnerability to poor mental health outcomes. Therefore, efforts to meet the psychological or mental health needs of persons impacted by disasters are important components of disaster readiness, response, and recovery (Spurlock, Brown, & Rami, 2009).

It is also well known that natural disasters can create a tremendous burden on existing health-care systems, negatively impacting the capacity to deliver timely and critical services to those in need. Lack of timely access to health care is noted

W. R. Spurlock · S. C. Brown (✉)
Southern University and A & M College, Baton Rouge, LA, USA
e-mail: WANDA_SPURLOCK@subr.edu

© Springer Nature Switzerland AG 2021

K. E. Cherry, A. Gibson (eds.), *The Intersection of Trauma and Disaster Behavioral Health*, https://doi.org/10.1007/978-3-030-51525-6_12

as one of the leading causes of mortality following disasters, leading to negative health outcomes. In fact, interruption of medical care has been cited as a significant contributor to mortality and morbidity in the aftermath of natural disasters (Kishore et al., 2018).

On the front line of disaster care are nurse practitioners (NPs), who have a long history, dating back to the 1960s of providing health care to vulnerable populations in urban and rural areas. Due to the shortage of primary care physicians, coupled with the increase in an aging and ethnically diverse populations, there is a demand for NPs who are skilled at assessing, diagnosing, and managing acute and chronic health problems (DeNisco & Barker, 2016). NPs play a significant role in improving access to primary health care by filling these gaps in health professional shortage and medically underserved areas, where timely access to health-care services is limited.

Vulnerable Populations and Disasters

All populations are not equally impacted by a disaster. Moreover, disasters have a profound impact on the health and well-being of individuals, families, and communities, especially vulnerable populations who bear a disproportionate burden of chronic illnesses, such as hypertension and diabetes. Vulnerable populations also bear a disproportionate burden of poor health outcomes and mortality when faced with disasters (Chesnay & Anderson, 2016; Davis, Wilson, Brock-Martin, Glover, & Svendsen, 2010; Smith, Rice, Danna, & Bennett, 2013) due to their marginalized sociocultural status, limited access to economic resources, and personal characteristics such as race, ethnicity, age, and gender. Additionally, the health and health-care problems of vulnerable populations are known to intersect with social factors such as poverty and inadequate education, which can serve as precursors to poor mental health outcomes.

Nurse Practitioners Filling the Gap in Disaster Care

Globally, nurses are recognized for their significant contributions to disaster response. Advanced practice registered nurses (APRNs), i.e., NPs, have master's or doctoral degrees and pass national examinations to become board certified in specific population foci. In response to disasters, NPs fill a critical role in "surge" capacity. Surge capacity is the ability of the health-care system to rapidly expand to meet an overwhelming demand for medical care in the event of bioterrorism and other wide-scale public health emergencies, including disasters (National Association of Public Hospitals and Health Systems, 2007). Moreover, according to the National Advisory Council on Nurse Education and Practice (2009), "the effectiveness of the healthcare system's response to a public health emergency is largely dependent on the surge capacity of the nurse workforce" (p. 2).

The availability of health-care providers, especially primary care physicians and mental health professionals, can significantly vary based on geographic locations. According to the American Association of Nurse Practitioners (AANP) (2019), there are over 270,000 NPs licensed in the United States. Recognized as expert clinicians, NPs provide a comprehensive, holistic approach to initiating and managing treatment plans, including prescribing medications, with an emphasis on health promotion and disease prevention. In the United States, millions of patients choose NPs as their health-care providers, with annual visits totaling more than 1.06 billion (AANP, 2019). NPs deliver health care in all states and the District of Columbia and practice under the rules and regulations of the state in which they are licensed.

Family Nurse Practitioners

The largest percentage of NPs, 66.9%, are certified as family nurse practitioners (FNPs) and practice across diverse health-care sites including community-based, acute and long-term care, and academic settings. FNPs provide a broad range of family-focused health-care services to patients across the life span: from infants to adolescents to adults and the elderly (AANP, 2019). Awareness of the importance of their role in providing primary care services following a natural disaster has become more apparent, considering historic weather-related disasters such as Hurricanes Katrina, Harvey, and Maria that have impacted individuals, families, and entire communities. FNPs can attend to the physical health needs of people with mental disorders as well as the mental health needs of those suffering from chronic diseases. FNPs also have the expertise to develop viable culturally relevant patient- and family-centered treatment plans and possess the competencies to foster interdisciplinary collaboration to prevent fragmentation in care, which often occurs following a disaster.

Phases of Disasters

Disaster management occurs in phases beginning with preparedness. FNPs make vast contributions to all phases of disaster management. Divided into three stages, preparedness, relief response, and recovery, "disaster management encompasses the efforts to deal with hazards and the disasters that they produce" (Powers, 2010, p. 8). Discussion of these phases is presented next.

Preparedness

Disaster preparedness refers to strategies and steps taken prior to the occurrence of a disaster. The goal of this phase is to minimize or mitigate potential loss and devastation as a result of the disaster. Spain, Clements, DeRanieri, and Holt (2012) have simply stated, "Disaster preparedness maximizes safe conditions, decreases vulnerability, and minimizes risk to individuals when they are confronted with a hazard" (p. 43). Although planning for disasters has been a priority of federal, state, and local entities for many years, these efforts often fall short of considering the overall preexisting health of the community and how this will intersect with disaster planning efforts, especially for vulnerable populations. It is important to recognize that the preexisting health of local communities will impact its capacity to recover following a disaster (see Mosby, Chap. 13, this volume). Therefore, successful community planning efforts must consider the social, economic, and cultural factors that heighten a preexisting state of vulnerability before, during, and after a disaster. As primary care providers, FNPs are knowledgeable of the connection between social conditions and environmental factors that shape health, health care, and health outcomes. Data are available at the local, state, and national levels that serve as excellent resources to undergird preparedness efforts. Nurses are a part of organizational and system-wide disaster preparedness teams. As part of these teams, they facilitate disaster preparedness in acute care facilities and long-term care and community-based sites, tailoring the disaster-preparedness plans to meet the specific needs of the clients in these various health-care settings.

America's health rankings The annual report, America's Health Rankings (United Health Foundation [UHF], 2018), is the longest running annual assessment of the nation's health on a state-by-state basis. Data contained in this report can be used by policy makers, public health officials, health-care providers, and community leaders to develop tailored, state-specific approach to disaster planning, response, and recovery efforts. For example, increases in mortality and chronic diseases, such as obesity, were highlighted in the 2018 AHR's report. Additionally, a rising suicide rate, drug deaths, and frequent mental distress validate mental health concerns across the nation. Moreover, self-reported mental distress, measured by the Behavioral Risk Factor Surveillance System (BRFSS), increased 7% in the past 2 years (UHF, 2018).

The top five highest ranked states for overall health in 2018 were Hawaii (#1), Massachusetts (#2), Connecticut (#3), Vermont (#4), and Utah (#5), respectively. The five lowest ranking states for overall health (poorest health) were Louisiana (50th), Mississippi (49th), Alabama (48th), Oklahoma (47th), and Arkansas (46th). It is important to note that three of the states with the lowest overall health rankings are coastal states, also referred to as Gulf States that are more vulnerable to weather-related events such as hurricanes. Therefore, states with the poorest health rankings should be a top priority for disaster planning, especially for allocation of health-care resources.

Know the state of local health when preparing for a disaster While data on the nation's health on a state-by-state basis is available through the *America's Health Rankings* annual report, most states also publish their own *Health Report Card*, describing the overall health of its citizens, including behavioral health issues. These reports also provide data on health issues confronting specific regions of the state, i.e., specific counties and parishes. A large majority of FNPs practice in community-based clinics and are knowledgeable of the health issues that influence the communities in which they serve. Therefore, they are positioned to serve as leaders in disaster planning efforts.

Relief Response

According to Powers (2010), the emphasis of this stage is the delivery of health care to meet the immediate needs of the community. Response time is always a major factor in this phase of disaster management. Initially, response capacity is strained due to an increase in the volume of persons seeking treatment of acute illnesses and injuries, brought about by the disaster. However, the overall community's capability and readiness to respond is determined not only by its ability to respond to medical needs but also by its ability to efficiently respond to mental health needs. The timely delivery of mental health services, especially in rural and remote areas, is even more challenging.

FNPs strive to safeguard the health and well-being of communities in which they practice. People of all ages can be impacted by disasters. More specifically, FNPs can perform assessments such as history and physical examinations; screen for pre-existing chronic conditions or newly diagnosed conditions such as diabetes, hypertension, depression and anxiety, and acute infections; implement treatment plans; and initiate referrals that are culturally relevant and age appropriate, thus validating the significant contributions that FNPs provide in the immediate response to a disaster.

As states and local communities engage in efforts aimed at meeting the mental health-care needs of its citizens following a disaster, use of innovative treatment modalities, such as telehealth, is essential. Telehealth includes the use of video-based conferences, electronic information, and telecommunication to support long-distance clinical care (Balestra, 2018). This technological advancement has vast potential to fill the gap in the delivery of mental health services, ensuring that high-risk patients receive timely assessment, treatment, and follow-up. In addition, FNPs serve as members of interdisciplinary mobile medical crisis teams that are integral to addressing immediate behavioral health needs following a disaster. Spanning the continuum of relief response, FNPs are also engaged in emergency triage, often volunteer to staff shelters, and participate in faith-based disaster relief initiatives.

Recovery

While disasters can affect the health of communities, the health of communities can also affect disaster relief and recovery. Primary health care, a comprehensive approach to health and well-being, is centered on the needs and preferences of individuals, families, and communities. Furthermore, primary health care addresses the "whole" person, throughout the life span. When necessary, it includes case management and coordinated referrals, ensuring that physical, mental, and social health and well-being are addressed.

In the recovery phase, communities may be left to manage on their own as outside assistance, disaster aid, and other resources tend to diminish over time. According to Freedy and Simpson (2007), the utilization of primary health-care services usually increases for 12 or more months following major disasters, validating the need for electronic health records (EHR) that can be accessed by other health-care providers involved in the delivery of health care. The use of EHR facilitates timely delivery of services, prevents duplication of diagnostic tests, and reduces the likelihood of medical error while promoting continuity of care (Spurlock et al., 2009).

People of all ages can be impacted by disasters, and exposure to these events can result in a wide range of psychological and behavioral responses, which can vary by age groups. For example, psychological and behavioral responses that more commonly occur in adults include anger, anxiety, sleep disturbances, fear, irritability, and increase in use of alcohol and tobacco (see Waters and Copeland, Chap. 17, this volume). Preschool-aged children more often experience separation anxiety; regressive, hyperactive, or withdrawn behaviors; and nightmares. Psychological and behavioral responses in school-age children usually manifest with difficulties in concentration, sleep disturbances, somatic complaints, and safety concerns, whereas adolescent responses more closely resemble that of adults (Townsend & Morgan, 2018).

Based on the severity of the disaster, it is essential that continuity of care is maintained for individuals affected by disasters as they adapt to life changes while attempting to return to a sense of normalcy. It is during this time of disaster recovery that FNPs are essential in connecting individuals and families to more long-term community-based resources and services, such as grief counseling, support groups, and stress management and self-care strategies.

Conclusion and Future Directions

Although disasters are unpredictable, they are inevitable. As the most trusted profession, nurses continue to comprise the largest segment of the nation's health-care workforce. Advanced practice nurses, such as FNPs, fill a critical gap in the delivery of primary health-care services to individuals whose access to health care is limited.

Likewise, FNPs play an essential role during all phases of disaster management. Due to their advanced clinical reasoning, diagnostic skills, and ability to treat individuals across the life span, FNPs are uniquely positioned to meet the holistic health-care needs of individuals, families, and communities when disaster strikes.

It is likely that disasters will continue to increase in frequency and intensity. However, many nurses still feel ill-prepared to respond when disaster strikes. Therefore, nurses should possess specific disaster-based competencies and credentials such as the National Healthcare Disaster Certification awarded through the American Nurses Credentialing Center. Nurses should also register for the Emergency System for Advance Registration of Volunteer Health Professionals (ESAR-VHP), administered on a state level, so that they can respond quickly when disaster strikes. By registering through ESAR-VHP, volunteers' identities, credentials, accreditations, and hospital privileges are all verified in advance, saving valuable time in emergency situations. Additionally, some states have adopted Enhanced Nurse Licensure Compact (eNLC) legislation providing one multistate license that allows nurses to quickly cross state borders and provide critical services following a disaster. Unfortunately, to date, not all states have passed this legislation, a noteworthy obstacle that awaits future advocacy efforts and sustained attention by nurse leaders and legislators.

References

American Association of Nurse Practitioners. (2019). *NP fact sheet*. Retrieved from https://www.aanp.org/about/all-about-nps/np-fact-sheet

Balestra, M. (2018). Telehealth and legal implications for nurse practitioners. *The Journal of Nurse Practitioners, 14*(1), 33–39.

Chesnay, M., & Anderson, B. (2016). *Caring for the vulnerable: Perspectives in nursing theory, practice, and research*. Burlington, MA: Jones and Barlett Learning.

Davis, J. R., Wilson, S., Brock-Martin, A., Glover, S., & Svendsen, E. R. (2010). The impact of disasters on populations with health and health care disparities. *Disaster Medicine Public Health Preparedness, 4*(1), 30–38.

DeNisco, S. M., & Barker, A. M. (2016). *Advanced practice nursing: Essential knowledge for the profession* (4th ed.). Boston: Jones and Bartlett.

Field, C. C., Barros, V. R., Dokken, D. J., Mach, K. J., Mastrandrea, M. D., Bilir, T. E., et al. (2014). Summary for policymakers. In *Climate change 2014: Impacts, adaptation, and vulnerability. Part A: Global and sectoral aspects. Contribution of working group II to the fifth assessment report of the Intergovernmental Panel on Climate Change* (pp. 1–32). Cambridge, UK/New York: Cambridge University Press.

Freedy, J., & Simpson, W. (2007). Disaster related physical and mental health: A role for the family physician. *American Family Physician, 75*, 401–405.

Kishore, N., Marques, D., Mahumud, A., Kiang, M., Rodriquez, I., Fuller, A., et al. (2018). Mortality in Puerto Rico after hurricane Maria. *The New England Journal of Medicine, 379*, 162. https://doi.org/10.1056/NEJMsa1803972. https://www.nejm.org/doi/full/10.1056/NEJMsa1803972

National Association of Public Hospitals and Health Systems (2007). *Research brief: Hospital staffing and surge capacity during a disaster event*. Retrieved from http://essentialhospitals.org/wp-content/uploads/2014/10/May2007_Research_Brief.pdf

National Advisory Council on Nurse Education and Practice. (2009). *Challenges faced by the nurse workforce in a changing environment. Part I: Surge capacity: Educating the nursing workforce for emergency and disaster preparedness*. Retrieved from https://www.hrsa.gov/advisorycommittees/bhpradvisory/nacnep/Reports/seventhreport.pdf

Powers, R. (2010). Introduction to disasters and disaster nursing. In R. Powers & E. Daily (Eds.). *International disaster nursing* (pp. 1–11). New York: Cambridge

Smith, E. L., Rice, K. L., Danna, D., & Bennett, M. J. (2013). Providing culturally competent care during disasters: Strategies for nurses. *The Journal of Continuing Education in Nursing, 44*(4), 151–152.

Spain, K. M., Clements, P. T., DeRanieri, J. T., & Holt, K. (2012). When disaster happens: Emergency preparedness for nurse practitioners. *The Journal for Nurse Practitioners, 8*(1), 38–44.

Spurlock, W., Brown, S., & Rami, J. (2009). Delivering primary health care to hurricane evacuees. *American Journal of Nursing, 109*(8), 50–53.

Townsend, M., & Morgan, K. (2018). *Psychiatric mental health nursing: Concepts of care in evidence-based practice* (8th ed.). Philadelphia: F.A. Davis Company.

United Health Foundation. (2018). *America's health rankings*. Retrieved from https://www.americashealthrankings.org/learn/reports/2018-annual-report

Chapter 13
Frameworks of Recovery: Health Caught at the Intersection of Housing, Education, and Employment Opportunities After Hurricane Katrina

Kim Mosby

Introduction

Living through disasters and traumatic events can affect mental health. Research shows some individuals describe themselves as more resilient following a disaster because they develop skills and knowledge that will help them cope in the future. However, in some situations, negative mental health outcomes are associated with experiencing traumatic disaster-related events (Bonanno, Brewin, Kaniasty, & Greca, 2010; Kessler et al., 2008; Ursano, Fullerton, & Terhakopian, 2008). Common negative mental health impacts from disasters include depression, anxiety, and post-traumatic stress disorder (PTSD). Other negative outcomes include headaches, feelings of anger, increased stress, and developing substance abuse problems to cope with disaster-related stress (see Waters & Copeland, Chap. 17, this volume). While these experiences are common, many survivors do not receive counseling or treatment to help them recover from the trauma they experienced (Lindsey, Chambers, Pohle, Beall, & Lucksted, 2013; Nicholls, Picou, Curtis, & Lowman, 2015).

Following a disaster, many factors influence health outcomes. The type of disaster can have an impact on the psychological health of affected communities. Fritz (1961) and Quarantelli and Dynes (1977) found "therapeutic" communities form after natural disasters when community members come together to help each other recover physically, which may lessen psychological impacts. In contrast, technological disasters often foster a "corrosive" community, where victims and other groups assign blame and expect compensation (Freudenburg, 1997; Furedi, 2007; Picou, Marshall, & Gill, 2004). After technological disasters, victims are often involved in long-term legal struggles to fight for compensation, which forces them to relive the disaster multiple times, slows down their recovery, and negatively affects their mental and emotional health (Picou et al., 2004). Ursano et al. (2008)

K. Mosby (✉)
Coastal Sustainability Studio, Louisiana State University, Baton Rouge, LA, USA
e-mail: kmosby4@lsu.edu

© Springer Nature Switzerland AG 2021

K. E. Cherry, A. Gibson (eds.), *The Intersection of Trauma and Disaster Behavioral Health*, https://doi.org/10.1007/978-3-030-51525-6_13

suggest that psychological responses happen in phases, where social cohesion is the immediate response. Depending on how local, regional, state, and federal agencies conduct disaster response and recovery, the corrosive community reaction may emerge as a second phase, especially among minorities who may not receive the same level of assistance as their white counterparts (Ursano et al., 2008).

In addition to the type of disaster, experiencing displacement from familiar surroundings or disruption in social networks also affects the mental and emotional health of individuals recovering from a disaster. Research suggests exposure to a disaster alone does not increase negative mental health outcomes (Fergusson, Horwood, Boden, & Mulder, 2014; Galea et al., 2002). Such studies found that participants had to experience trauma related to the disaster in order to develop negative mental health consequences, such as PTSD, depression, anxiety, or substance dependency. Disaster-related trauma may include physical injury or stress related to difficulty finding shelter, food, and clothing (Jacob, Mawson, Marinelle, & Guignard, 2008; Orengo-Aguayo, Stewart, de Arellano, Suárez-Kindy, & Young, 2019; Van Griensven et al., 2006).

Loss of familiar surroundings and normal social connections due to displacement can also exacerbate mental health issues. For instance, Jacob et al. (2008) and Van Griensven et al. (2006) found displaced participants had a higher prevalence of mental health symptoms than those who were not displaced. Van Griensven et al. (2006) suggested that although symptoms decreased between the 8-week and 9-month post-event surveys, levels of PTSD, anxiety, and depression symptoms remained higher in the displaced population than in the non-displaced population. These studies show that even though displacement is often depicted as a solution to meet immediate needs following disasters, it can also be associated with negative mental health outcomes. Displacement disrupts social networks, removes survivors from their pre-disaster social and physical environment, and prevents them from reestablishing their normal routines (Adams, Van Hattum, & English, 2009; Jacob et al., 2008; Van Griensven et al., 2006). Adams et al. (2009) argued that survivors who return to communities that experienced drastic changes due to recovery efforts may still feel displaced because their pre-disaster community and way of life is permanently altered.

Social networks also play important roles during times of stress and trauma because they offer the capacity for people to respond positively to disaster stress (Bankoff, 2014; Bonanno et al., 2010). Engaging with one's social network can foster feelings of connectedness, which may lessen the negative mental health effects of disaster-related trauma. Studies show that social networks are an important factor in coping successfully and increasing individual resiliency to future disaster events (Bonanno et al., 2010; Jones, Gupta, Murphy, & Norris, 2011). Jacob et al. (2008) found that being separated from loved ones and familiar settings operated as a greater stressor than physical injury. Having community and social support reduces rates of depression, anxiety, and other mental health challenges following exposure to disaster trauma (Bonanno et al., 2010; Fergusson et al., 2014; Jones et al., 2011). Bankoff (2014) found that participating in religious and cultural practices may help survivors cope with disaster-related stress and trauma (see Marks, Hatch, Lu, & Cherry, 2015, for further discussion).

While social networks facilitate coping and resilience, disaster response can exacerbate mental health issues if government programs further disrupt survivors' lives. In disasters where housing is destroyed, shelters at churches, schools, and other civic buildings tend to house people for a few days after the event. Government disaster response efforts create temporary solutions that provide housing for several months, such as travel trailers arranged into temporary trailer parks. However, these solutions often exacerbate mental and behavior health because temporary housing locations tend to be isolated and far away from familiar surroundings (Edwards, 1998; Lu & Singelmann, 2009). Edwards (1998) found temporary trailer parks increase anxiety, depression, isolation, displacement, etc.

Research suggests that housing instability, displacement, disruption of social networks, and loss of employment can lead to negative mental health outcomes (Galea et al., 2007; Lu, 2011; Lu & Singelmann, 2009; Mortensen, Wilson, & Ho, 2009). Mortensen et al. (2009) found displacement to Houston resulted in employment and income loss, which decreased mental and physical health. However, Lu and Singelmann (2009) found those living in Houston apartments reported less anxiety than those living in FEMA trailer parks in Louisiana because Houston apartments offered greater stability in less isolated areas. Lu (2011) connected housing instability and loss of employment to increased stress which negatively impacted participant health. Galea et al. (2007) found the top stressors that led to the high rates of anxiety and PTSD in participants were physical illness or injury as well as sleeping in shelters and having difficulty accessing adequate food and clothing, which the authors categorized as extreme physical adversity. Further, studies found that connection to community involvement and social networks mediated negative mental health impacts, such as depression and anxiety (Lu, 2011; Lu & Singelmann, 2009; Ursano et al., 2008). Longitudinal studies found these negative mental health outcomes lingered overtime rather than returning to pre-storm levels (Galea et al., 2007; Greenough et al., 2001; Kessler et al., 2008). However, these studies ended a few years after Hurricane Katrina. They did not track participants to examine longitudinal mental and emotional health in long-term displacement in light of recovery policies that changed housing, education, employment, and healthcare options in New Orleans after Hurricane Katrina.

The purpose of this chapter is to explore how disaster recovery and post-Katrina policy changes affected the mental and emotional health of African Americans displaced from New Orleans. The following section covers the theoretical perspectives that guided the research. The next section presents the research design and methods used for data collection and analysis that lead to a discussion of the findings. The chapter concludes with recommendations to improve mental health outcomes following future disasters.

Theoretical Perspectives

The present research draws upon several theoretical frameworks to investigate how individual decision-making regarding the process of return impacted the mental and emotional health of African Americans displaced from New Orleans by Hurricane Katrina and the 2005 levee failure. These theories include critical theory, critical race theory (CRT), intersectionality, and social vulnerability theory. Critical theory and CRT deconstruct assumptions about social institutions in search for inequalities and injustices (Bronner, 2011; Creswell, 2007; Delgado & Stefancic, 2001). Agger states, "Critical theorists attempt to develop a mode of consciousness and cognition that breaks the identity of reality and rationality, viewing social facts not as inevitable constraints on human freedom but as pieces of history that can be changed" (1991, p. 109). Critical theories can influence research design as a methodology or substantively (Agger, 1991; Creswell, 2007). The present research incorporates critical theory and critical race theory substantively because the author conceptualizes the process of return as constrained by race, class, and gender differences. The project's goal was to deconstruct assumptions of return and ground the experience in structural inequalities and in race, gender, and class differences.

Drawing on Crenshaw's (1991) concept of political intersectionality, the author examined how housing, education, employment, and healthcare policies overlapped to influence the decision to return after Hurricane Katrina. Brodkin Sacks (1994) identified these social institutions as key to upward mobility in the United States and as arenas in which the federal government supported systemic discrimination against African Americans. Tomlinson (2013), Spade (2013), and Verloo (2013) recognize important points of intersection are determined by power relations within society. Like Spade (2013) and Cho, Crenshaw, and McCall (2013), the author rejects a single-axis approach to understanding the processes of displacement, return, and recovery where survivors found themselves at the intersection of several changing institutional forces. Bringing marginalized voices to the decision-making table provides the opportunity for the disenfranchised to discuss how possible solutions may interact with their specific context to further, rather than alleviate, discrimination and inequality (Cho et al., 2013; Verloo, 2013; Young, 1990).

The present research examined nuances that influenced individual agency within the context of structural constraints on recovery and return (Bourdieu, 1990; Giddens, 1984). The analysis revealed the barriers to return as identified by African Americans to illuminate possible causes of the disparate return rates among African American and white residents that previous research reported (see Henderson et al., 2015). Through the experiences of African Americans and the unique standpoint that emerged following the political nature of Hurricane Katrina's recovery, this study also aimed to critically examine structural policy decisions and their impacts as well as media representations of recovery. The next section focuses on disaster theory and illustrates how the critical perspective influences the theoretical frameworks used to understand the causes and impacts of disasters.

How societies frame disasters may determine how governments, communities, and individuals prepare and respond to disaster situations (Dynes, 1993; Furedi,

2007; Kroll-Smith & Couch, 1991; Quarantelli, 2000). Classifying disasters as "Acts of God" fostered a fatalistic perspective because this view locates disasters outside of human influence and control (Quarantelli, 2000, p. 5). The shift to an "Act of Nature" approach emphasized the ability of engineering to reduce risk by creating safer buildings and protection systems (2000, p. 5). This perspective introduces limited human agency to lessen the impact of disasters through improving physical structures (2000). As technological accidents, as opposed to natural weather phenomena, began to cause more disasters and as sociologists undertook empirical studies of disasters, society began to view disasters as "Acts of Men and Women" (2000, p. 3–4). Situating disasters as social constructions reveals the possibility of human action to mitigate and minimize the impact of disasters (2000).

From this shift in framing disasters emerged the social vulnerability theory of disaster (Phillips, Thomas, Fothergill, & Blinn-Pike, 2010; Wisner, Blaikie, Cannon, & Davis, 2004). Social vulnerability theory recognizes that power relations in society determine how people experience disasters. Marginalized and disadvantaged populations due to race, gender, class, age, ability, etc., are more likely to suffer negative consequences during and after disasters because they have the least resources to respond and often are in more vulnerable positions than their counterparts before the disaster (Wisner et al., 2004). Phillips et al. (2010) suggest that while disaster scholars have generally accepted the role social inequity plays in disasters, popular culture and media representations continue to portray disasters as "Acts of God" because it shifts blame from the structures of power to processes outside of human influence. Thus, mainstream media and popular discourse tend not to challenge the balance of power after disasters.

Research Study: An Example

This chapter draws on a larger study that explored how African Americans frame challenges to return to New Orleans after Hurricane Katrina (Mosby, 2017). The questions that guided the project were as follows: How do post-disaster challenges affect behavioral health outcomes and decisions to return? What strategies are used to overcome emergent challenges? What resources do African Americans need to return to their communities after a disaster? How do participants and the media portray the influence of race, class, and gender on the process of return?

Research Design and Methods

This study employed a qualitative approach to explore the decision-making process of African Americans displaced from New Orleans by Hurricane Katrina. Qualitative research involves an in-depth examination of social phenomena rather than a statistical description (Babbie, 2004; Creswell, 2007). This project sought to uncover the nuanced decision-making process of displaced African Americans and how they framed obstacles to return. Furthermore, this author hoped to gain a deeper understanding of the relationship between individual agency and structural changes following Hurricane Katrina (Bourdieu, 1990; Giddens, 1984).

In this chapter, the author compared three different cases of post-Katrina experiences: (1) residents that returned after Hurricane Katrina; (2) residents permanently displaced to Houston, TX; and (3) residents that returned to New Orleans for more than 1 year before relocating to Houston. The author conducted in-depth, semi-structured interviews with open-ended questions with 17 residents displaced by Hurricane Katrina and a content analysis of media articles published in the *New York Times* (*NYT*) and the *Times-Picayune* (*TP*). The articles offer a local and a national perspective on planning processes and policy changes after Hurricane Katrina. Due to its influence on national politics and its history for acknowledging social differences, *NYT* was chosen as a national media narrative that helped individuals outside of the metro area make sense of New Orleans recovery after the disaster. Initial interviews were conducted in person (2011–2012) with follow-up phone interviews with 14 of the original participants in 2016. Most interviews were conducted individually and lasted 60–90 minutes. One interview lasted 2 hours because three family members were interviewed at the same time. Two other participants were interviewed together and required a second interview to complete the initial interview process. All follow-up interviews were conducted individually. Pseudonyms are used throughout this chapter to protect confidentiality.

Sampling and Data

The project sampled at multiple levels. The author selected Houston because it sheltered the largest number of African American evacuees after the storm and housed the largest federal disaster assistance program (Bliss, 2015; Weber & Peek, 2012). To select participants, the author used snowball and maximum variation sampling to increase the possibility for different perspectives and experiences to emerge (Creswell, 2007; Seidman, 2006). Saturation and sufficiency determined the sample size (Seidman, 2006; Yin, 2009). The study included 17 participants over 18 years old in 2011 who self-identify as African Americans. Table 13.1 presents a summary of the methodological features and sociodemographic characteristics of the sample.

To sample media articles, the author performed a database search and limited the search results by date: from August 29, 2005, to August 1, 2016. Keywords for the search included New Orleans, Hurricane Katrina, recovery, Road Home, housing, school, education, income, employment, jobs, health, and return. The author created a database of potential articles from the keyword search and used purposive sampling to select articles by date and headline themes. The author narrowed the sample until removing additional articles would result in holes. Table 13.2 presents a summary of the characteristics of the content analysis sample.

Table 13.1 Methodological features and sociodemographic characteristics by case assignment

Total sample/ recruitment	Respondents	Age[a]	Gender
Maximum variation, snowball sampling	17	18 years or older	8 women 9 men
Sample by group			
Case 1: Returned to New Orleans, LA	5	14–56 years	2 women 3 men
Case 2: Displaced to Houston, TX	9	13–38 years	6 women 3 men
Case 3: Returned to New Orleans, LA[b] and later relocated to Houston, TX	3	30–42 years	3 men

Notes:
[a]For the total sample, age reflects participants' chronological age at the time of the interview (between 2011 and 2012), while age by case membership reflects chronological age in 2005.
[b]Respondents resided in post-Katrina New Orleans for at least 12 months before relocating to Houston, TX.

Table 13.2 Characteristics of the content analysis sample

Source	Search engine	Articles in database	Articles in sample
Times-Picayune	NewsBank	462	59
New York Times	LexisNexis	587	56

Analysis

All interviews were transcribed verbatim and double-checked for discrepancies between the audio recording and the written transcript. After transcription, the author coded the interviews and the news articles with deductive, inductive, and in vivo coding (Creswell, 2007). Deductive codes that emerged from the literature included: recovery planning, recovery policy, uneven recovery, recovery outcomes, continued vulnerability, barriers to rebuild, barriers to return, assistance, race, class, and gender. Inductive coding allowed codes to emerge during analysis (2007). In vivo codes were labeled with direct quotes from the data, such as "not what it was" (Coffey & Atkinson, 1996; Creswell, 2007).

After coding, the author analyzed the codes for meaning and combined them to create themes. To analyze the data, this project used a holistic rather than an embedded approach (Creswell, 2007). Embedded approaches selectively analyze specific facets where holistic approaches analyze the case as a whole (2007). Rather than limiting analysis to institutional obstacles presented by housing, education, healthcare, and employment policy changes, the author analyzed each case in its entirety, including physical, emotional, and psychological obstacles. A deeper understanding of the needs and experiences of displaced African Americans emerged because the participants' framework and identification of barriers directed the analysis process.

Findings

During the analysis process, the codes merged to form the theme: policy compounds vulnerability. The findings showed post-disaster policy decisions that affected housing, education, healthcare, and employment opportunities failed to help individuals and the community as a whole recover. These policy changes increased the vulnerable status of residents by decreasing their access to healthcare as well as increasing the difficulty of finding housing, employment, and schools and maintaining their connectedness with their social networks in post-Katrina New Orleans. This agrees with prior research that indicates recovery policies often exacerbate rather than reduce social vulnerabilities (Chhotray & Few, 2012; Ingram, Franco, Rumbaitis-del Rio, & Khazai, 2006). Participants framed policy decisions as compounding their vulnerability by describing the negative mental and emotional impacts they experienced while coping with post-Katrina changes. For example, Lemont Cummings described how recovery influenced his emotional state and his decision to relocate to Houston: *"When I came back, the changes in the city…and everything that all the hardships caused really…left an emotional stain on me where I feel like if I go back to New Orleans anytime soon it will be too soon."*

The media samples provided examples of how policy compounded vulnerability in the fields of housing, education, and employment. Two *TP* and six *NYT* sampled articles focused specifically on how healthcare policy compounded vulnerability after Hurricane Katrina. During recovery, city and state officials chose to build a new facility instead of repairing Charity Hospital, a state-run teaching hospital that served the city's poor, uninsured, and mentally ill before the disaster. It took 10 years for the new medical complex that replaced Charity Hospital to open. The sampled authors found this compounded the vulnerability of returned residents, who encountered limited healthcare options for a decade.

Both media samples viewed policy decisions that left the city without healthcare facilities for 10 years as compounding the vulnerability of all returned residents. In the *TP*, Hunter and Pope (2006) wrote:

> The lack of resources is more acutely felt in the realm of mental health care, and patients with a history of mental illness are most at risk. Mental health officials and law enforcement agencies in the New Orleans area have reported a sharp increase in the number of requests to pick up mentally ill patients, a rise in the number of people who resist violently, or both… Dr. Jeffrey Rouse, the deputy psychiatric coroner for Orleans Parish… said the rise in violent resistance since Katrina is a direct result of the all but total obliteration of the mental health system in the New Orleans area.

The samples used expert opinions to highlight the negative consequences of post-Katrina healthcare shortages, which included higher rates of relying on emergency rooms rather than primary care physicians, emergency room wait times increased from hours to days, and developing substance abuse and additional psychological issues (such as depression, anxiety, or PTSD). In the *NYT*, Chen (2012) discussed the ability of government decisions in the short and long term to impact mental health outcomes following a disaster and criticized post-Katrina response and recovery measures for contributing to high levels of mental health issues among survivors still struggling to meet their basic needs. The sampled authors showed the slow pace of rebuilding healthcare facilities exacerbated health issues for all residents by limiting access to care.

Previous research shows some individuals develop substance abuse issues if they use drugs, alcohol, or nicotine as coping mechanisms for dealing with anxiety, depression, or stress following traumatic events (Bonanno et al., 2010; also see Waters & Copeland, Chap. 17, this volume). Other individuals turn to faith-based practices or social organizations (Bankoff, 2014; Jenkins, 2012). While these coping mechanisms emerged here, participants also described two unique strategies to cope with stress and displacement. These strategies included moving repeatedly and relyin3.g on women in their social networks.

Participants in Houston and New Orleans discussed moving repeatedly to cope with the trauma and adjustment of being displaced. When participants returned to New Orleans, they found poor housing options and experienced a sense of displacement (Adams et al., 2009). For example, Joseph Johnson stated:

> If I was to be put back in New Orleans now, it would just feel like I'm in Houston a little bit because... none of my family's there... My friends aren't there... Everything has changed so much... The people were the life of New Orleans. You drive through New Orleans now; it's like it's so dead... It's like torture... It's like waking up...in a dream, a nightmare, and ya know it's like you're the only person there... It's like *I am Legend*, ya know. You're the only one, and it's so deserted.

Participants in Houston also moved repeatedly in order to meet their physical and mental health needs. They tended to move less frequently than participants in New Orleans because disaster assistance required those in Houston to enter into leases (Lu & Singelmann, 2009). When leases expired, respondents would move to another part of Houston trying to find peace since they could not return to New Orleans.

All Case 3 participants relocated to Houston because they could not secure long-term employment and financial stability in post-Katrina New Orleans. Marcus Booker explained:

> My story is that for two years, three years after Katrina and maybe even more, maybe four years. I don't think I've lived in one place longer than 6 or 7 months. I was constantly, constantly moving... just trying to find a situation to be at peace. That's kinda why I moved to Houston. Let me go over here and see. I know I'll find a place to live. I know I'll find some work, and I know I can just be in one place for an extended period of time, and even if I didn't like it, I can just relax.

Moving to Houston provided the opportunity to find steady employment and higher wages. Participants framed the move to Houston as a way to find a place to settle so they could experience peace instead of the constant stress and anxiety brought on by unemployment and poor housing options. Although participants described moving as a way to cope, research suggests housing instability exacerbates negative mental health outcomes following disasters (Lu, 2011; Lu & Singelmann, 2009).

Relying on the women in their social networks also helped participants cope with the stress and increased demands they experienced after Hurricane Katrina (Roberto, Kamo, & Henderson, 2009). In some households, wives became the breadwinners because their pre-storm jobs existed after the storm while their husbands' jobs did not return. In other situations, girlfriends and mothers provided stable housing and emotional support to help their loved ones deal with the emotional trauma brought on by trying to survive in displacement or in a partially functioning city. In other situations, displaced aunts and cousins took in the children in their network to reduce expenses for struggling parents. Almost every participant described the ways their family and social networks chose separation as a means to meet their needs as well as the stress and depression they experienced as a result of that separation (Jacob et al., 2008).

Participants framed their health as caught in the intersection of housing, education, and employment issues. They attributed problems with anxiety, depression, PTSD, high blood pressure, anger, fatigue, and headaches to post-storm challenges finding housing, schools, and jobs. Seeing the devastation linger in their neighborhoods for years also caused emotional distress, which fostered a corrosive community (Freudenburg, 1997; Picou et al., 2004). In the following quote, Lemont Cummings expressed his anger and his view of the role race played in the recovery trajectory of New Orleans:

> That's another part that just kinda infuriates me. It does. I think that a big reason why they haven't done what they could've done because $7 billion is a lot of money to rebuild. …They don't want the black population back. They don't want them to come back, and they know if they rebuild that area they would come back. …I think it's all government driven to keep those people from coming back.

Participants in each case expressed anger, resentment, and statements that the city did not care about its residents because recovery efforts did not repair African American neighborhoods enabling return (White, Philpot, Wylie, & McGowen, 2007). Many Case 2 participants stayed displaced when faced with the slow pace of recovery in New Orleans as Ella Taylor described:

> My aunt…left and went back home and I was just here with my kids. …I was like I want to go home, and then every time I would visit it was like I was in a foreign land. I didn't know anybody. …I can't bring my kids back to this. School systems was to the kaputz, and I was like I'll just stay here in Houston. But I'll get homesick and want to go home, so we would travel all the time. …I would always say I want to move back home, and then when I go home…everything is still looking like the hurricane hit. You still have cars turned over, and the Lower 9th Ward…it's like it's a foreign land. …Canal Street, French Quarters all that is back to up and running, but everything in the outside and the surround areas is just still damaged. …I was like I can't do it. I can't do it. But every time I would come back to Houston, I would keep chronically depressed, sick to my stomach, can't eat for a couple of days, and stay locked up in my room in the dark and cry. I just wanted to go back home.

Although participants chose to prolong their displacement, establishing a life in Houston presented additional challenges. Barbara Davis described the emotional impact of the barriers she encountered:

> It took me two years to find a job…One person that I interviewed with had the courage to tell me, 'It's going to be hard for you to find a job because they've already told us not to hire people from New Orleans.' …That was at a temp agency. …I just felt like violated. I just felt like I was judged. …I've always been a person that worked. I'm not a person that lived off of the government… I wanted to work. I wanted to be able. I'm an independent person,

and I felt like I wanted my independence back, and people weren't allowing me the opportunity to get that back.

In Houston, discrimination, challenges accessing resources, and being separated from social networks increased feelings of depression, anxiety, and anger (Mosby, 2017).

A cycle of deteriorating health also emerged in which caring for family members increased stress that deteriorated the health of caretakers. The storm affected family members differently. Some respondents reported that their elderly parents and grandparents experienced a deterioration of their physical and mental health due to displacement. As family members suffered from depression, dementia, and a deterioration of physical health, other family members emerged as caretakers. However, the stress from caring for these family members negatively impacted the mental, emotional, and physical health of the caretakers. Respondents who did not take care of other family members generally reported an easier time acclimating to post-Katrina life. Those who had multiple children or parents that did not recover mentally after the storm reported the stress caused high blood pressure, anxiety, and feelings of helplessness and anger. Respondents coped with the added pressure by turning to alcohol and food, which further deteriorated their physical health.

Negative impacts to behavioral health in this study were long-lasting. Previous research suggests that in the months and years that follow a disaster, rates of mental health disorders decrease and eventually return to normal (Bonanno et al., 2010; Van Griensven et al., 2006). However, participants stated they experienced negative impacts to their mental health for many years after the storm. Some displaced participants said they still struggle with depression due to displacement and experience intense anxiety when it rains. For example, Ella Taylor stated:

It took a long time to adjust...probably about 3 to 4 years... If it was a hard rain, my kids would all be together. They want to sleep...in the bed with me. ...My kids wouldn't sit down in the tub to take a bath. They would stand up 'cause they were scared of the water. ...I have a friend right now, still to this day...water can't touch her feet 'cause she actually went through the five days staying in New Orleans... She will panic if water touches her feet. ...If it's raining outside and she step in a puddle of water, she'll freak out. ...We haven't had any counseling for this. We're still dealing with this to this day. Even I still feel angry sometimes. I still feel, ya know, resentment, and I'm still adjusting to being here, ya know. I still get depressed sometimes.

Feelings of anger about the pace and progress of recovery also continue to linger more than a decade later.

In the follow-up interviews conducted in 2016, participants recognized their mental health continued to suffer due to their experiences after Hurricane Katrina. Although they recognized the need for mental healthcare, almost all of them said they did not receive professional help to overcome these challenges nor did they discuss their traumatic experiences with family or friends (Mortensen et al., 2009). Most chose not to seek treatment because of social stigmas in society and their culture (Lindsey et al., 2013). Only two participants sought mental healthcare between 2011 and 2016, one for himself and the other for her daughter, who was 6 in 2005 and 14 at the time of treatment. Both were diagnosed as bipolar although neither had a family history with mental disorders. Both participants viewed the emergence of the disorder as a result of dealing with too much stress following Hurricane Katrina.

Conclusions and Recommendations

Research shows that experiencing symptoms of mental health disorders often happens after a disaster (see Cherry, Calamia, Birch, & Moles, Chap. 16, this volume). The present findings suggest that post-disaster policy can exacerbate the trauma associated with disasters when government policy makes returning and rebuilding difficult. To reduce post-disaster trauma, federal, state, and local governments should minimize policy changes after a disaster (Adams et al., 2009). People often want to return to what was. As such, officials should not use a disaster as an opportunity to recreate an entire way of life in the name of improving a city's quality of life. Drastic changes, such as those implemented in New Orleans after Hurricane Katrina, can lead to feelings of displacement even for residents that return (2009). Instead post-disaster policy should facilitate the ability of households, social networks, and communities to stay together, especially if displacement is necessary (Jacob et al., 2008). In addition, policies should help foster a sense of community in temporary housing sites to help alleviate the negative mental health effects of disaster trauma and displacement (Fergusson et al., 2014; Lu, 2011; Lu & Singelmann, 2009).

While seeking to minimize displacement and separation, recovery policies should also provide more funding for mental health treatment. Mental health services should be available for longer than 11 months post-event (Lystad, 1990). This is especially important when recovery is slow and takes multiple years because these situations tend to create corrosive communities rather than therapeutic communities (Ursano et al., 2008). In these types of disasters, residents are more likely to experience negative mental health issues for longer periods of time. Services should be provided for more than a year following the event because people in the midst of a protracted recovery will continue to need mental healthcare until recovery is finished, maybe longer (Greenough et al., 2001; Ursano et al., 2008).

However, simply providing funding for therapy will not be effective because of the social stigmas related to mental healthcare in the United States (Lindsey et al., 2013). Although the participants in this study recognized their need for mental healthcare, they refused to seek treatment due to fear of being stigmatized. To overcome these challenges, two different approaches are needed. First, social programs should be created to normalize seeking mental health services. A public education campaign could be created that uses the media to inform people about normal post-disaster responses, which includes depression, anxiety, post-traumatic stress disorder, anger, fatigue, and headaches (Lystad, 1990). This may help people realize they are not alone in what they are feeling, nor is there something wrong specifically with them. The campaign could also educate residents about what resources are available and where to seek help for mental healthcare following disaster or non-disaster-related trauma (1990). Additionally, doctors, teachers, and other disaster relief workers should be taught about normal mental and behavioral health responses to disasters and encouraged to recommend mental health resources to survivors of disaster-related trauma even if the primary purpose of the visit was not for mental or behavioral health concerns.

Second, policies should help community centers, faith-based organizations, and other social organizations create spaces to address mental health concerns (Nicholls et al., 2015). This could happen in many forms. It may be providing funding for meetings to be held in these locations where residents typically gather. Funding could allow professionals to host meetings for residents to talk about recovery stresses, seek support from neighbors dealing with similar challenges, and bond with their community members. Encouraging people to engage in religious and cultural activities can instill a sense of normality and peace into their lives during a tumultuous time (Bankoff, 2014; Jenkins, 2012). Furthermore, fostering a sense of community and social cohesion can reduce negative mental health disorders (Fergusson et al., 2014; Jacob et al., 2008).

Outside of increasing funding for mental health services and providing alternative means for people to access the support and care they need, aid to help families following a disaster should also be expanded. As families struggled to find adequate resources after the storm, those participants caring for ailing family members experienced more stress and reported more issues with their mental health than those participants who only had to fend for themselves (Thompson, Norris, & Hanacek, 1993). To alleviate this, recovery policy should offer services that help residents care for their ailing family members. Providing home healthcare services to help care for elderly relatives and those who live in rural settings with limited community resources would lessen the stress on caretakers, who typically experience an increase in responsibilities after disasters (Mosby, Birch, Moles, & Cherry, in press).

In closing, reducing negative behavioral health consequences associated with disaster-related trauma is possible through purposeful policy interventions. First, disaster response measures need to prioritize keeping families and social networks together during response and recovery. Second, citizens need to be informed about normal psychological responses to experiencing trauma and where to seek help. Third, mental health services need to be available for longer periods of time and offered in locations and formats where people already gather. Finally, additional services should be made available for individuals that have to care for family members that experience extreme negative consequences following a disaster. Implementing these changes to current approaches to disaster recovery assistance can help facilitate better mental health outcomes following disasters.

Acknowledgments This research was made possible by the generous support of the University of New Orleans and the Gulf Research Program of the National Academies of Sciences, Engineering, and Medicine and the Robert Wood Johnson Foundation (under award number 2000008299).

References

Adams, V., Van Hattum, T., & English, D. (2009). Chronic disaster syndrome: Displacement, disaster capitalism, and the eviction of the poor from New Orleans. *American Ethnologist, 36*(4), 615–636.

Agger, B. (1991). Critical theory, poststructuralism, postmodernism: Their sociological relevance. *Annual Review of Sociology, 17*, 105–131.

Babbie, E. (2004). *The practice of social research* (10th ed.). Belmont, CA: Wadsworth, Thomson Learning, Inc..

Bankoff, G. (2014). Disaster medicine in Southeast Asia. In T. Harper & S. S. Amrith (Eds.), *Histories of health in Southeast Asia: Perspectives on the long twentieth century*. Bloomington, IN: Indiana University Press.

Bliss, L. (2015, August 25). *10 years later, there's so much we don't know about where Katrina survivors ended up: An incomplete picture of displacement and return in New Orleans, in maps*. CityLab. Washington, D. C. Retrieved from http://www.citylab.com/politics/2015/08/10-years-later-theres-still-a-lot-we-dont-know-about-where-katrina-survivors-ended-up/401216/

Bonanno, G. A., Brewin, C. R., Kaniasty, K., & Greca, A. M. L. (2010). Weighing the costs of disaster: Consequences, risks, and resilience in individuals, families, and communities. *Psychological Science in the Public Interest, 11*(1), 1–49.

Bourdieu, P. (1990). *The logic of practice*. Stanford, CA: Stanford University Press.

Brodkin Sacks, K. (1994). How did Jews become white folks? In S. Gregory & R. Sanjek (Eds.), *Race* (pp. 78–102). New Brunswick, NJ: Rutgers University Press.

Bronner, S. E. (2011). *Critical theory: A very short introduction*. Cary, NC: Oxford University Press.

Chen, P. W. (2012, November 2). Easing the trauma after the storm. *New York Times* Blogs.

Chhotray, V., & Few, R. (2012). Post-disaster recovery and ongoing vulnerability: Ten years after the super-cyclone of 1999 in Orissa, India. *Global Environmental Change, 22*(3), 695–702.

Cho, S., Crenshaw, K. W., & McCall, L. (2013). Toward a field of intersectional studies: Theory, applications, and praxis. *Signs: Journal of Women in Culture and Society, 38*(4), 785–810.

Coffey, A., & Atkinson, P. (1996). *Making sense of qualitative data: Complementary research strategies*. Thousand Oaks, CA: Sage Publications.

Crenshaw, K. W. (1991). Mapping the margins: Intersectionality, identity politics, and violence against women of color. *Stanford Law Review, 43*(6), 1241–1299.

Creswell, J. W. (2007). *Qualitative inquiry & research design: Choosing among five approaches* (2nd ed.). Thousand Oaks, CA: SAGE Publications, Inc.

Delgado, R., & Stefancic, J. (2001). *Critical race theory: An introduction*. New York: New York University Press.

Dynes, R. R. (1993). Disaster reduction: The importance of adequate assumptions about social organization. *Sociological Spectrum, 13*, 175–192.

Edwards, M. L. K. (1998). An interdisciplinary perspective on disasters and stress: The promise of an ecological framework. *Sociological Forum, 13*(1), 115–132.

Fergusson, D. M., Horwood, L. J., Boden, J. M., & Mulder, R. T. (2014). Impact of a major disaster on the mental health of a well-studied cohort. *JAMA Psychiatry, 71*(9), 1025–1031.

Freudenburg, W. R. (1997). Contamination, corrosion and the social order: An overview. *Current Sociology, 45*(3), 19–39.

Fritz, C. E. (1961). Disaster. In R. K. Merton & R. A. Nisbet (Eds.), *Contemporary social problems* (pp. 651–694). New York: Harcourt, Brace, & World.

Furedi, F. (2007). The changing meaning of disaster. *Area, 39*(4), 482–489.

Galea, S., Ahern, J., Resnick, H., Kilpartick, D., Bucuvalas, M., Gold, J., et al. (2002). Psychological sequelae of the September 11 terrorist attacks in New York City. *New England Journal of Medicine, 346*, 982–987.

Galea, S., Brewin, C. R., Gruber, M., Jones, R. T., King, D. W., King, L. A., et al. (2007). Exposure to hurricane-related stressors and mental illness after Hurricane Katrina. *Archives of General Psychiatry, 64*(12), 1427–1434.

Giddens, A. (1984). *The constitution of society: Outline of the theory of structuration*. Cambridge, MA: Polity Press.
Greenough, G., McGeehin, M., Bernard, S. M., Trtanj, J., Riad, J., & Engelberg, D. (2001). The potential impacts of climate variability and change on health impacts of extreme weather events in the United States. *Environmental Health Perspectives, 109*(suppl 2), 191–198.
Henderson, M., Davis, B., Climek, M., Thouta, S., Hagan, G., Williamson, C., et al. (2015). *Views of recovery ten years after Katrina and Rita: A survey of residents of the city of New Orleans and residents throughout Louisiana*. Baton Rouge, LA: Louisiana State University Public Policy Research Lab.
Hunter, M., & Pope, J. (2006, July 3). Stressed? No, we're OK, really, we are – Actually, we're on the edge, mental health experts say. *Times-Picayune*, p. 01.
Ingram, J. C., Franco, G., Rumbaitis-del Rio, C., & Khazai, B. (2006). Post-disaster recovery dilemmas: Challenges in balancing short-term and long-term needs for vulnerability reduction. *Environmental Science and Policy, 9*(7–8), 607–613.
Jacob, B., Mawson, A. R., Marinelle, P., & Guignard, J. C. (2008). Disaster mythology and fact: Hurricane Katrina and social attachment. *Public Health Reports, 123*(5), 555–566.
Jenkins, P. (2012). After the flood: Faith in the diaspora. In L. Weber & L. Peek (Eds.), *Displaced* (pp. 218–229). Austin, TX: University of Texas Press.
Jones, E. C., Gupta, S. N., Murphy, A. D., & Norris, F. H. (2011). Inequality, socioeconomic status, and social support in post-disaster mental health in Mexico. *Human Organization, 70*, 33–43.
Kessler, R. C., Galea, S., Gruber, M. J., Sampson, N. A., Ursano, R. J., & Wessely, S. (2008). Trends in mental illness and suicidality after Hurricane Katrina. *Molecular Psychiatry, 13*, 374–384.
Kroll-Smith, S., & Couch, S. R. (1991). What is a disaster? An ecological-symbolic approach to resolving the definitional debate. *International Journal of Mass Emergencies and Disasters, 9*(3), 355–366.
Lindsey, M. A., Chambers, K., Pohle, C., Beall, P., & Lucksted, A. (2013). Understanding the behavioral determinants of mental health service use by urban, under-resourced Black youth: Adolescent and caregiver perspectives. *Journal of Child and Family Studies, 22*(1), 107–121.
Lu, A. (2011). Stress and physical health deterioration in the aftermath of Hurricanes Katrina and Rita. *Sociological Perspectives, 54*(2), 229–250.
Lu, A., & Singelmann, J. (2009). Location matters? A comparison of placement conditions for hurricane evacuees in Houston and Louisiana. *International Journal of Mass Emergencies and Disasters, 27*(3), 191–217.
Lystad, M. (1990). United States programs in disaster mental health. *International Journal of Mental Health, 19*(1), 80–88.
Marks, L. D., Hatch, T. G., Lu, Y., & Cherry, K. E. (2015). Families and faith-based communities after a disaster: Successes and failures in the wakes of Hurricanes Katrina and Rita. In K. E. Cherry (Ed.), *Traumatic stress and long-term recovery: Coping with disasters and other negative life events* (pp. 247–270). New York: Springer.
Mortensen, K., Wilson, R. K., & Ho, V. (2009). Physical and mental health status of Hurricane Katrina evacuees in Houston in 2005 and 2006. *Journal of Health Care for the Poor and Underserved, 20*(2), 524–538.
Mosby, K. (2017). *Frameworks of Recovery: Exploring the Intersection of Policy & Decision-Making Processes After Hurricane Katrina* (Doctoral Dissertation, University of New Orleans). Retrieved from https://scholarworks.uno.edu/td/2421
Mosby, K., Birch, T., Moles, A., & Cherry, K. E. (in press). Disasters. In L. W. Kaye (Ed.), *Handbook of rural aging*. New York: Routledge, Taylor & Francis.
Nicholls, K., Picou, J. S., Curtis, J., & Lowman, J. A. (2015). The utility of community health workers in disaster preparedness, recovery, and resiliency. *Journal of Applied Social Science, 9*(2), 191–202.

Orengo-Aguayo, R., Stewart, R. W., de Arellano, M. A., Suárez-Kindy, J. L., & Young, J. (2019). Disaster exposure and mental health among Puerto Rican youths after Hurricane Maria. *JAMA Network Open, 2*(4), e192619.

Phillips, B. D., Thomas, D. S. K., Fothergill, A., & Blinn-Pike, L. (Eds.). (2010). *Social vulnerability to disasters*. Boca Raton, FL: CRC Press.

Picou, J. S., Marshall, B. K., & Gill, D. A. (2004). Disaster, litigation, and the corrosive community. *Social Forces, 82*(4), 1493–1522.

Quarantelli, E. L. (2000). *Disaster planning, emergency management and civil protection: The historical development of organized efforts to plan for and to respond to disasters (Preliminary Paper No. PP301)*. Newark, DE: Disaster Research Center, University of Delaware.

Quarantelli, E. L., & Dynes, R. R. (1977). Response to social crisis and disaster. *Annual Review of Sociology, 3*, 23–49.

Roberto, K., Kamo, Y., & Henderson, T. (2009). Encounters with Katrina: Dynamics of older adults' social support networks. In K. E. Cherry (Ed.), *Lifespan perspectives on natural disasters* (pp. 133–152). New York: Springer Publishing Company.

Seidman, I. (2006). *Interviewing as qualitative research: A guide for researchers in education and the social sciences* (3rd ed.). New York: Teachers College Press.

Spade, D. (2013). Intersectional resistance and law reform. *Signs: Journal of Women in Culture and Society, 38*(4), 1031–1055.

Thompson, M. P., Norris, F. H., & Hanacek, B. (1993). Age-differences in the psychological consequences of Hurricane Hugo. *Psychology and Aging, 8*(4), 606–616.

Tomlinson, B. (2013). To tell the truth and not get trapped: Desire, distance, and intersectionality at the scene of argument. *Signs: Journal of Women in Culture and Society, 38*(4), 993–1017.

Ursano, R. J., Fullerton, C. S., & Terhakopian, A. (2008). Disasters and health: Distress, disorders, and disaster behaviors in communities, neighborhoods, and nations. *Social Research, 75*, 1015–1028.

Van Griensven, F., Chakkraband, M. S., Thienkrua, W., Pengjuntr, W., Cardozo, B. L., Tantipiwatanaskul, P., et al. (2006). Mental health problems among adults in tsunami-affected areas in southern Thailand. *JAMA, 296*(5), 537–548.

Verloo, M. (2013). Intersectional and cross-movement politics and policies: Reflections on current practices and debates. *Signs: Journal of Women in Culture and Society, 38*(4), 893–915.

Weber, L., & Peek, L. A. (2012). *Displaced: Life in the Katrina diaspora*. Austin, TX: University of Texas Press.

White, I. K., Philpot, T. S., Wylie, K., & McGowen, E. (2007). Feeling the pain of my people: Hurricane Katrina, racial inequality, and the psyche of Black America. *Journal of Black Studies, 37*(4), 523–538.

Wisner, B., Blaikie, P. M., Cannon, T., & Davis, I. (2004). *At risk: Natural hazards, people's vulnerability, and disasters* (2nd ed.). London and New York: Routledge.

Yin, R. K. (2009). *Case study research: Design and methods* (4th ed.). Los Angeles, CA: SAGE Publications, Inc.

Young, I. M. (1990). *Justice and the politics of difference*. Princeton, NJ: Princeton University Press.

Part III
Community Practice

Chapter 14
Disasters and the Built Environment: Understanding the Potential and Pitfalls of Urban Planning in Post-Disaster Recovery

Traci Birch, Isaac Henry, and Marla Nelson

Introduction

The physical characteristics of the natural and built environment impact peoples' experiences of places and their attachments to them (Hayden, 1995; Stokols & Shumaker, 1981; Tuan, 1974; Ujang, 2012). But just as places consist of more than the space they occupy, the affective bond between people and place goes beyond the physical environment to include the meanings or symbols attributed to places and the social relationships and networks embedded within them (Altman & Low, 1992; Amundsen, 2012; Madgin, Bradley, & Hastings, 2016; Tester, Ruel, Anderson, Reitzes, & Oakley, 2011; Ujang, 2012). In some instances, social attachments to places may be even stronger than the attachments to physical characteristics (Hidalgo & Hernandez, 2001; see Nemeth, Kurianski, & Onishi, this volume, Chap. 18). Regardless of the type of attachment, people-place bonding is a significant component of individual and community health and well-being (Marshall & Bishop, 2015).

Place loss that comes in the wake of a disaster – natural or human caused – can be traumatizing. Due to the sudden and often violent nature of disasters, place loss can also bring grief and threats to identity. The experience of the disaster itself is upsetting, and the sudden loss of the built environment and social networks that people depend greatly upon is both shocking and stress inducing (Brown & Perkins,

T. Birch (✉)
LSU School of Architecture, Coastal Sustainability Studio, Louisiana State University, Baton Rouge, LA, USA
e-mail: tbirch@lsu.edu

I. Henry
Department of Planning, Roanoke County, VA, USA

M. Nelson
Department of Planning and Urban Studies, University of New Orleans, New Orleans, LA, USA

© Springer Nature Switzerland AG 2021

K. E. Cherry, A. Gibson (eds.), *The Intersection of Trauma and Disaster Behavioral Health*, https://doi.org/10.1007/978-3-030-51525-6_14

1992; Cox & Perry, 2011; Diaz, 2013; Erikson, 1976; Fussell, 2012; Kaniasty & Norris, 2004; Osofsky, Osofsky, Kronenberg, Brennan, & Hansel, 2009; Scannell, Cox, Fletcher, & Heykoop, 2016). It is within this type of traumatized state that people must often work to rebuild their homes and neighborhoods and reestablish a sense of meaning, belonging, and connection to that reconstructed place.

In this chapter, we focus on the role of urban planning in rebuilding communities in the wake of a catastrophic disaster. Broadly defined, urban planning is a future-oriented discipline dedicated to improving cities and regions and the lives of the people that live and work within them. Taking an integrative, interdisciplinary approach, urban planning examines how the built environment influences human behavior and interaction. Regardless of the strength of a community's pre-disaster planning, choices about disaster recovery and moving forward in the face of realized vulnerability can be overwhelming. Urban planning within the recovery context is similar to day-to-day planning processes but with the enormous challenges of time compression and the loss of key infrastructural and socioeconomic networks (Olshansky, 2006). Planners must work with residents to restore the built environment in a way that helps residents reshape and rebuild their attachment to place while simultaneously encouraging development patterns that reduce future risk. In the next section of this chapter, we examine peoples' attachment to place and the trauma experienced through place loss. From there, we focus on recovery planning and its transformative potential to strengthen individual and community capacity to overcome loss and recover from disasters. In the following section, we identify challenges to post-disaster planning, including the perennial issues of mistrust and the messiness of participatory processes. This chapter closes with a brief discussion on rebuilding after disaster to expand opportunities and restore well-being for all residents of disaster-impacted communities.

Place, Place Attachment, and Loss

Efforts to understand the connections – emotional, psychological, and social – that people have to *place* are common across varied disciplines such as urban studies, geography, sociology, and environmental psychology. The roots of place scholarship lie with the geographers Tuan (1975, 1977) and Relph (1976). Relph (1976) first asserted that place is meaningful in that it combines location, shared experience, and social relations to form the underpinning components of human identity through time. For decades since, environmental psychologists and geographers have examined relationships to place to better understand how place meaning and place attachment shape attitudes and behaviors (Altman & Low, 1992; Manzo, 2003; Proshansky, 1978; Tuan, 1977). At the same time, community psychologists and sociologists developed a rich scholarship embracing ecological and social systems principles to inform community development contexts including social capital, sense of place, and community participation (Flora & Flora, 1996). Starting with the urban upheavals of the 1960s, urban scholars such as Jacobs (1961), Gans (1968), and Appleyard (1981) – in their now classic works – studied people's perceptions of the places they lived and their on-the-ground experiences of place to inform the urban planning process. Drawing on the connections made by these urban scholars, Manzo and Perkins (2006) examine how place meaning and attachment can galvanize residents to participate in local planning efforts and work together to protect or improve their community.

While the literature addressing place is complex, diverse, and multidisciplinary, it is also highly segmented, despite calls for more coordination across disciplines (Manzo & Perkins, 2006; Trentelman, 2009). The colloquial "commonsense" nature of the word "place" makes it a difficult concept to understand in a nuanced way (Cresswell, 2004, p. 127–128). Attempts to define "place" are usually either incredibly broad (e.g., Agnew, 1987), highly abstract (e.g., Tuan, 1977), or intentionally narrow and specific to the purpose of a particular study (e.g., Friedmann, 2010). Further, Marshall and Bishop (2015) note there is a distinct difference in the way urban planners see placemaking as interventions to improve the quality of places and enhance people-place connections, and other disciplines that focus more on the process of people-place bonding. There is one thing, though, that these divergent attempts to define place have in common: they all include people's attachment to places as *part of the places themselves.*

It is also widely held that place attachment provides stability, familiarity, and security and thus is positively associated with health and well-being (Altman & Low, 1992; Amundsen, 2012, 2015; Fullilove, 1996; Hess, Malilay, & Parkinson, 2008; Rollero & De Piccoli, 2010; Williams & Patterson, 2008). If places are underpinning components of human identity, central to individual and community well-being, then the loss of place can be truly devastating. Individual and community identity are tied to the quality of its physical, social, and emotional characteristics. When a disaster strikes, it can erase those unique features. With the place rendered unrecognizable, community identity feels lost (Butterfield, 2014). Scholars often use forms of the word "trauma" to describe how humans react to loss of place (Atherton, 2017; Baum, 1999; Erikson, 1976; Fullilove, 2004; Kropf & Jones, 2014; Malone, Pomeroy, & Jones, 2011). They hold that the traumatic experience of place loss causes those affected to enter a phase of "bereavement," "mourning," or "grief" (Cox & Perry, 2011; Diaz, 2013; Eisenbruch, 1991; Erikson, 1976; Malone et al., 2011; Marris, 1974; Scannell et al., 2016). Not surprisingly, they often describe loss of place as akin to loss of identity or loss of self (Albrecht et al., 2007; Burley, 2010; Dugan, 2007; Eisenbruch, 1991; Fullilove, 1996; Glavovic, 2008; Malone et al., 2011). According to Fullilove (2004), individuals experiencing place loss undergo *root shock*, a "traumatic stress reaction to the destruction of all or part of one's emotional ecosystem" (p. 11).

Post-Disaster Planning and Its Transformative Potential

Planning for recovery and reconstruction after a disaster is no small task. The destruction and upheaval caused by disasters create conditions of pervasive uncertainty and information scarcity, which compound the difficulty of planning (Nelson, Ehrenfeucht, & Laska, 2007; Pelling, 2003). Post-disaster, planners face a tension in the conflicting need for both speed and deliberation in developing a recovery plan (Olshansky, 2006). Planners and local officials working in post-disaster situations will face intense pressure from residents and economic interests to restore some sense of normalcy (Haas, Trainer, Bowden, & Bolin, 1977; Kates, Colten, Laska, & Leatherman, 2006; Kates, Erikson, Pijawka, & Bowden, 1977). This entails making quick decisions to reduce uncertainty and guide individual decision-making to keep businesses alive, minimize population loss, house people whose homes were destroyed, and minimize the overall sense of loss experienced by the community (Boyd, 2014; Olshansky, 2005, 2006).

At the same time, planners are also expected to implement improvements to reduce and manage future risk, so that the same degree of destruction and upheaval will not be repeated (Beatley, 1998; Boyd, 2014; Ehrenfeucht & Nelson, 2011; Glavovic, 2008; Kates, 1977; Mileti, 1999; Olshansky, 2005; Schwab, 2014). While many researchers identify the time immediately following a disaster as a window of opportunity to implement significant changes (Schwab, Topping, Eadie, Deyle, & Smith, 1998), residents and other stakeholders often resist widespread change, seeking instead to hurriedly recreate the city as it was before the disaster (Olshansky, 2006). Planners thus face a unique tension in timing between the need to act quickly to get the city up and running and the need to develop an inclusive, deliberative recovery process to determine to best ways to reduce risk that are supported by the community.

Despite the formative challenges planners often face, planning can play a crucial role in the rebuilding and recovery of place, and hence the restoration of individual and community well-being. Under well-designed planning processes, a community can come together often across racial, class, or geographic lines to rebuild and become closer and stronger than ever before (Ganapati, 2009; Haas et al., 1977; Mileti, 1999; Norris, Baker, Murphy, & Kaniasty, 2005; Oliver-Smith, 1979; Prince, 1920). In involving those who have lost their places in planning for their reconstruction, planners can also provide the space for individuals and communities to work through their bereavement (see Cherry, 2020, for discussion of grief in connection with disaster-related losses), a prerequisite to recovery. Marris (1974) argues that when one experiences loss, including loss of place, "the working out of grief is the central, most urgent task" (p. 149). For the bereaved, he believes it is hard to find meaning and motivation until grief has run its course. If grief goes unacknowledged, if the bereaved are expected to immediately pick right up and move on, "the suppressed tensions will in the end prove more profoundly disruptive than the social conflicts which relieve them" (Marris, 1974, p. 103). Marris (1974, p. 82) further contends that the working out of grief depends upon the articulation of "conflicting impulses" – the desire to return to the past and the desire to forget about it entirely and embrace the future. Mastery of grief occurs when the bereaved are able to appreciate what was truly important about the relationship that was lost and "rehabilitate" this meaning for ongoing use (Marris, 1974, p. 34).

Other scholars have noted planning processes' immense potential for creating the space for the "conflicting impulses" described by Marris (1974, p. 82) to work themselves out (Baum, 1998; Baum, 1999; Erfan, 2017; Laurian & Inch, 2018).

Baum (1999, p. 11-12) argues that planning processes can create a "transitional space, between past and future," where communities can mourn their losses, find new meaning in the past, and contemplate the future. Erfan (2013, 2017) argues that the planning profession is uniquely situated to provide space for the reconciliation of past and future in traumatized communities. In her time working with a First Nations community on Vancouver Island in British Columbia, Erfan (2013, 2017) observed that community members were more likely to discuss their grief in planning meetings than in meetings that existed for the sole purpose of discussing grief and trauma. Planning meetings, Erfan notes, present an "ideal forum" for opening up about trauma and working through grief in ways that are not "too scary, too self-indulgent or too irrelevant to the material realities of everyday life" (2013, p. 272). Following Sandercock (1998), Erfan (2013) espouses the benefits of "therapeutic planning," described as "emotionally engaged planning, which intends to support a process of healing and reconstruction of meaning" (p. 37).

The Challenge of Achieving Planning's Potential: The Difficulty of Public Engagement

Disaster, although devastating, can create opportunities through the recovery planning process to advance a shared vision of a safer community. The creation of this vision is accomplished through robust public engagement, which is often required by law due to its use of public funds. Public engagement is first and foremost about listening. A well-designed, deliberate engagement process can build hope, trust, and confidence in government and community leaders and provide opportunities to acknowledge and address long-standing community challenges (Butterfield, 2014). However, actively and meaningfully engaging residents, particularly post-disaster, is a difficult task. Residents are often distressed and can be immobilized by the loss and the complexity of issues they face (Butterfield, 2014). And if mismanaged, recovery planning and public engagement can perpetuate long-standing power imbalances. As noted by Arnstein (1969) in her seminal work *A Ladder of Citizen Participation*, "there is a critical difference between going through the empty ritual of participation and having the real power needed to affect the outcome of the process" (p. 216). Meaningful public engagement is also about a redistribution of power that goes beyond allowing the powerful to claim that all sides were considered while maintaining the status quo. Recovery planning without true public engagement and empowerment has the potential to make those recovering feel distrustful of the process, unsure of what is happening, why, and when.

Post-disaster planning processes that involve the public also have the advantage of being able to draw upon local knowledge (Fischer, 2000). Rarely is it possible to amass data at a grain fine enough to understand nuances at the neighborhood or block level. However, residents of a community are often great knowledge-bearers, possessing far more detailed understandings about the social, environmental, and economic conditions of their neighborhood than those from outside (McCrea,

Walton, & Leonard, 2015). If engaged in a true knowledge-sharing process, residents can provide valuable location-specific information with planners in a way that meaningfully addresses place attachment, risk reduction, and future well-being (Oulahen & Doberstein, 2012; Sadiqi, Coffey, & Trigunarsyah, 2011).

The aftermath of a disaster is not the ideal time to initiate planning. As noted by Olshansky et al. (2008, p. 278), "the best preparation for recovery planning is to have active planning processes beforehand, including networks of well-established community organizations, lines of communication, and a variety of planning documents and tools." While this is not always possible, positioning recovery planning so that it aligns with community needs and supports community-wide recovery is critical. Planning processes that involve the public are more conducive to producing sustainable communities – in the environmental, social, and economic senses – than those that don't (Beatley, 1998; Fullilove, 1996; Godschalk, Kaiser, & Berke, 1998). Yet planners must incorporate the public in recovery planning processes from the very beginning to avoid formulating unsupported proposals (Alexander, 2004; Nelson et al., 2007; Olshansky et al., 2008).

People traumatized by the experience of a disaster – the violence and suddenness of the event, the evacuation process, and the ensuing uncertainty about the future of place – are likely to resist proposals that call for great change or are developed by experts behind closed doors (Ehrenfeucht & Nelson, 2011; Ganapati, 2009; Nelson et al., 2007). Nelson et al. (2007, p. 45) describe how the opportunity to rebuild New Orleans "safer and smarter" after Hurricane Katrina was lost when the initial planning efforts to restrict redevelopment and shrink the footprint of the city neglected to consult the public and were met with widespread resistance. In response to the backlash from residents of flooded neighborhoods, the mayor, who was in the midst of a reelection campaign, allowed all residents to rebuild in place, albeit with heightened vulnerability. The authors argue that in the aftermath of Hurricane Katrina, only transparent planning processes that involved the public from the beginning would have stood a chance of building support for any major rearrangement of the built environment.

It is important to understand the elements of a robust planning process and the need for local knowledge to answer the following questions (Butterfield, 2014):

- Why are we here and what problems do we need to solve both as a result of the disaster and historical inequities? (Clearly define problem or need.)
- What information do we need from the public to determine how to solve the problem? (Create a multidimensional system of stakeholder engagement tied to specific defined problems/outcomes.)
- What does the public need to fully identify, understand, and address the problem? (Determine how this can be a process of community healing.)

Due to resident mistrust and the messy nature of recovery planning, it is necessary that institutional support be strong (Irazabal & Neville, 2007; Nelson et al., 2007; Olshansky et al., 2008; Sandercock, 1998) and that professional planning expertise be asserted throughout the process (Chandrasekhar, Zhang, & Xiao, 2014; Fischer, 2000; Glavovic, 2008; Helling, 1998; Olshansky, 2006). In the immediate aftermath of a disaster, institutional support and planning expertise are needed to quickly restore basic infrastructure and soften the blow of the disaster (Ganapati, 2009; Irazabal & Neville, 2007; Walsh, 2007). During the actual planning process, involvement of professional planners is necessary to ensure that the solutions produced are implementable. Citizen participants, overtaken by nostalgia (Baum, 1999), may propose unattainable recovery alternatives. At some point, difficult decisions will have to be made that will leave some segments of the population dissatisfied. A poorly conceived participation process that lacks a recognized organizational structure risks dedicating staff time and money to a process that has no clear authority, purpose, or end (Butterfield, 2014; Nelson et al., 2007). Nelson et al. (2007) describe how the lack of a clear organizational structure for recovery planning in New Orleans after Hurricane Katrina led to a drawn-out process with convoluted outcomes (see Mosby, this volume, Chap. 13).

Though it is essential for the public to play an active and meaningful role in the post-disaster planning process, planning expertise is also necessary to ensure that the path of recovery creates a place worth living in. In the wake of large-scale disasters, communities are likely to "reconstitute in way that is vastly different from the pre-event era" (Kropf & Jones, 2014, p. 285). Urban design is a powerful tool in post-disaster recovery; in these situations, planners can use their expertise to design spaces that enhance place attachment and facilitate new opportunities for face-to-face interaction and community building (Ganapati, 2009). Planners can help a community to "abstract what was fundamentally important" about the pre-disaster place and "rehabilitate it" into something new and beautiful (Marris, 1974, p. 34).

Stories about communities working to recover from large-scale disasters are often celebrated as tales of collective redemption. In practice, however, these processes are messy and rife with challenges. For starters, planners who facilitate post-disaster participatory planning processes must wear many hats – expert, mediator, negotiator, organizer (Forester, 1999). This reality divides attention and can be exhausting, particularly when planners are themselves also impacted residents of the community. There will be pressure to complete the planning process quickly, as an extended planning process undermines the prospect of a quick recovery. While it is generally thought that those who experience place loss are eager to participate in recovery planning (Ganapati, 2009; Trioni, 2015), participants may also view planners as opponents in this process (Baum, 1999). There is often widespread distrust of planning institutions among disaster-affected populations. Past transgressions (Olshansky et al., 2008) may elevate current distrust. Further, distrust may be heightened if government disaster response is perceived as lacking (Ganapati, 2009). Marris (1974) describes how groups experiencing communal loss often need to identify themselves in opposition to institutional agents of change for the negotiation of change to begin. To people uncertain of the future of the place they consider their home, planners may be seen as the enemy due to their association with change. Baum (1999) argues that citizen participants may engage planners in open conflict as a means of managing anxiety about change. Planners, he argues, should not suppress this conflict but rather use it as an opportunity to help people discuss their fears – to come to grips with the past and begin to conceptualize the future. However, the tension between speed and deliberation comes back into play here, as too much conflict between planners and residents can prolong the planning processes to the point that its effectiveness is diminished.

Organizing equitable participation in process is difficult, as those with money and education are far more likely to participate than those without (Chandrasekhar et al., 2014; Glavovic, 2008). This inequity in participation is dramatized further in post-disaster situations, as disasters tend to exacerbate previous inequalities (see Pelling, 2003; Gotham & Powers, 2015; Mosby, this volume, Chap. 13). Once the planning process has begun, there are many competing voices to be heard, and it often proves difficult to build consensus (Kropf & Jones, 2014; Murakami & Wood, 2014; Oulahen & Doberstein, 2012). And as the dust settles, divisions within the community often resurface stronger than before, as the needs of one's self and family may conflict with the needs of the community as a whole (Oliver-Smith, 1979, 1996). Mounting physical and emotional fatigue combined with continued scarcity of resources creates fertile ground for interpersonal strife (Norris et al., 2005). The appearance of substantive forms of aid from outside the community often serves to exacerbate, not mend, these conflicts, as it is inevitable that some will feel that resources have been unfairly distributed (Mayer, Running, & Bergstrand, 2015; Morris, Grattan, Mayer, & Blackburn, 2013; Oliver-Smith, 1979).

Given these tensions, post-disaster participatory planning initiatives rarely meet all of their lofty goals, widening the gap between theory and practice (Davidson, Johnson, Lizarralde, Dikmen, & Sliwinski, 2007). Planners must learn to set realistic goals when undertaking these processes and work to artfully manage the tension between "speed and deliberation" described by Olshansky (2006). If properly done, planners can create a process to envision desirable futures and deliberate on alternatives within a range of plausible and advantageous outcomes.

Participatory decision-making is critical. But planners also need to recognize the limitations of participatory processes in affecting decisions beyond their control. Gotham and Greenberg (2014) illustrate this in the design of federal disaster recovery programs implemented in New York after 9/11 and New Orleans after Hurricane Katrina. Uneven recovery aid exacerbated preexisting socioeconomic disparities, resulting in inequitable recovery outcomes. Despite highly engaged residents and grassroots activists in both New York and New Orleans, displacement and insufficient recovery resources weakened already lagging neighborhoods and increased their vulnerability to future crises (Gotham & Greenberg, 2014; Mosby, this volume, Chap. 13).

Conclusion

The disorientation experienced by survivors in the immediate aftermath of disasters does not evaporate quickly. Large-scale disasters tend to leave prolonged periods of uncertainty in their wake, as places and the social relationships embedded in them cannot readily be reestablished (Kytola, Cherry, Marks, & Hatch, 2015; Marris, 1974, p. 57). The uncertainty that necessarily follows large-scale, place-altering disasters can lead to a multiplying of the traumatic, grief-inducing, and identity-threatening effects of living through a disaster (Goto, Wilson, Kahana, & Slane, 2006; Kates, 1977; Kropf & Jones, 2014; Shalev & Ursano, 2003; Trioni, 2015; Wisner, Blaikie, Cannon, & Davis, 2003). Communities whose places are destroyed or radically altered by disasters deal with the side effects for years, if not generations, after the actual disaster (e.g., Cox & Perry, 2011; Morris et al., 2013). However, this is not to say that those who lose their places to disasters must necessarily begin an endless downward spiral of despair. The potential for rehabilitation does exist, and there are opportunities to be found. Leach (2005) in his observations of post-9/11 New York argues that collective loss can serve to reinforce collective identity. This process of reconstituting and reinforcing collective identity can be brought about through public involvement in post-disaster planning processes.

In working together to enable residents to plan for the reconstruction of places they have lost, planners can help residents begin to reshape and rebuild their attachment to place (Anguelovski, 2013). While researchers have investigated place attachment largely in terms of emotional bonds to the residence or neighborhood (Manzo, 2003), place meanings and attachments exist at multiple levels (Brown & Perkins, 1992; Qian, Zhu, & Liu, 2011). In some cases, the significance of the city or region is as salient or more salient than home and the neighborhood (Cuba & Hummon, 1993; Hidalgo & Hernandez, 2001; Lewicka, 2008) and that attachment to place at the city level positively influences more dimensions of well-being than neighborhood attachment (Rollero & De Piccoli, 2010). In attempts to restore normalcy, residents may be more concerned with returning to the same city or region than with returning to the exact same neighborhood. In post-Hurricane Katrina New Orleans, flooding had altered residents' attachments to their homes and neighborhoods. Given the complete destruction of the physical and social landscape, residents had to adjust their expectations of what they could return to (Mosby, this volume, Chap. 13; Nelson, 2014). Realizing that their homes and neighborhoods would never be restored to a pre-disaster state, many focused on the importance of returning to New Orleans (Nelson, 2014).

Planners need to acknowledge the loss and grief residents face (Cherry, 2020), recognize the transformative potential of planning to create a space to work through their grief and envision their future, and strive to overcome the challenges of participation to establish an open and inclusive process to deliberate fair and conceivable outcomes. Planners also need to recognize the limits of participation, identifying decisions beyond resident control and supporting residents in calling for broader, more inclusive post-disaster visions to influence the pace and pattern of recovery and expand opportunities and restore well-being for all residents of disaster-impacted communities.

References

Agnew, J. (1987). *Place and politics: The geographical mediation of state and society*. Boston and London: Allen and Unwin.

Albrecht, G., Sartore, G. M., Connor, L., Higginbotham, N., Freeman, S., Kelly, B., et al. (2007). Solastalgia: The distress caused by environmental change. *Australasian Psychiatry, 15*(suppl 1), S95–S98.

Alexander, D. (2004, April). *Planning for post-disaster reconstruction*. Paper presented at I-Rec 2004 International Conference: Improving Post-Disaster Reconstruction in Developing Countries. Montreal. Retrieved from http://www.grif.umontreal.ca/pages/papers2004/Paper%20-%20Alexander%20D.pdf.

Altman, I., & Low, S. (1992). *Place attachment*. New York: Springer.

Amundsen, H. (2012). Illusions of resilience? An analysis of community responses to change in northern Norway. *Ecology and Society, 17*(4). Retrieved from https://rs.resalliance.org/vol17/iss4/art46/

Amundsen, H. (2015). Place attachment as a driver of adaptation in coastal communities in Northern Norway. *Local Environment, 20*(3), 257–276.

Anguelovski, I. (2013). From environmental trauma to safe haven: Place attachment and place remaking in three marginalized neighborhoods of Barcelona, Boston, and Havana. *City & Community, 12*(3), 211–237.

Appleyard, D. (1981). *Livable streets*. Berkeley, CA: University of California Press.

Arnstein, S. R. (1969). A ladder of citizen participation. *JAIP, 35*(4), 216–224.

Atherton, T. L. (2017). Mitigating distrust: Trauma and redevelopment in Eastwick, Philadelphia. Retrieved from: http://udspace.udel.edu/handle/19716/23534

Baum, H. S. (1998). Ethical behavior is extraordinary behavior; it's the same as all other behavior: A case study in community planning. *Journal of the American Planning Association, 64*(4), 411–423.

Baum, H. S. (1999). Forgetting to plan. *Journal of Planning Education and Research, 19*(1), 2–14.

Beatley, T. (1998). The vision of sustainable communities. In R. J. Burby (Ed.), *Cooperating with nature* (pp. 233–262). Washington, D.C.: Joseph Henry Press.

Boyd, A. (2014). Long term recovery planning: Goals and policies. In *PAS report 576: Planning for post-disaster recovery: Next generation* (chapter 5). Retrieved from https://www.fema.gov/media-library-data/1425503479190-22edb246b925ba41104b7d38eddc207f/APA_PAS_576.pdf

Brown, B. B., & Perkins, D. D. (1992). Disruptions in place attachment. In I. Altman & S. Low (Eds.), *Place attachment* (pp. 279–304). New York: Plenum Press.

Burley, D. M. (2010). *Losing ground: Identity and land loss in coastal Louisiana*. Jackson, MS: University Press of Mississippi.

Butterfield, C. (2014). *Planning for post-disaster recovery briefing papers: Public engagement in recovery planning*. Report for the American Planning Association. Retrieved from https://www.planning.org/research/postdisaster/

Chandrasekhar, D., Zhang, Y., & Xiao, Y. (2014). Nontraditional participation in disaster recovery planning: Cases from China, India, and the United States. *Journal of the American Planning Association, 80*(4), 373–384.

Cherry, K. E. (2020). *The other side of suffering: Finding a path to peace after tragedy*. New York, NY: Oxford University Press.

Cox, R. S., & Perry, K. M. E. (2011). Like a fish out of water: Reconsidering disaster recovery and the role of place and social capital in community disaster resilience. *American Journal of Community Psychology, 48*, 395–411.

Cresswell, T. (2004). Defining place. In M. Hawley & A. Fitzsimmons (Eds.), *Critical encounters with texts: Finding a place to stand*. Boston: Pearson.

Cuba, L., & Hummon, D. M. (1993). A place to call home: Identification with dwelling, community and region. *The Sociological Quarterly, 34*, 111–131.

Davidson, C. H., Johnson, C., Lizarralde, G., Dikmen, N., & Sliwinski, A. (2007). Truths and myths about community participation in post-disaster housing projects. *Habitat International, 31*, 100–115.

Diaz, J. O. P. (2013). Recovery: Re-establishing place and community resilience. *Global Journal of Community Psychology Practice, 4*(3), 1–10.

Dugan, B. (2007). Loss of identity in disaster: How do you say goodbye to home? *Perspectives in Psychiatric Care, 43*(1), 41–46.

Ehrenfeucht, R., & Nelson, M. (2011). Planning, population loss and equity in New Orleans after Hurricane Katrina. *Planning, Practice & Research, 26*(2), 129–146.

Eisenbruch, M. (1991). *From post-traumatic stress disorder to cultural bereavement: Diagnosis of Southeast Asian refugees* [author manuscript]. Retrieved from http://www.eisenbruch.com/about_us/Eisenbruch%20-%20From%20PTSD%20to%20cultural%20bereavement.pdf

Erfan, A. (2013). *An experiment in therapeutic planning: Learning with the Gwa'sala-'Nakwaxda'xw first nations* (unpublished doctoral dissertation). The University of British Columbia, Vancouver, BC.

Erfan, A. (2017). Confronting collective traumas: An exploration of therapeutic planning. *Planning Theory & Practice, 18*(1), 34–50.

Erikson, K. T. (1976). *Everything in its path: Destruction of community in the Buffalo Creek flood.* New York: Simon and Schuster.

Fischer, F. (2000). *Citizens, experts, and the environment.* Durham, NC: Duke University Press.

Flora, C., & Flora, J. (1996). Creating social capital. In W. Vitek & W. Jackson (Eds.), *Rooted in the land: Essays on community and place* (pp. 217–219). New Haven, CT: Yale University Press.

Forester, J. (1999). *The deliberative practitioner: Encouraging participatory planning processes.* Cambridge, MA: The MIT Press.

Friedmann, J. (2010). Place and place-making in cities: A global perspective. *Planning Theory & Practice, 11*(2), 149–165.

Fullilove, M. T. (1996). Psychiatric implications of displacement: Contributions from the psychology of place. *American Journal of Psychiatry, 153*, 1516–1523.

Fullilove, M. T. (2004). *Root shock: How tearing up city neighborhoods hurts America, and what we can do about it.* New York: Random House.

Fussell, E. (2012). Help from family, friends, and strangers during hurricane Katrina: Finding the limits of social networks. In L. Weber & L. Peek (Eds.), *Displaced: Life in the Katrina diaspora* (pp. 150–166). Austin, TX: University of Texas Press.

Ganapati, N. E. (2009). Rising from the rubble: Emergence of place-based social capital in Gölcük, Turkey. *International Journal of Mass Emergencies and Disasters, 27*(2), 127–166.

Gans, H. (1968). *People and plans: Essays on urban problems and solutions.* New York: Basic Books.

Glavovic, B. C. (2008). Sustainable coastal communities in the age of coastal storms: Reconceptualising coastal planning as 'new' naval architecture. *Journal of Coastal Conservation, 12*(3), 125–134.

Godschalk, D. R., Kaiser, E. J., & Berke, P. R. (1998). Integrating heard mitigation and local land use planning. In R. J. Burby (Ed.), *Cooperating with nature* (pp. 85–118). Washington, D.C.: Joseph Henry Press.

Gotham, K., & Powers, J. (2015). Building resilience: Social capital in disaster recovery. *Contemporary Sociology: A Journal of Reviews, 44*(1), 30–31.

Gotham, K. F., & Greenberg, M. (2014). *Crisis cities: Disaster and redevelopment in New York and New Orleans.* New York: Oxford University Press.

Goto, T., Wilson, J. P., Kahana, B., & Slane, S. (2006). The Miyake Island volcano disaster in Japan: Loss, uncertainty, and relocation as predictors of PTSD and depression. *Journal of Applied Social Psychology, 36*(8), 2001–2026.

Haas, J. E., Trainer, P. B., Bowden, M. J., & Bolin, R. (1977). Reconstruction issues in perspective. In J. E. Hass, R. W. Kates, & M. J. Bowden (Eds.), *Reconstruction following disaster* (pp. 25–68). Cambridge, MA: MIT Press.

Hayden, D. (1995). *The power of place: Urban landscapes as public history*. Cambridge, MA: The MIT Press.
Helling, A. (1998). Collaborative visioning: Proceed with caution!: Results from evaluating Atlanta's Vision 2020 project. *Journal of the American Planning Association, 64*(3), 335–349.
Hess, J. J., Malilay, J. N., & Parkinson, A. J. (2008). Climate change: The importance of place. *American Journal of Preventative Medicine, 35*(5), 468–478. Retrieved from: https://www.ajpmonline.org/article/S0749-3797(08)00689-2/fulltext
Hidalgo, M. C., & Hernandez, B. (2001). Place attachment: Conceptual and empirical questions. *Journal of Environmental Psychology, 21*, 273–281.
Irazabal, C., & Neville, J. (2007). Neighbourhoods in the lead: Grassroots planning for social transformation in post-Katrina New Orleans? *Planning Practice and Research, 2007*, 1–30.
Jacobs, J. (1961). *The death and life of great American cities*. New York: Random House.
Kaniasty, K., & Norris, F. H. (2004). Social support in the aftermath of disasters, catastrophes, and acts of terrorism: Altruistic, overwhelmed, uncertain, antagonistic, and patriotic communities. In R. J. Ursano, A. E. Norwood, & C. S. Fullerton (Eds.), *Bioterrorism: Psychological and public health interventions* (pp. 200–229). Cambridge, UK: Cambridge University Press.
Kates, R. W. (1977). Major insights: A summary and recommendations. In J. E. Hass, R. W. Kates, & M. J. Bowden (Eds.), *Reconstruction following disaster* (pp. 261–293). Cambridge, MA: MIT Press.
Kates, R. W., Colten, C. E., Laska, S., & Leatherman, S. P. (2006). Reconstruction of New Orleans after Hurricane Katrina: A research perspective. *PNAS, 103*(40), 14653–14660.
Kates, R. W., Erikson, N. J., Pijawka, D., & Bowden, M. J. (1977). Alternative pasts and futures. In J. E. Haas, R. W. Kates, & M. J. Bowden (Eds.), *Reconstruction following disaster* (pp. 207–259). Cambridge, MA: MIT Press.
Kropf, N. P., & Jones, B. L. (2014). When public tragedies happen: Community practice approaches in grief, loss, and recovery. *Journal of Community Practice, 22*(3), 281–298.
Kytola, K. L., Cherry, K. E., Marks, L. D., & Hatch, T. G. (2015). When neighborhoods are destroyed by disaster: Relocate or return and rebuild? In K. E. Cherry (Ed.), *Traumatic stress and long-term recovery: Coping with disasters and other negative life events* (pp. 211–229). New York: Springer.
Laurian, L., & Inch, A. (2018). On time and planning: Opening futures by cultivating a "Sense of Now". *Journal of Planning Literature, 34*(3), 267–285.
Leach, N. (2005). 9/11. In M. Crinson (Ed.), *Urban memory: History and amnesia in the modern city* (pp. 75–92). New York: Routledge.
Lewicka, M. (2008). Place attachment, place identity, and place memory: Restoring the forgotten city past. *Journal of Environmental Psychology, 28*, 209–231.
Madgin, R., Bradley, L., & Hastings, A. (2016). Connecting physical and social dimensions of place attachment: What can we learn from attachment to urban recreational spaces? *Journal of Housing and the Built Environment, 31*(4), 677–693.
Malone, P. A., Pomeroy, E. C., & Jones, B. L. (2011). Disoriented grief: A lens through which to view the experience of Katrina evacuees. *Journal of Social Work in End-of-Life & Palliative Care, 2*(3), 241–262.
Manzo, L. C. (2003). Beyond house and haven: Toward a revisioning of emotional relationships with places. *Journal of Environmental Psychology, 23*(1), 47–61.
Manzo, L., & Perkins, D. (2006). Finding common ground: The importance of place attachment to community participation and planning. *Journal of Planning Literature, 20*(4), 335–350.
Marris, P. (1974). *Loss and change*. London: Routledge & Kegan Paul.
Marshall, N., & Bishop, K. (2015). Is there any correlation between place-making and place attachment? A challenge for interdisciplinary research. *The International Journal of Interdisciplinary Social and Community Studies, 9*(3–4), 1–10.
Mayer, B., Running, K., & Bergstrand, K. (2015). Compensation and community corrosion: Perceived inequalities, social comparisons, and competition following the Deepwater Horizon oil spill. *Sociological Forum, 30*(2), 369–390.

McCrea, R., Walton, A., & Leonard, R. (2015). Developing a model of community wellbeing and resilience in response to change. *Social Indices Research, 129*, 195–214.
Mileti, D. S. (1999). *Disasters by design: A reassessment of natural hazards in the United States*. Washington, D.C.: John Henry Press.
Morris, J. G., Grattan, L. M., Mayer, B. M., & Blackburn, J. K. (2013). Psychological responses and resilience of people and communities impacted by the Deepwater Horizon oil spill. *Transactions of the American Clinical and Climatological Association, 124*, 191–201.
Murakami, K., & Wood, D. W. (2014). Planning innovation and post-disaster reconstruction: The Case of Tohoku, Japan. *Planning Theory & Practice, 15*(2), 237–265.
Nelson, M. (2014). Using land swaps to concentrate redevelopment and expand resettlement options in post-Hurricane Katrina New Orleans. *Journal of the American Planning Association, 80*(4), 426–437.
Nelson, M., Ehrenfeucht, R., & Laska, S. (2007). Planning, plans, and people: Professional expertise, local knowledge, and governmental action in post-hurricane Katrina New Orleans. *Cityscape, 9*(3), 23–52.
Norris, F. H., Baker, C. K., Murphy, A. D., & Kaniasty, K. (2005). Social support mobilization and deterioration after Mexico's 1999 flood: Effects of context, gender, and time. *American Journal of Community Psychology, 36*(1–2), 15–28.
Oliver-Smith, A. (1979). Post disaster consensus and conflict in a traditional society: The 1970 avalanche of Yungay, Peru. *Mass Emergencies, 4*, 39–52.
Oliver-Smith, A. (1996). Anthropological research on hazards and disasters. *Annual Review of Anthropology, 25*, 303–328.
Olshansky, R. B. (2005, October). *How do communities recover from disaster? A review of current knowledge and an agenda for future research*. Symposium presented at 46th Annual Conference of the Association of Collegiate Schools of Planning, Kansas City.
Olshansky, R. B. (2006). Planning after hurricane Katrina. *Journal of the American Planning Association, 72*(2), 147–153.
Olshansky, R. B., Johnson, L. A., Horne, J., & Nee, B. (2008). Longer view: Planning for the rebuilding of New Orleans. *Journal of the American Planning Association, 74*(3), 273–287.
Osofsky, H. J., Osofsky, J. D., Kronenberg, M., Brennan, A., & Hansel, T. C. (2009). Posttraumatic stress symptoms in children after hurricane Katrina: Predicting the need for mental health services. *American Journal of Orthopsychiatry, 79*(2), 212–220.
Oulahen, G., & Doberstein, B. (2012). Citizen participation in post-disaster flood hazard mitigation planning in Peterborough, Ontario, Canada. *Risk, Hazards & Crisis in Public Policy, 3*(1), 1–26.
Pelling, M. (2003). *The vulnerability of cities: Natural disasters and social resilience*. Sterling, VA: Earthscan Publications.
Prince, S. (1920). Catastrophe and social change. *Studies in History, Economics, and Public Law, 94*(1), 118–140.
Proshansky, H. (1978). The city and self-identity. *Environment and Behavior, 10*(2), 147–169.
Qian, J., Zhu, H., & Liu, Y. (2011). Investigating urban migrants sense of place through a multiscalar perspective. *Journal of Environmental Psychology, 31*(2), 170–183.
Relph, E. (1976). *Place and placelessness*. London: Pion Limited.
Rollero, C., & De Piccoli, N. (2010). Does place attachment affect social well-being? *Revue Europeene De Psychologie Apliquee, 60*, 33–238.
Sadiqi, Z., Coffey, V., & Trigunarsyah, B. (2011). Post-disaster housing reconstruction: Challenges for community participation. In *Proceedings of the International Conference on Building Resilience: Interdisciplinary approaches to disaster risk reduction, and the development of sustainable communities*, Heritance Kandalama, Sri Lanka.
Sandercock, L. (1998). *Cosmopolis II: Mongrel cities in the 21st century*. London: Continuum.
Scannell, L., Cox, R. S., Fletcher, S., & Heykoop, C. (2016). 'That was the last time I saw my house': The importance of place attachment among children and youth in disaster contexts. *American Journal of Community Psychology, 58*, 158–173.

Schwab, J., Topping, K., Eadie, C., Deyle, R., & Smith, R. (1998). *Planning for post-disaster recovery and reconstruction* (PAS report no. 483/484). Chicago, IL: American Planning Association.

Schwab, J. C. (2014). Next steps in creating resilient communities. In *PAS Report 576: Planning for post-disaster recovery: Next generation* (chapter 8). Retrieved from https://www.fema.gov/media-library-data/1425503479190-22edb246b925ba41104b7d38eddc207f/APA_PAS_576.pdf

Shalev, A., & Ursano, R. (2003). *Mapping the multidimensional picture of acute responses to traumatic stress*. Oxford, UK: Oxford University Press.

Stokols, N., & Shumaker, S. A. (1981). People in places: A transactional view of settings. In J. Harvey (Ed.), *Cognition, social behavior, and the environment* (pp. 441–488). Hillsdale, NJ: Erlbaum.

Tester, G., Ruel, E., Anderson, A., Reitzes, D. C., & Oakley, D. (2011). Sense of place among Atlanta public housing residents. *Journal of Urban Health, 88*(3), 436–453. Retrieved from https://europepmc.org/articles/pmc3126922

Trentelman, C. (2009). Place attachment and community attachment: A primer grounded in the lived experience of a community. *Sociologist, Society & Natural Resources, 22*(3), 191–210.

Trioni, M. (2015). Disastrous publics: Counter-enactments in participatory experiments. *Science, Technology, & Human Values, 40*(4), 564–587.

Tuan, Y. (1974). *Topophilia: A study of environmental perception, attitudes, and values*. New York: Columbia University Press.

Tuan, Y. (1975). Place: An experiential perspective. *Geographical Review, 65*(2), 151–165.

Tuan, Y. (1977). *Space and place: The perspective of experience*. Minneapolis, MN: University of Minnesota Press.

Ujang, N. (2012). Place attachment and continuity of urban place identity. *Procedia – Social and Behavioral Sciences, 49*(1), 156–167.

Walsh, F. (2007). Traumatic loss and major disasters: Strengthening family and community resilience. *Family Process, 46*, 207–227.

Williams, D. R., & Patterson, M. E. (2008). Place, leisure, and well-being. In J. Eyles & A. Williams (Eds.), *Sense of place, health, and quality of life* (pp. 105–119). Aldershot, UK: Ashgate.

Wisner, B., Blaikie, P., Cannon, T., & Davis, I. (2003). *At risk: Natural hazards, people's vulnerability, and disasters* (2nd ed.). New York: Routledge.

Chapter 15
Community Vulnerabilities and Wellbeing after Disaster

Aimee Moles, Traci Birch, Yi Ling Chan, Dahyung Yang, Haojie Zhu, and Katie E. Cherry

Introduction

When disasters strike, disparities in community health and wellbeing become glaringly apparent. In few places has this been better demonstrated than south Louisiana, where the impacts of coastal land loss, anthropogenic hazards, and extreme weather occur at regular intervals. Communities with abundant resources require less assistance and are able to recover more quickly (Mosby, Birch, Moles, & Cherry, 2020). Less-prepared communities may be unable to access needed social, emotional, and financial resources and often have long-term negative consequences as a result (Cutter & Emrich, 2006). How can these disparities be addressed so that all communities can adapt and maintain wellbeing after a natural disaster? A key challenge for researchers and policymakers is to understand the role that the complex socio-economic and built environments play in disaster recovery for individuals and communities (see Kuriansky, Chap. 19, this volume). Developing quantitative methods to measure community members' needs, as well as modeling vulnerabilities within coastal and inland communities, can lead to best practice recommendations for

A. Moles (✉)
Social Research and Evaluation Center, Louisiana State University, Baton Rouge, LA, USA
e-mail: amoles@lsu.edu

T. Birch
LSU School of Architecture, Coastal Sustainability Studio, Louisiana State University, Baton Rouge, LA, USA

Y. L. Chan · H. Zhu
Department of Geography and Anthropology, Louisiana State University, Baton Rouge, LA, USA

D. Yang
Korea National Institute of Forest Science, Seoul, South Korea

K. E. Cherry
Department of Psychology, Louisiana State University, Baton Rouge, LA, USA

© Springer Nature Switzerland AG 2021

K. E. Cherry, A. Gibson (eds.), *The Intersection of Trauma and Disaster Behavioral Health*, https://doi.org/10.1007/978-3-030-51525-6_15

adaptation. Furthermore, a quantitative approach to these issues can enable appropriate long-range plans to ensure the best possible physical and emotional health for resilient communities (Cutter, 2014).

In this chapter, we examine the spatial, social, economic, and environmental conditions that are assumed to effect wellbeing and the ability of people living in vulnerable places to recover from the impact of hazards. We begin with an overview of Bronfenbrenner's (1979) ecological systems theory, which has been adapted to study the psychological impact of natural disasters (Kilmer & Gil-Rivas, 2010; Kytola, Cherry, Marks, & Hatch, 2015; Weems & Overstreet, 2009). An ecological systems approach, among other contextual theories, provides a useful framework for thinking about individuals nested within the broader social contexts of family, community, and cultural traditions and heritage, as discussed next. In the following section, we describe our community wellbeing index, which was based on archival data collected from multiple sources in a three-parish (county) region that was most severely impacted by the 2016 flooding in south Louisiana (see Cherry, Calamia, and Elliott, Chap. 8, this volume). In the last section, we focus on policy implications for disaster-affected communities in the years after natural disaster.

Theoretical Approach and Literature

Wellbeing is not just the absence of disease or illness. It is a complex and self-identified combination of physical, mental, emotional, political, and socioeconomic conditions that contribute to one's ability to cope with the normal stresses of life, work productively, and contribute to community (McCrea, Walton, & Leonard, 2015; Wiseman & Brasher, 2008). The circumstances that people are born into, grow up with, and live in greatly impact wellbeing and overall ability to cope and adapt over time for both individuals and communities as a whole. For the purposes of this work, we focus specifically on community wellbeing, which includes the broad range of self-identified cultural and social values that guide community development (Cox, Frere, West, & Wiseman, 2010). Coastal areas such as those in this study are subject to events and situations, which have the capacity to erode this sense of wellbeing along with the physical environment, leading to social disruption, decreased resilience, and increased vulnerability (Picou, Marshall, & Gill, 2004).

It is due to this complexity that we chose Bronfenbrenner's ecological systems theory as a starting point (see Banford Witting and Bagley, Chap. 5, this volume). Bronfenbrenner's (1977, 1979) ecological systems theory provides an integrative framework for studying factors that affect adaptation and wellbeing after a disaster (Kilmer & Gil-Rivas, 2010; Kytola et al., 2015; Weems & Overstreet, 2009). In particular, the assumption is that people function within multiple nested contexts or ecologies that vary in proximity to the individual. Proximal ecologies include family, school, and peers, among other influences close to a person. Distal ecologies include the farthest sources of influence, such as government, sociocultural values, and beliefs. Proximal and distal ecologies are assumed to exert bidirectional influence, where changes in one ecology may influence another as well as an individual's development (Kytola et al., 2015).

Bronfenbrenner's original formulation of ecological systems theory, which emphasized multiple nested ecologies at increasing levels of abstraction, later gave rise to his bioecological model with greater focus on characteristics of the individual over time as a critical determinant of development (Bronfenbrenner & Morris, 2006). An ecological framework for disaster research, which emphasizes proximal and distal factors affecting wellbeing in individuals and communities, has motivated research on topics as diverse as childhood wellness and community resilience after disaster (Pfefferbaum, Pfefferbaum, & Norris, 2010), to the emotional consequences of destruction and loss after Hurricane Katrina for disaster-exposed youth in New Orleans (Weems & Banks, 2015; Weems & Overstreet, 2009), and coping behaviors and wellbeing among Katrina-displaced older adults (Kamo, Henderson, & Roberto, 2011). Here we adopt an ecological systems perspective to guide our work on the development of an index of community wellbeing after the August 2016 flood in southeast Louisiana.

The Current Project: A Community Wellbeing Index

The geographic area chosen for the present research was based on the pattern of existing challenges common to coupled inland-coastal regions across the Gulf Coast and the southeastern US coastal region (see Fig. 15.1). In particular, the advent of climate change, disturbances of the natural environment by human activities, severe weather impacts, and an aging infrastructure are all testing the adaptive capacity of these coastal areas and the surrounding inland regions. In the present research, the geographic area for the index encompasses the Louisiana parishes (counties) of Ascension, East Baton Rouge, and Livingston, which are 302 mi^2, 472 mi^2, and 653 mi^2, respectively. The region is characterized by low elevation and slope that make the area susceptible to frequent flooding (Jacobsen, 2017). In fact, large water bodies and proximity to water are the defining characteristics of the region. The study area is bound by the Mississippi River to the west, and a large estuary system comprising Lakes Maurepas and Pontchartrain to the south. The Amite River and its largest tributary, the Comite River, divide the study area roughly in half. While the Mississippi River is contained within the US Corps of Engineers levee system, the Amite and Comite Rivers do not have similar controls to manage periodic flooding across these large floodplains (Gulf Engineers and Consultants, 2015). In addition, the proximity of the combined Maurepas/Pontchartrain estuary means eastern portions of the study area are also prone to storm surge flooding from tropical storm events.

Due to low elevations and proximity to water, many of these areas remained undeveloped or were used as agricultural lands for generations – either through necessity or regulation (Harland Bartholomew & Associates, 1946). However,

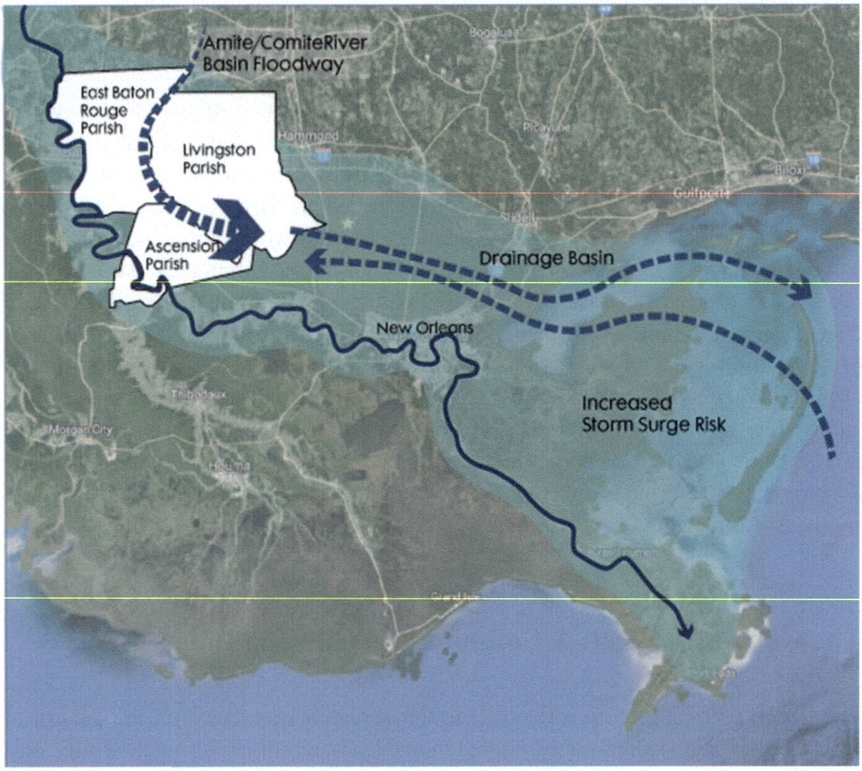

Fig. 15.1 Geographic area with shared coastal environment. (Courtesy of the LSU Coastal Sustainability Studio, 2017)

between 1980 and 2016 the region expanded and added 231,000 additional people, an increase of nearly 50% (U.S. Census Bureau, 1990, 2019). Ascension and Livingston Parishes saw their populations more than double as the region grew outward away from the City of Baton Rouge toward the Amite River floodplain (U.S. Census Bureau, 2019). Unlike many coastal areas of the state, which have implemented strict building regulations to elevate and reduce flooding, much of the growth in the study area targeted here has occurred as if it is not in a floodplain. This is a critical issue, because residential development is primarily slab-on-grade construction, meaning that there are few conservation provisions in place to maintain wetlands, and the addition of infrastructure and impervious surfaces have exacerbated flooding issues due to altered hydrologic regimes (Hardy, Mitchell, & Grueskin, 2018). Indeed, this was the case in 2016 when catastrophic flooding happened across south Louisiana, resulting in 13 deaths (see Cherry, Calamia, and Elliott, Chap. 8, this volume). Over 109,000 homes were directly impacted, which forced residents of flooded communities to find alternative housing (Terrell, 2016).

East Baton Rouge Parish is the largest parish in the state, with an estimated population of 440,956 (U.S. Census Bureau, 2019). The parish is home to the state's capital city of Baton Rouge, state government offices, and the central business district, and as a result is the oldest and most urbanized of the three parishes in the study area. Southeast of Baton Rouge is Ascension Parish with the incorporated cities of Donaldsonville and Gonzales. While historically agricultural, Ascension Parish has become one of the fastest growing residential parishes in the state. Directly to the east of the City of Baton Rouge, Livingston Parish contains the fast-growing cities of Denham Springs and Walker, which have also become commuter suburbs for the region. Further east, Livingston and Ascension maintain their rural character, but are still experiencing growth. Important comparisons can also be made among these locations due to their unique histories and their varied demographic characteristics.

While the effects of climate change vary across coastal and inland regions, these places and processes are not disconnected. Oftentimes climate impacts most closely associated with the coast, such as sea level rise and intensifying storms, are felt far inland. The Louisiana Capital region is not typically considered part of the coastal zone, except for the furthest eastern reaches. However, loss of coastal wetlands along estuarine shorelines, increased intensity of storm surge due to tropical storms, and high-volume rain events have all served to increase water in the landscape. In addition to migration from Baton Rouge to the suburbs, these parishes are also expected to serve as receiver communities as residents move inland from threatened coastal locations (Intergovernmental Panel on Climate Change, 2015; Peduzzi, Dao, Herold, & Mouton, 2009). However, if extreme weather events such as those that caused the August 2016 floods occur with higher frequency, risk for inland residents will begin to resemble that for coastal locations. Without careful planning and recognition of increased risk due to climate change, continued development in the floodplain will increase vulnerability to floods and other natural hazards (see Birch, Henry, and Nelson, Chap. 14, this volume). The multiple, dynamic processes of increased suburban sprawl, coastal migration, and flooding coupled with sea-level rise, land loss, and increased weather intensity make this region an important area for continued study and research.

The Current Project

In this project, the authors focused on the conceptualization and measurement of community wellbeing across a coupled coastal-inland region, which has been impacted by a range of slow-onset and sudden natural disasters. As previously noted, wellbeing is a complex combination of physical, mental, emotional, political and socio-economic conditions. As such, there cannot be just one indicator with which wellbeing can be measured (see also Cherry, Calamia, Birch, and Moles, Chap. 16, this volume). Consequently, we used a composite indicator made up of a combination of variables to create an overall measure of wellbeing.

This present index builds on the prolific work of social science researchers, including Colten, Kates, and Laska (2008), Cutter, Burton, and Emrich (2010), and

Burton (2015), who have primarily focused on the measurement of social vulnerabilities and disaster resilience through the development of composite indices. Composite indicators are a common way to communicate complex, observed relationships and phenomena to inform important stakeholders and policymakers (Cutter et al., 2010; Cutter, Boruff, & Shirley, 2003; Saisana & Cartright, 2007). There are numerous indices seeking to measure social processes and, in some way, capture facets of wellbeing. Possibly the best-known index measuring social conditions across a large geographic area is the Social Vulnerability Index (SOVI) (Cutter et al., 2003). Others include The Disaster Risk Index (Peduzzi et al., 2009) and The Prevalent Vulnerability Index (Cardona, 2005). Other indices such as the Environmental Vulnerability Index (Kaly, Pratt, & Mitchell, 2004) have focused on the vulnerability of the natural environment, which is an important aspect of this work. What the authors found in starting the present work is that while there are many examples, none included adequate measures of place-based risk factors such as development in floodplains, repetitive flood impacts, access to parks, or other factors that address quality of life and wellbeing. While the index reported here was developed for a small section of southeast Louisiana, we believe that it can easily be scaled up and replicated across similarly vulnerable coastal-inland regions' geographies. It is unique in its focus specifically on how to measure wellbeing in relationship to natural disasters.

Components of Wellbeing

As noted, a review of literature encouraged an ecological systems approach as the basis for the selection of variables and their categorization. Our variables were chosen according to three important considerations: (1) relevance to wellbeing based on social science literature, (2) scalability to the census tract level, and (3) availability of reliable, publicly available data. In addition to traditional quantitative sources, such as US Census data, proxy measures were employed for place-specific elements such as community health, access to open space, and density of roadways, among others (Cutter et al., 2008a; Schneiderbauer & Ehrlich, 2006). Out of the 100+ variables identified as potentially relevant, 70 met all three considerations and were included in the initial index. Variables were available in geographically diverse formats, including census tract, zip code, and land cover type, which were analyzed using ArcGIS (version 10.6) geo-mapping software and aggregated to the census tract level (DeMers, 2001). Indicators were standardized to percentages, densities, and per capita functions (see Cutter et al., 2003). Subsequently, each variable was rescaled to an identical range between 0 and 1; a value of 1 indicating the highest relative level of wellbeing, and 0 indicated the lowest. To guard against variables with significantly high intercorrelations, variables with Pearson's R-values greater than 0.70 were identified and eliminated. This resulted in 49 remaining variables, which were examined collectively and categorized into four classifications for a more nuanced examination of wellbeing through the ecological systems lens.

Table 15.1 contains a complete list of variables, their sources, and relevant literature citations. These classifications include community stress, economic health, environmental health (with natural environment and built environment as subcategories), and public safety. For each classification, a composite score was created by adding the values for each indicator in that census tract together and dividing by the number of variables. Similarly, each census tract received a composite score by combining the score for each classification and dividing the total by the number of categories.

The community stress classification includes personal demographic factors commonly attributed to social vulnerability and resilience. These demographics include measures such as increased ethnic/minority populations, lower education levels, increased elderly populations, and people with disabilities, the presence of which have been associated with decreased community resilience (Cutter et al., 2010). While aggregated to the community level, each of these indicators is associated with individual community members. In addition, this classification includes community-based protective factors, such as the number of civic and social advocacy organizations and access to childcare programs, which have also been associated with increased community resilience and wellbeing (Cutter et al., 2010). A higher density of these organizations and programs can indicate a stronger sense of community and provide stronger networks to allow a community to recover after a natural disaster. Social stressors such as social isolation can negatively impact individuals. In fact, there is evidence that distance to social networks can influence individual health (Corburn, 2015; Friel et al., 2011).

Economic health was designed to measure the community's fiscal stability, resource diversity and equity, and the exposure of economic assets. Key indicators include gross income, homeownership rates, and employment rates. Proxies for equity include percentage of the population spending more than 30% of their income on housing, and percentage of households with more than 1.5 people/bedroom. This classification includes community-based economic factors such as the number of traditional and payday lending institutions, each of which are proxies for access to financial tools and overall community resilience (Economic Innovation Group, 2017; Queste & Lauwe, 2006). This classification also included hazard-related financial stress related to a lack of flood insurance, which was of particular interest in the region after the 2016 flooding. It is known that lower income residents are disproportionately affected by disasters (Flanagan, Gregory, Hallisey, Heitgerd, & Lewis, 2011). With reduced access to financial resources, people are often unable to evacuate during an event or adequately recover afterwards. Household expenses often represent a larger portion of a low-income individual's total assets and for this reason, it is proportionately more expensive to replace or repair after a flood (Flanagan et al., 2011; Tierney, 2006). Families with a higher income level have greater safety nets and are able to withstand losses that may occur due to a flood. These greater financial resources enable them to "bounce back" and recover more efficiently than someone who cannot afford to repair a home, obtain alternate housing, or purchase new clothing after a flood (Cutter et al., 2003). However, even with safety nets, if residents do not have flood insurance their ability to bounce back and return to their lives is compromised (see Cherry et al., Chap. 16, this volume).

Table 15.1 Composite list of variables in the index

	Variable	Sources	Justification
Community stress			
1	% population that is not elderly	2016 ACS	Cutter et al. (2010)
2	% population that doesn't speak English as a second language	2016 ACS	Cutter et al. (2010)
3	% population without a disability	2016 ACS	Cutter et al. (2010)
4	% population that is not a minority	2016 ACS	Tobin (1999)
5	% population with at least a high school diploma	2016 ACS	U.S. Indian Ocean Tsunami Warning System Program (2007)
6	% of population that are <18	2016 ACS	Cutter et al. (2010)
7	Child care programs per 1000 population	Louisiana Department of Education	H. John Heinzlll Center (2002)
8	Ratio % college degree to % no high school diploma	2016 ACS	Cutter et al. (2010)
9	Religious organizations per 1000 population	Reference USA	Cutter et al. (2010)
10	Social advocacy organizations per 1000 population	Reference USA	Cutter et al. (2010)
11	% population with vehicle access	2016 ACS	Cutter et al. (2010)
12	% population with telephone access	2016 ACS	Cutter et al. (2010)
13	Community services (Historic Sites, Libraries, Museums) per 1000 population	Reference USA	Lochner, Kawachia, and Kennedy (1999)
14	Vacancy rate	2016 ACS	Economic Innovation Group (2017)
Economic health			
15	% of households spending >30% of income on housing	2016 ACS	Cox et al. (2010)
16	% homeownership	2016 ACS	Cutter et al. (2010)
17	% working-age population that is employed	2016 ACS	Cutter et al. (2010)
18	Per capita household income	2016 ACS	Tobin (1999)
19	Lending institutions per 1000 population	Reference USA	Queste and Lauwe (2006)
20	Payday loans per 1000 population	Reference USA	Economic Innovation Group (2017)
21	% workforce employed in professional occupations	2016 ACS	Cumming et al. (2005)
22	Hazard-related financial stress – flood insurance	Louisiana Watershed Resilience Study (FEMA)	Cox et al. (2010)

(continued)

Table 15.1 (continued)

Environmental health			
Built environment			
23	% population living in high-intensity urban areas (ratio of road density/ population density)	2016 ACS/DOT	Geis and Kutzmark (1995)
24	% of population with more than 1.5 occupants per room	2016 ACS	Economic Innovation Group (2017)
25	Density of single-family detached homes	2016 ACS	Economic Innovation Group (2017)
26	% land area that is developed open space	2011 NLCD	Geis and Kutzmark (1995)
27	Sidewalks	ADA compliant sidewalks from EBR Parish GIS	Corburn (2015)
28	Impervious surfaces	2011 NLCD	Lundgren and Jonsson (2012)
29	Dedicated cycling trails	Bike BR	Cox et al. (2010)
Natural environment			
30	% land area that does not contain erodible soils	USDA NCRS	Bradley and Grainger (2004)
31	% land area not in an inundation zone (100-year flood plain)	2016 FEMA	Cutter et al. (2008a)
32	% land area that is nondeveloped forest	2011 NLCD	Cutter et al. (2008a)
33	% land area with no wetland decline	2011 NLCD	Cutter et al. (2008a)
34	% land area with no land-cover/land-use change, 1992-2001	2001/2011 NLCD	United Nations Department of Economic and Social Affairs (2007)
35	% land area that is arable cultivated land	2011 NLCD	United Nations Department of Economic and Social Affairs (2007)
36	Access to parks (Access to areas of public open space, Appearance of public areas)	EBR Parish GIS, Ascension Parish GIS, and hand digitization	Cox et al. (2010)
37	Tree canopy	2011 NLCD	Corburn (2015)
Public Safety			
39	Presence of plumbing facilities	2016 ACS	Economic Innovation Group (2017)
40	Presence of complete kitchen facilities	2016 ACS	Economic Innovation Group (2017)
41	Doctors and medical professionals per 1000 population	2016 ACS	Cutter et al. (2010)

(continued)

Table 15.1 (continued)

42	% households covered by National Flood Insurance Program policies	2017 FEMA	Cutter et al. (2010)
43	% workforce employed in emergency services (firefighting, law enforcement, protection)	2016 ACS	Cutter et al. (2008b)
44	% fire, police, emergency relief services, and temporary shelters outside of hazard zones	Department of Homeland Security	U.S. Indian Ocean Tsunami Warning System Program (2007)
45	Principal arterial miles	2016 DOT	Cutter et al. (2010)
46	Flood depths	2016 FEMA	Suarez, Anderson, Mahal, and Lakshmanan (2005)
47	Violent crime rates per FBI uniform crime report definitions	2016_2017 crime counts from parish and city police	
48	Non-violent crime rates per FBI uniform crime report definitions	2016_2017 crime counts from parish and city police	
49	Death rates per Louisiana Department of Public Health	2015_2017 deaths from Louisiana Dept of Health	

CDC Centers for Disease Control and Prevention, *DOT* Department of Transportation, *LDH* Louisiana Department of Health, *RWJF* Robert Johnson Foundation Program

Environmental health considerations include indicators related to both built and natural environments. Built environment variables include factors such as percentage of impervious surfaces, road density, and bike path access, each of which serve as proxies for flood exposure and public health impacts. Natural environment considerations include density of forests, existing wetlands, and soil quality, each of which are indicators of reduced flood exposure. In addition to their importance to flooding, built and natural environment indicators play a role in psychological wellbeing (see Birch et al., Chap. 14, this volume). Urban sprawl, characterized by low-population densities on the edges of cities increases reliance on vehicles for transportation and makes walking and biking to destinations difficult (Arthurson, Lawless, & Hammet, 2016; New South Wales Health, 2009). Additionally, the distance from the urban core puts community members at a disadvantage during natural disasters, where rescuers may have to travel far to provide aid. A robust network of pedestrian sidewalks can promote physical activity and increase the chances of social interaction, improving wellbeing (Corburn, 2015). Furthermore, access to green spaces in parks can increase these social interactions, providing opportunities for exercise and access to nature. Green spaces and high densities of vegetation provide important ecosystem services, such as reducing the impacts of floods and other natural disasters.

The public safety classification was designed to capture relationships between people's communities and their overall public health. This classification proved the most difficult to populate, as there were gaps in data availability due to privacy concerns and the scale of available data. For example, psychosocial support services and prevalence of obesity, alcoholism and depression is only available at the parish (county) level, and is much less available in suburban communities versus urbanized areas. Ultimately, proxies for public health and safety, such as violent and nonviolent crime rates, death rates, plumbing and kitchen facilities, and access to doctors and medical professionals were used for measurement. We also included hazard-specific variables such as density of emergency services and shelters, and the percentage of households covered by flood insurance policies to better understand community ability to respond in the event of a disaster.

Results

After locating multiple data sources and processing these data geographically, we analyzed the resultant maps for both baseline levels of community wellbeing and changes in community wellbeing post flood using 2017 data if available. We studied both composite index and classification index results to see where there were risk and protective factors apparent, and to determine whether results fit our initial expectations. We expected areas with flood damage to show marked changes post flood, and that flood-adjacent communities could also experience change due to the movement of evacuating residents, housing shortages, loss of infrastructure, etc. The initial pre-flood composite index map shown in Fig. 15.2 is the integration of all 49 variables and demonstrates the variability from census tract to census tract over the entire three-parish area. Lighter areas had comparatively higher pre-flood composite scores and darker areas had comparatively lower pre-flood composite scores.

Pre-flood composite scores ranged from 0.32 to 0.51. Some patterns are apparent, many of which were expected. As can be seen in Fig. 15.2, inner-city urban areas, especially in northern East Baton Rouge Parish, have a lower wellbeing baseline due to high poverty and crime rates, as well as low densities of community amenities and limited financial resources. A portion of the rural parishes of Ascension and Livingston parishes also struggle due to a lack of bike and pedestrian infrastructure, limited emergency facilities, and a low number of community service providers despite high numbers of those in poverty or who are elderly. Higher wellbeing baselines are found in the central portion of the study area, which is generally more suburban in nature. Poverty levels as a whole are moderate in these communities, homeownership is high, and crime rates low. Similar variations appear across the region.

In the postflood composite index map (Fig. 15.3) we layered flood depths over the composite scores to look for patterns of change in wellbeing related to flood exposure. Because flooding was not an aspect of the preflood index, we did not include it

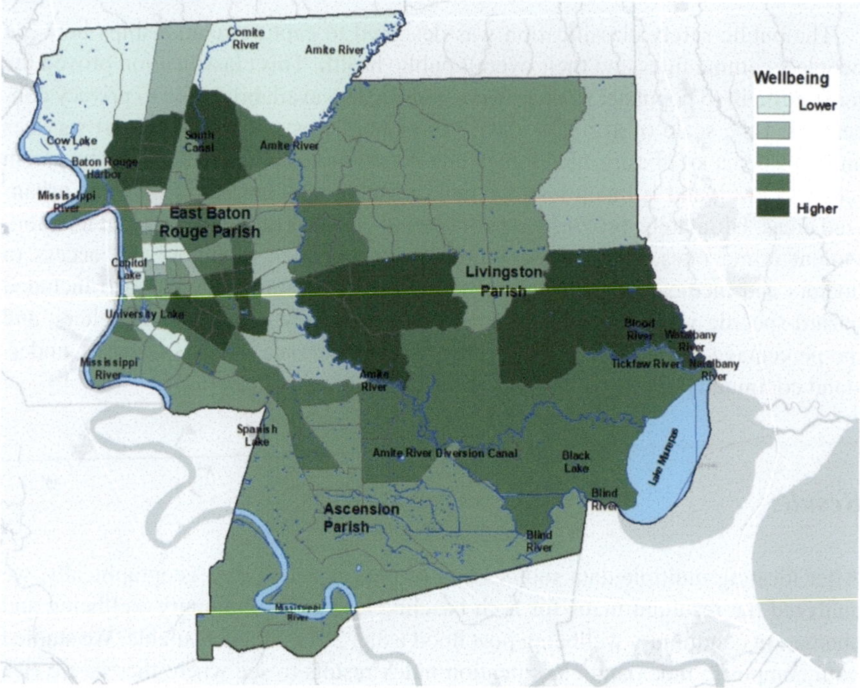

Fig. 15.2 Study area with pre-flood integrated variables

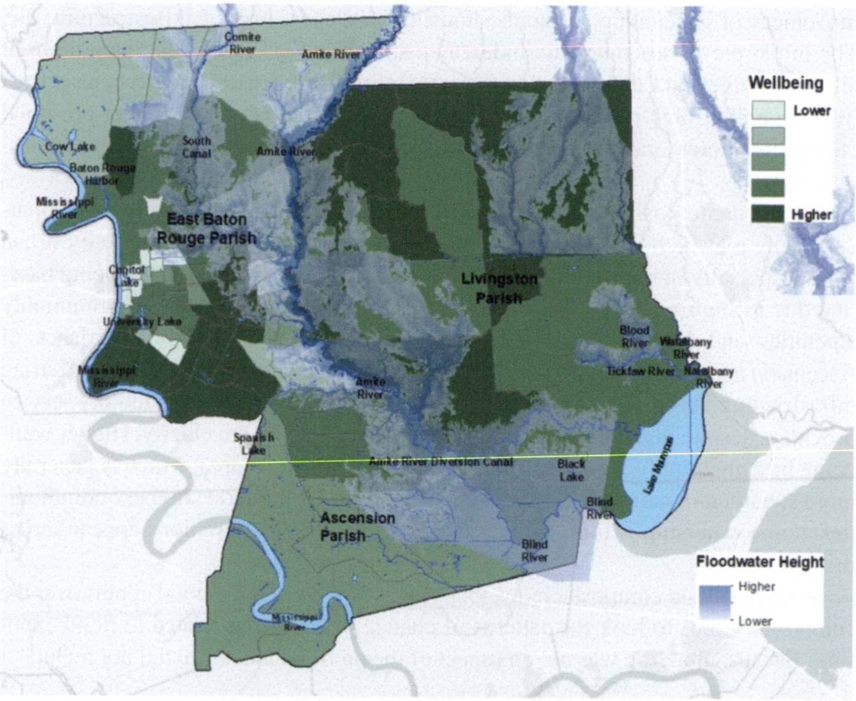

Fig. 15.3 Study area with postflood integrated variables

in the postflood index, but instead used it to visually inspect changes in the maps. Interestingly, postflood composite scores ranged from 0.30 to 0.57, similar to the preflood results. As can be seen in Fig. 15.3, the flood layer cuts through the center of the map along the Amite River as well as into North Baton Rouge along the Comite River. While mapping the area using index scores did help us visualize well-being, census tract boundaries limit the ability to see both the bigger picture and also the more nuanced details. The authors employed the process of regionalization to get around the impasse of census tract boundaries, which allows a closer examination of the individual communities within these groupings. Regionalization is a procedure that allows polygons (such as census tracts) to be combined into meaningful groups (Openshaw, 1977). While there are many algorithms for regionalization, all methods maximize both within-cluster similarities and between-cluster differences. Using GeoDa 1.12, an open source software package that conducts spatial data analysis and geo visualization, we conducted a regionalization procedure using the max-p algorithm (Duque et al., 2012). The max-p algorithm is particularly suited for choosing spatial regions that are both maximally different from each other and internally homogenous. The max-p algorithm requires a minimal threshold for a predefined attribute, in our case population. Each cluster in this analysis contains 10% of the study area's residents (6933 people), as this proved to be the optimal number for dividing the region into groups of similar census tracts with respect to our index. The resulting analysis produces both a cluster map defining each region with a different number and color, and a chart with each region's Z scores for all five index classifications. This allowed us to compare regions across the study areas based on the composite index and each of the five index classifications (see Fig. 15.4). For clarity, Fig. 15.5 displays the wellbeing clusters and their corresponding Z-scores to demonstrate where each group of communities excels or struggles based on how far above or below the study area median a particular cluster falls.

Geographic clusters are represented by the numbers 1–8 in Fig. 15.5 below the corresponding group of index values. Cluster one census tracts are located along the Mississippi River and historic core of the region, which also contains some of the most urbanized and densely populated communities. Pre-flood, this group of census tracts falls slightly below the median in all five index classifications, with economic health being the most noticeable challenge. Similarly, Cluster three is geographically adjacent to Cluster one, and has some similar density and urbanization characteristics, but is more residential in character. This area encompasses some of the original commuter suburbs developed in the 1950s, and has seen significant disinvestment over recent decades. These conditions manifest in the uniformly low scores for each of the index classifications in this cluster. Clusters two and four, on the other hand, are uniformly high in all five categories of the index. Somewhat separated physically, Cluster two is in a more rural setting with a number of small commuter suburbs and historic towns located outside of the urbanized core, cluster four is relatively dense, but suburban in character. Adjacent areas that have seen population loss and disinvestment, these areas have experienced stability and growth over recent decades, which is reflected in the higher-than-average index classification scores. Less uniform in nature are clusters five, six, seven, and eight. Cluster

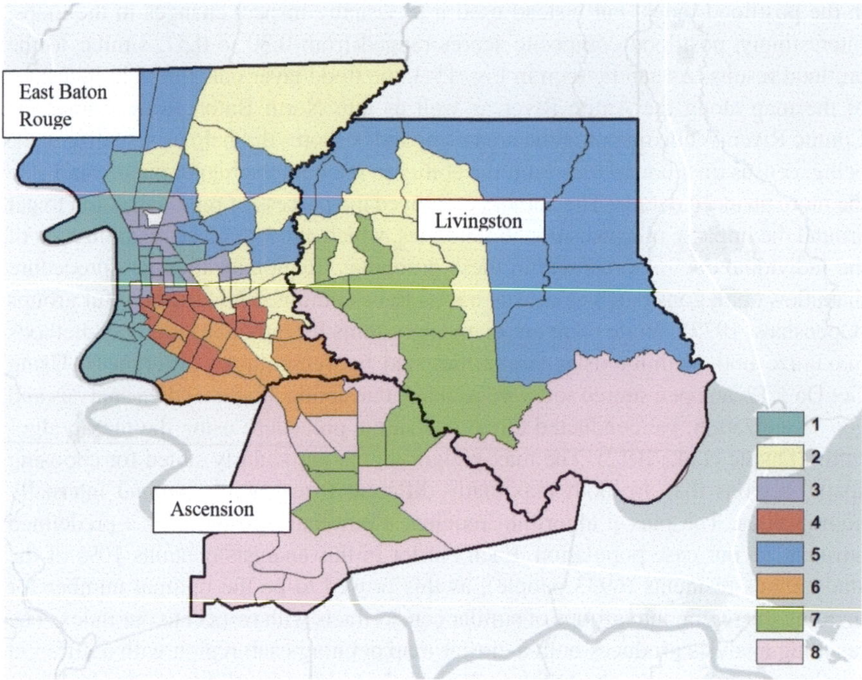

Fig. 15.4 Five index classifications with preflood integrated variables

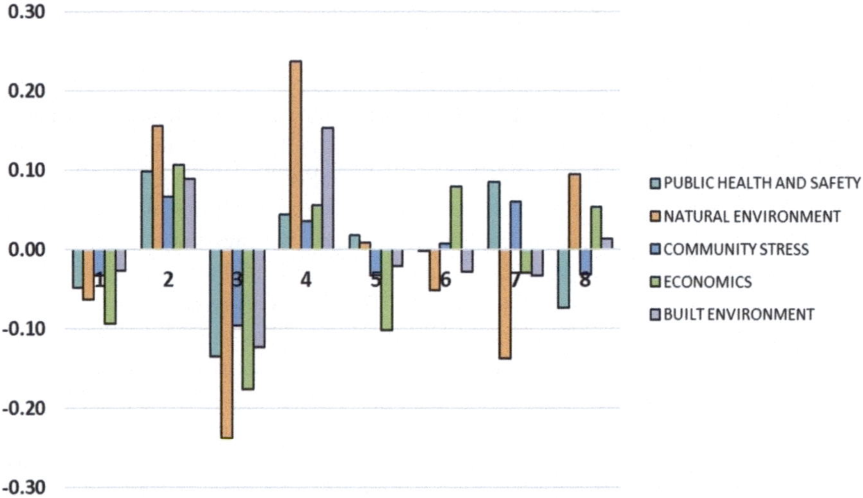

Fig. 15.5 Wellbeing clusters with preflood integrated variables

five tracts are not geographically contiguous, though all are in the northern portion of the study area. These tracts are lower in economic wellbeing and have higher community stress levels when compared to other communities in the study area. Housing costs as a percentage of income were higher than the median Z score in cluster five in 2015 as were levels of unemployment. Vehicle ownership in this non-urban location is lower than the median, all of which contribute to higher community stress scores. Cluster six tracts span the southern edge of East Baton Rouge Parish and portions of the towns of Prairieville and Gonzales in Ascension Parish. This grouping is at the median in community stress and public health and safety and above the median in economics. However, natural and built environments are challenges. These tracts border the Mississippi River and hold many other natural water sources, including bayous and wetlands. Wetland decline, erodible soil, and impervious surfaces as well as higher-than-median percentages of land use/land change contribute to the lower index scores. Cluster seven lies in the eastern portion of the study area, and straddles the Amite River through Livingston and Ascension Parishes. While community stress levels, public health, and safety scores were higher than the median, this cluster faces challenges economically as well as with aspects of the natural environment and the built environment. Poverty is higher than the median for this area as is unemployment, which negatively impacts the index score for these tracts. This cluster is situated in the federally determined 100-year flood zone and additionally has a higher-than-median percentage of erodible soils which impacts the index score for natural environment aspects. The built environment has limited bike paths or sidewalks for community use. Finally, Cluster eight contains most of Ascension Parish, excluding Prairieville and Gonzales, which were included in Cluster seven. This group of tracts performed above the median in all components of the index except for public health and safety and community stress. This area contains the towns of Donaldsonville, Sorrento, and St. Amant. Index scores in this area were impacted by a higher-than-typical number of homes without adequate plumbing and kitchen facilities as well as a lower number of medical professionals, shelters, and social welfare organizations. Violent crime rates per 1000 were higher than the median for the study area.

With a functional picture of the region's strengths and challenges before the flooding in 2016 (see Figs. 15.4 and 15.5), the next step was to examine these same areas for postflood changes. We took into consideration the communities most impacted by water inundation in 2016, as well as the indirect impact on other locations surrounding the flooding. Returning to the same clusters with data obtained from 2017 after the flooding, we produced two additional figures. Figure 15.6 compares composite index scores from the original preflood data with the postflood data. Economic health improved overall, but none of the other four components improved. In fact, community stress showed the biggest drop in scores indicating that these aspects of wellbeing were noticeably impacted (Fig. 15.6).

To permit a more nuanced analysis of these wellbeing indicators, we examined the individual component scores for each cluster using a bar graph to show how the clusters compare to each other, relative to the median for the entire geographic area

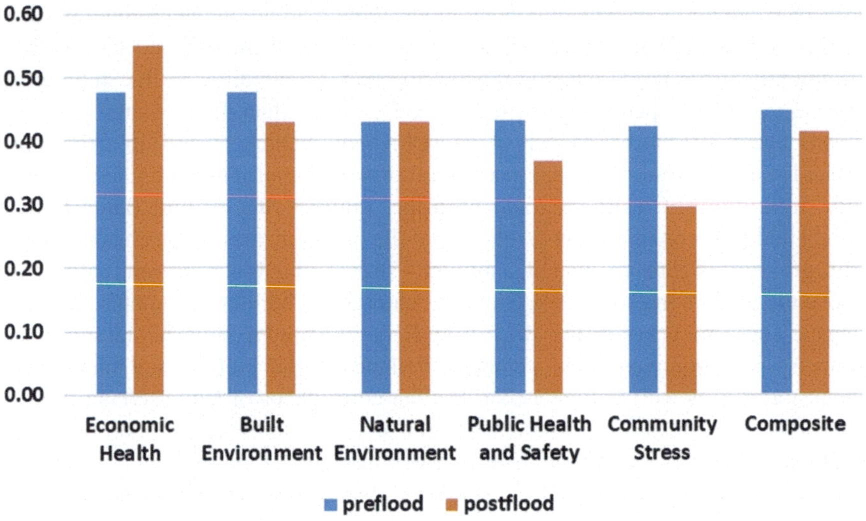

Fig. 15.6 Composite Index Scores pre- and postflood

(see Fig. 15.7). As in the previous figures, numbers 1–8 represent the original clusters.

As can be seen in Fig. 15.7, there is a much more varied distribution of strengths and weaknesses for each cluster. We examined these data for drivers of change demonstrated by the index scores for each of the geographic groupings. We found a complex mix of population shifts, new multifamily housing in several locations in the study area, and a continued outgrowth into Ascension and Livingston Parishes. Flooded areas experienced drops in population while communities just outside of these locations experienced both growth and crowding of individual homes, possibly because of families taking up temporary residence with friends and relatives as their homes were repaired. Additionally, some flooded locations lost community services, childcare, and religious organizations. Many of these losses occurred in communities highly dependent on these services. Pulling apart the complex threads of these changes and their resulting impact will require continued investigation. Ideally, input from community residents should be sought to help explain what has occurred based on personal experience. Ongoing data collection will assist with understanding more long-term trends and help to inform our understanding of community wellbeing in the face of changing climates.

Conclusion and Future Directions

In this chapter, we presented an index to measure community wellbeing before and after the 2016 flooding in south Louisiana. This index adds to existing work on proximal and distal variables that may impact wellbeing in individuals and

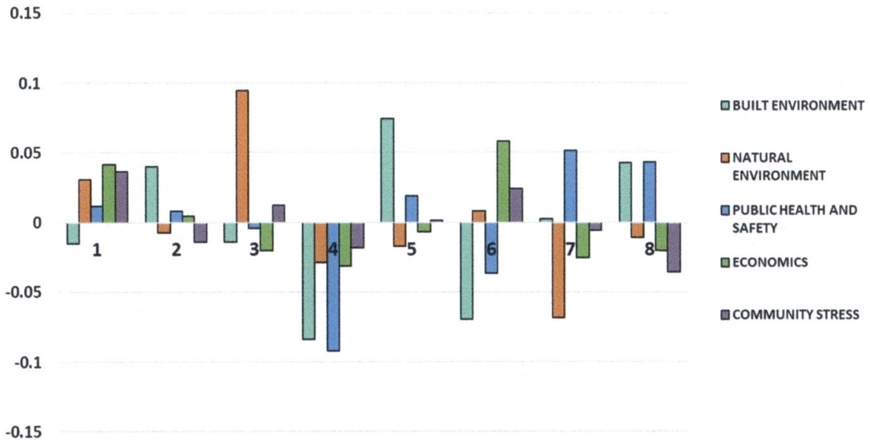

Fig. 15.7 Wellbeing clusters with preflood integrated variables

communities (Pfefferbaum et al., 2010), consistent with an ecological systems theory framework for disaster research (Kytola et al., 2015; Weems & Overstreet, 2009). Defining the forces driving the shifts we observed and quantifying them in a meaningful way are important measurement challenges necessary to inform changes from the preflood baseline and allow for specific interventions to shore up vulnerabilities and decrease risk from future disasters. In closing, we note the difficulties inherent in constructing an empirically based index of wellbeing, as well as policy implications of this work.

Challenges

In this research, our greatest challenges related to data availability. Ascension and Livingston Parish have fewer comprehensive datasets available in comparison to East Baton Rouge Parish. This difference is likely due to East Baton Rouge Parish's higher population and a more robust set of resources for data collection, such as the local research institutions. Thus, to successfully create an index across the three-parish area, we could not use the datasets that existed for just East Baton Rouge Parish.

One noteworthy limitation of this study concerned the lack of resources for health data collection at the census tract level. The health variables have a low standard deviation due to the larger scale we had to use for these. This indicates that all areas tend to be close to the mean so that those variables become less useful. Additionally, the composite indexes had to be developed with different numbers of variables, therefore we had to be cautious about overweighting any particular variable. Despite these difficulties, our index provides a starting point for measuring resilience and recovery. During the early stages of this project, we discovered that there were no readily available indicators of mental health (see Cherry et al., Chap. 16, this volume); therefore, the present index does not include mental health data. Continued refinement of the index and the addition of mental health data are important goals that await future research.

Policy Implications

The policy choices made by and for communities may have far-reaching consequences for community wellbeing due to the complex interplay of socioeconomic, cultural, and environmental factors. In reality, there are issues related to biological susceptibility, human behavior, unpredictable changes in the financial markets and other outside influences. The unknowns surrounding climate change add to the complexities of creating wise policies for the future given these uncertainties. As a result, the need for solid research upon which policies can be based is critical.

Whether a community can demonstrate adequate resilience following a disaster is effectively a function of pre-disaster wellbeing. Because many types of disasters cannot be prevented, practitioners are instead tasked with the need to predict and mitigate vulnerabilities while at the same time building and maintaining strong, equitable communities. Investing in strategies that build social and ecological capacity requires recognizing that population health functions as a system rather than multiple disconnected parts. As the authors developed this index, the importance of these interconnections became very apparent and complex in nature. The geographic setting, mix of community types, and historical background related to significant weather events provided us with a number of insights into community wellbeing as it related to place (see Birch et al., Chap. 14, this volume). Of importance was also the need to quantify community characteristics such as proximity to open space that are often overlooked because they don't present in typical forms (e.g. U.S. Census data), but are critical to community wellbeing and risk reduction.

The research reported here presented few surprises in that Louisiana policymakers are aware of the issues they face related to flooding and coastal erosion as well as the need for changes in building practices. As the maps were surveyed and the cluster analysis completed, the authors were reassured in that this index was compatible with our perceptions of the surrounding landscapes and populations. Our next steps are to measure the postflood environment again (perhaps with mental health data) and with updated natural environment data to supply important information about recovery and resilience as well as how these issues relate to the measures in our index. In the meantime, more flooding can certainly not be ruled out. The geographic areas in our study present a challenge for planners and policymakers due to the unique landscape and mix of urban, suburban, and rural communities tied together by a rapidly changing natural environment and a shared watershed. While community leaders often wish to see themselves as separate as economic, cultural, and geographic entities, they have no choice but to recognize their dependence on each other. This has become most apparent as recent natural disasters have required cooperative efforts for recovery following severe damage to property and supplies of goods and services as well as forced migration from affected areas into less-affected ones.

In addition to recognition of the interconnectedness of the physical environment, the communities in our study share traumatic memories of hurricanes, floods, oil spills, and other events that have negatively impacted social/emotional wellbeing through forced migration and communal loss of resources (Cherry, Marks, Adamek, & Lyon, 2015). These traumas are powerful predictors of community wellbeing (Sampson, 2003). Abramovitz and Albrecht (2013) define this as a collective loss which may be experienced quite differently depending on geographic location. Neighborhoods with multiple stressors such as high rates of poverty, crime, and ecological inequality are more severely impacted by the same experiences (Abramovitz & Albrecht, 2013) and the loss of those already limited resources significantly impacts the ongoing ability of the community to function effectively (Hobfoll & Lilly, 1993). The parishes under investigation here have struggled to recover from disasters and to absorb the in-migration of large numbers of displaced people and the subsequent disruptions following Hurricane Katrina and again post-flood of August 2016 (Bergeron, 2005). These migrations following natural disasters not only threaten the migrating group, but also impact the ecosystem of the receiving community by increasing traffic, crime, crowding and other stressors. For some communities, the collective memory of loss is also a potential asset. The awareness of risk, skills for recovering both physically and psychologically, and the capacity to thrive in uncertain and unpredictable environments may become part of the community culture (Magis, 2010; Robertson & Blinder, 2016). Prior to weather or other emergency events, for example, we can be aware of inequality of place and its effects on population health, as well as negative impacts on safety such as crowded highways or gaps in transportation systems, communities without personal vehicles or multiple older or disabled individuals.

Moving forward, the index provides a picture of community capacity, parish capacity, and coastal Louisiana as a whole, which is useful knowledge for planning purposes. That is, resilience building can be site specific and may include building of necessary partnerships as well as protocols for meeting the physical and emotional needs of community members. Immediate response to emergency situations can be informed by awareness of necessary evacuation routes, locations for interim shelters, healthcare, and supply stations that will be closest to those most in need of them.

In closing, this index underscores the value of taking an ecological systems perspective on disaster impact, highlighting the need for further study of individuals nested within multiple ecologies (Kilmer & Gil-Rivas, 2010; Kytola et al., 2015; Weems & Overstreet, 2009). This index provides a unique contribution to the disaster science literature guided by an ecological systems theory perspective. Addressing the needs of families and communities impacted by disasters is a key challenge for social scientists and urban planners. Disaster survivors' lived experiences also raise awareness of the need to prepare effectively for future disasters on local, state, and national levels.

Acknowledgments This research was supported by the Gulf Research Program of the National Academies of Sciences, Engineering, and Medicine and the Robert Wood Johnson Foundation (under award number 2000008299). This support is gratefully acknowledged.

Correspondence concerning this chapter should be addressed to Aimee Moles, Social Research and Evaluation Center, Louisiana State University, Baton Rouge, LA 70803-5501 (e-mail: amoles@lsu.2edu).

References

Abramovitz, M., & Albrecht, J. (2013). The community loss index: A new social indicator. *Social Service Review, 87*(4), 677–724. https://doi.org/10.1086/674112

Arthurson, K., Lawless, A., & Hammet, K. (2016). Urban planning and health: Revitalising the alliance. *Urban Policy and Research, 34*(1), 4–16.

Bergeron, A. (2005). Its population doubled, Baton Rouge is forced to adjust urban planning. *Architectural Record, 193*(10), 46. Retrieved from http://libezp.lib.lsu.edu/login?url=http://search.ebscohost.com/login.aspx?direct=true&db=a9h&AN=18821172&site=ehost-live&scope=site

Bradley, D., & Grainger, A. (2004). Social resilience as a controlling influence on desertification in Senegal. *Land Degradation and Development, 15*, 451–470.

Bronfenbrenner, U. (1977). Toward an experimental ecology of human development. *American Psychologist, 32*, 513–531.

Bronfenbrenner, U. (1979). *The ecology of human development.* Cambridge, MA: Harvard University Press.

Bronfenbrenner, U., & Morris, P. A. (2006). The bioecological model of human development. In R. M. Lerner & W. R. Dumon (Eds.), *Handbook of child psychology Vo. 1. Theoretical models of human development* (6th ed., pp. 793–828). Hoboken, NJ: Wiley.

Burton, C. (2015). A validation of metrics for community resilience to natural hazards and disasters using the recovery from Hurricane Katrina as a case study. *Annals of the Association of American Geographers, 105*(1), 67–86.

Cardona, O. D. (2005). *Indicators of disaster risk and risk management. Summary report.* Washington, DC: Inter-American Development Bank.

Cherry, K. E., Marks, L. D., Adamek, R., & Lyon, B. A. (2015). Younger and older coastal fishers face catastrophic loss after Hurricane Katrina. In K. E. Cherry (Ed.), *Traumatic stress and long-term recovery: Coping with disasters and other negative life events* (pp. 327–348). New York: Springer.

Colten, C. E., Kates, R. W., & Laska, S. B. (2008). *Community resilience: Lessons from New Orleans and Hurricane Katrina. CARRI Research Report III.* Oak Ridge, TN: Community and Regional Resilience Initiative.

Corburn, J. (2015). City planning as preventive medicine. *Preventive Medicine, 77*, 48–51.

Cox, D., Frere, M., West, S., & Wiseman, J. (2010). Developing and using local community well-being indicators: Learning from the experiences of community indicators in Victoria. *Australian Journal of Social Issues, 45*(1), 71–89.

Cutter, S. L. (2014). Commentary: Resilience to what? Resilience for whom? *The Geographical Journal, 182*, 110–113. https://doi.org/10.1111/geoj.12174

Cutter, S. L., Boruff, B. J., & Shirley, W. L. (2003). Social vulnerability to environmental hazards. *Social Science Quarterly, 84*(2), 242–261.

Cutter, S. L., Burton, C. G., & Emrich, C. T. (2010). Disaster resilience indicators for benchmarking baseline conditions. *Journal of Homeland Security and Emergency Management, 7*(1), 1–22.

Cutter, S. L., & Emrich, C. T. (2006). Moral hazard, social catastrophe: The changing face of vulnerability along the Hurricane Coasts. *The Annals of the American Academy of Political and Social Science, 604*(1), 102–112. https://doi.org/10.1177/0002716205285515

Cutter, S. L., Barnes, L., Berry, M., Burton, C., Evans, E., Tate, E., & Webb, J. (2008a). A place-based model for understanding community resilience to natural disasters. *Global Environmental Change, 18*(4), 598–606. https://doi.org/10.1016/j.gloenvcha.2008.07.013

Cutter, S. L., & Finch, C. (2008b). Temporal and spatial changes in social vulnerability to natural hazards. *Proceedings of the National Academy of Sciences, 105*(7), 2301. https://doi.org/10.1073/pnas.0710375105

Cumming, G., Barnes, G., Perz, S. G., Schmink, M., Sieving, K., Southworth, J., Binford, M., Holt, R. D., Stickler, C., Van Holt, T. (2005). An exploratory framework for the empirical measurement of resilience. Ecosystems, 8, 975–987. https://doi.org/10.1007/s10021-005-0129-z(2005)

DeMers, M. N. (2001). *GIS modeling in raster*. New York: Wiley.

Duque, J. C., Anselin, L., & Rey, S. J. (2012). THE MAX- P-REGIONS PROBLEM. Journal of Regional Science, 52(3), 397–419. https://doi.org/10.1111/j.1467-9787.2011.00743.x

Economic Innovation Group. (2017). *The 2017 Distressed Communities Index*. Retrieved from the Economic Innovation Group website: https://eig.org/wp-content/uploads/2017/09/2017-Distressed-Communities-Index.pdf?_cldee=Y25ld21hbkBjb21tdW5pdHlhY3Rpb25wYXJ0bmVyc2hpcC5jb20%3d&recipientid=contact-e8f510447af1e61181013863bb35fd80-bba74121a126460f9166cb45f5ab4a41&esid=c4f806de-fcb7-e711-8124-e0071b6a20a1

Flanagan, B. E., Gregory, E. W., Hallisey, E. J., Heitgerd, J. L., & Lewis, B. (2011). A social vulnerability index for disaster management. *Journal of Homeland Security and Emergency Management, 8*(1), 1–22.

Friel, S., Akerman, M., Hancock, T., Kumaresan, J., Marmot, M., Melin, T., et al. (2011). Addressing the social and environmental determinants of urban health equity: Evidence for action and research agenda. *Journal of Urban Health, 88*(5), 860–874.

Geis, D., & Kutzmark, T. (1995). Developing sustainable communities. *Public Management* (00333611), 77(8), 4.

Gulf Engineers and Consultants. (2015). *Amite River basin floodplain management plan*. Available online: http://www.amitebasin.org/ARBFMNov%202015.pdf. Last accessed 19 Feb 2018.

Hardy, S., Mitchell, D., & Grueskin, C. (2018). What's changed in Louisiana's housing codes since 2016's epic flooding? Not much. *The Advocate*. Retrieved from https://www.theadvocate.com/baton_rouge/news/article_a7cb87f4-9752-11e8-8db9-aff7b1343455.html

Harland Bartholomew and Associates. (1946). *The 25 year city-parish plan for metropolitan Baton Rouge Louisiana*. St. Louis, MO: Harland Bartholomew and Associates.

Hobfoll, S. E., & Lilly, R. S. (1993). Resource conservation as a strategy for community psychology. *Journal of Community Psychology, 21*(2), 128–148. https://doi.org/10.1002/1520-6629(199304)21:2<128::AID-JCOP2290210206>3.0.CO;2-5

Intergovernmental Panel on Climate Change. (2015). Social, economic, and ethical concepts and methods. In *Climate Change 2014: Mitigation of Climate Change: Working Group III Contribution to the IPCC Fifth Assessment Report* (pp. 207–282). Cambridge, UK: Cambridge University Press. https://doi.org/10.1017/CBO9781107415416.009

Jacobsen, B. (2017). *August 2016 flood preliminary report Amite River basin. Report prepared for the Amite River Basin Drainage and Water Conservation District*. Available online: http://www.amitebasin.org/2016Flood/August%202016%20Flood%20Preliminary%20Report.pdf. Last accessed 15 Jan 2020.

Kaly, U. L., Pratt, C. R., & Mitchell, J. (2004). *The environmental vulnerability index (EVI) 2004*. Retrieved from http://www.sopac.org/sopac/evi/Files/EVI2004TechnicaReport.pdf

Kamo, Y., Henderson, T. L., & Roberto, K. A. (2011). Displaced older adults' reactions to and coping with the aftermath of Hurricane Katrina. *Journal of Family Issues, 30*, 1346–1370.

Kilmer, R. P., & Gil-Rivas, V. (2010). Introduction: Attending to ecology. In R. P. Kilmer, V. Gil-Rivas, R. G. Tedeschi, & L. G. Calhoun (Eds.), *Meeting the needs of children, families, and communities post-disaster: Lessons learned from Hurricane Katrina and its aftermath* (pp. 3–24). Washington, DC: American Psychological Association.

Kytola, K. L., Cherry, K. E., Marks, L. D., & Hatch, T. G. (2015). When neighborhoods are destroyed by disaster: Relocate or return and rebuild? In K. E. Cherry (Ed.), *Traumatic stress and long-term recovery: Coping with disasters and other negative life events* (pp. 211–229). New York: Springer.

Lochner, K., Kawachia, I., & Kennedy, B. P. (1999). Social capital: A guide to its measurement. *Health and Place, 5*, 259–270.

Lundgren, L., & Jonsson, A. (2012). Assessment of social vulnerability. Centre for Climate Science and Policy Research. (Briefing no. 9) https://www.divaportal.org/smash/get/diva2:552075/FULLTEXT01.pdf

Magis, K. (2010). Community resilience: An indicator of social sustainability. *Society & Natural Resources, 23*(5), 401–416. https://doi.org/10.1080/08941920903305674

McCrea, R., Walton, A., & Leonard, R. (2015). Developing a model of community wellbeing and resilience in response to change. *Social Indices Research, 129*, 195–214.

Mosby, K., Birch, T., Moles, A., & Cherry, K. E. (2020). Disasters and older adults in rural communities. In L. W. Kaye (Ed.), Handbook of rural aging. New York: Routledge, Taylor and Francis. Manuscript in press.

New South Wales Health. (2009). *Healthy urban development checklist: A guide for health services when commenting on development policies, plans and proposals*. Sydney, Australia: North Ryde.

Openshaw, S. (1977). A geographical solution to scale and aggregation problems in region-building, partitioning and spatial modelling. *Transactions of the Institute of British Geographers (New Series), 2*, 459–472.

Peduzzi, P., Dao, H., Herold, C., & Mouton, F. (2009). Assessing global exposure and vulnerability towards natural hazards: The disaster risk index. *Natural Hazards & Earth System Sciences, 9*(4), 1149–1159. https://doi.org/10.5194/nhess-9-1149-2009

Pfefferbaum, B., Pfefferbaum, R. L., & Norris, F. (2010). Community resilience and wellness for children exposed to Hurricane Katrina. In R. P. Kilmer, V. Gil-Rivas, R. G. Tedeschi, & L. G. Calhoun (Eds.), *Meeting the needs of children, families, and communities post-disaster: Lessons learned from Hurricane Katrina and its aftermath* (pp. 265–288). Washington, DC: American Psychological Association.

Picou, S., Marshall, B., & Gill, D. (2004). Disaster, litigation, and the corrosive community. *Social Forces, 82*(4), 1493–1522.

Queste, A., & Lauwe, P. (2006). User needs: Why we need indicators. In H. John Heinz lii Center (2002) & J. Birkmann (Eds.), *Measuring vulnerability to natural hazards: Towards disaster resilient societies* (pp. 103–114). Tokyo: United Nations University Press.

Robertson, C. & Blinder, A. (2016, August 23). A grim expertise is shared in Baton Rouge. *New York Times* (pp. A9–A13). Retrieved from http://libezp.lib.lsu.edu/login?url=http://search.ebscohost.com/login.aspx?direct=true&db=a9h&AN=117608813&site=ehost-live&scope=site

Saisana, M., & Cartright, F. (2007). Composite indicators: Science or artifacts? In *2007 Biannual conference*. Prague, Czech Republic: European Survey Research Association.

Sampson, R. J. (2003). The neighborhood context of well being. *Perspectives in Biology and Medicine, 46*(3), S53–S64.

Schneiderbauer, S., & Ehrlich, D. (2006). Social levels and hazard (in)dependence in determining vulnerability. In J. Birkman (Ed.), *Measuring vulnerability to natural hazards: Towards disaster resilient societies* (pp. 78–102). Tokyo: United University Press.

Suarez, P., Anderson, W., Mahal, V., & Lakshmanan, T. R. (2005). Impacts of flooding and climate change on urban transportation: A systemwide performance assessment of the Boston metro area. *Transportation Research: Part D, 10*(3), 231–244. https://doi.org/10.1016/j.trd.2005.04.007

Terrell, D. (2016). The economic impact of The August 2016 Floods on the State of Louisiana (p. 22). *Louisiana Economic Development*. Retrieved from http://gov.louisiana.gov/assets/

docs/RestoreLA/SupportingDocs/Meeting-9-28-16/2016-August-Flood-Economic-Impact-Report_09-01-16.pdf
Tierney, K. (2006). Social inequality, hazards, and disasters. In R. J. Daniels, D. F. Kettl, & H. Kunreuther (Eds.), *On risk and disaster: Lessons from Hurricane Katrina* (pp. 109–128). Philadelphia: University of Pennsylvania Press.
Tobin, G. A. (1999). Sustainability and community resilience: The holy grail of hazards planning? *Environmental Hazards, 1*, 13–25.
U. S. Census Bureau. (1990). *Population of states and counties of the United States: 1790 to 1990*. Retrieved from https://www.census.gov/population/www/censusdata/pop1790-1990.html
U. S. Census Bureau. (2019). *Annual estimates of the resident population: April 1, 2010 to July 1, 2018*. Retrieved from https://factfinder.census.gov/faces/tableservices/jsf/pages/productview.xhtml?pid=PEP_2018_PEPANNRES&src=pt. Last accessed 7 Jan 2020.
U.S. Indian Ocean Tsunami Warning System Program. (2007). *How resilient is your coastal community. A guide for evaluating coastal community resilience to tsunamis and other coastal hazards*. Bangkok, Thailand: U.S. Agency for International Development.
United Nations Department of Economic and Social Affairs (UNDESA). (2007). *Indicators of sustainable development: Guidelines and methodologies*. New York: United Nations Press.
Weems, C. F., & Banks, D. M. (2015). Severe stress and anxiety disorders in adolescence: The long-term effects of disasters. In K. E. Cherry (Ed.), *Traumatic stress and long-term recovery: Coping with disasters and other negative life events* (pp. 177–194). New York: Springer.
Weems, C. F., & Overstreet, S. (2009). An ecological-needs-based perspective of adolescent and youth emotional development in the context of disaster: Lessons from Hurricane Katrina. In K. E. Cherry (Ed.), *Lifespan perspectives on natural disasters: Coping with Katrina, Rita and other storms* (pp. 27–44). New York: Springer.
Wiseman, J., & Brasher, K. (2008). Community wellbeing in an unwell world: Trends, challenges, and possibilities. *Journal of Public Health Policy, 29*, 353–366.

Chapter 16
Assessing Mental Health After a Disaster: Flood Exposure, Recovery Stressors, and Prior Flood Experience

Katie E. Cherry, Matthew R. Calamia, Traci Birch, and Aimee Moles

Introduction

Catastrophic disasters such as hurricanes, floods, tornadoes, wildfires, earthquakes, and tsunamis can strike remote geographic regions, as well as crowded metropolitan areas. Disasters can destroy entire communities, disrupting the lives of thousands of people. In the immediate aftermath of a natural disaster, the infrastructure of directly affected communities can be overwhelmed. With threats to public health and safety, people may be forced out of their ruined homes and neighborhoods. Disaster survivors who must leave the area may find themselves suddenly thrust into new cities and towns with many challenges to navigate, often without assistance or the benefit of planning or preparation (Cherry, 2020).

Ample evidence in the disaster science literature documents the mental health consequences of disasters within the first 2 years of these events (Neria, Galea, & Norris, 2009; Norris & Elrod, 2006). However, long-term threats to health in the years after a disaster have received less attention (Cherry, 2015; Cherry et al., 2015). This gap in the literature on long-term outcomes is problematic, given the frequency of catastrophic disasters and increased likelihood of these events to reoccur in certain geographic areas. For instance, consider the US Gulf Coast region, where coastal residents have experienced back-to-back natural and technological disasters for over a decade. The Atlantic Hurricanes Katrina and Rita brought unparalleled

K. E. Cherry (✉) · M. R. Calamia
Department of Psychology, Louisiana State University, Baton Rouge, LA, USA
e-mail: pskatie@lsu.edu

T. Birch
LSU School of Architecture, Coastal Sustainability Studio, Louisiana State University, Baton Rouge, LA, USA

A. Moles
Social Research and Evaluation Center, Louisiana State University, Baton Rouge, LA, USA

© Springer Nature Switzerland AG 2021

K. E. Cherry, A. Gibson (eds.), *The Intersection of Trauma and Disaster Behavioral Health*, https://doi.org/10.1007/978-3-030-51525-6_16

destruction and loss of life in 2005 (Cherry, 2009; Mosby, Chap. 13, this volume). Five years later, the 2010 Deepwater Horizon oil spill occurred off the coast of south Louisiana. This technological disaster has crippled the shrimping and oyster industries in Florida, Alabama, Mississippi, Louisiana, and Texas (Cherry et al., 2015; Cherry, Marks, Adamek, & Lyon, 2015; Osofsky & Osofsky, 2013). Hurricane Isaac, a tropical storm upgraded to a Category 1 hurricane at landfall, pummeled coastal towns on August 29, 2012, 7 years to the day after Katrina. In August of 2016, a devastating flood occurred across a 22-parish (county) region in south Louisiana, bringing with it damages estimated in billions of dollars (Moles et al., Chap. 15, this volume). Understanding the long-term impacts of disasters is key for mitigating adversity, yet post-disaster mental and behavioral health assessment remains a daunting challenge (Benight & McFarlane, 2007; Ferreira, Buttell, & Ferreira, 2015; Galea, Maxwell, & Norris, 2008).

In this chapter, we present our first findings from a community-wide assessment that was undertaken to measure the impacts of the August 2016 flood that struck south Louisiana. Our primary objective was to examine individual-level psychological health at least two years after this flood. We suspect that three variables are important to wellbeing among residents exposed to the 2016 flood: (1) current flood exposure, (2) recovery stress, and (3) prior flood experience. We defined current flood exposure as structural damage to homes and properties due to water entering the residence. Recovery stress, a closely related variable, refers to the process of rebuilding and repairing flood-damaged homes and properties. By necessity, those with flooded homes had extensive losses of material possessions. Those who chose to rebuild faced extensive home repairs, including but not limited to gutting (removing damaged interior walls and insulation), floor removal and repair, electrical work, and locating and engaging with contractors for reconstruction purposes. Prior disaster experience is also likely to play a role in post-disaster mental and behavioral health, yet few studies have addressed the psychological impact of multiple hurricanes and floods (Benight, 2004; Bukvic, Gohlke, Borate, & Suggs, 2018). In the present research, we asked survey respondents to indicate their prior disaster experience (including prior flood and/or hurricane exposure) to provide new evidence pertaining to this issue. Taken together, the results of the present assessment inform the role that flood exposure, recovery stress, and prior flood experience may play in shaping post-disaster wellbeing.

Assessment of Mental Health Symptoms in a Disaster Context

The present research was designed to assess the mental health consequences of exposure to a catastrophic flood using an online survey. Measuring the impact of catastrophic disasters presents many different challenges and obstacles for researchers to overcome (Cherry, Calamia, & Elliott, Chap. 8, this volume; Goldmann & Galea, 2014). In response to the 2016 flooding in Baton Rouge, LA, the authors created an online survey designed to assess flood impact across a multi-parish (county) region. The methodological features of this survey and first findings based on initial analyses of the mental health data are summarized in turn next. The chapter closes with implications and future directions for research.

Overview of Method

Participants and Procedure. Recruitment postcards seeking research volunteers were distributed broadly to respondents at public meetings held in connection with Inland from the Coast (IFC), a multidisciplinary research project that addresses three dimensions of ongoing flood recovery and resilience across the greater metropolitan Baton Rouge, LA region. These dimensions include environmental modeling, community wellbeing, and applied architectural design (see Moles et al., Chap. 15, this volume for description). Additionally, recruitment cards were disseminated widely to community-dwelling individuals and groups across a three-parish (county) area that was directly impacted by the 2016 flooding. Participants could scan a QR code on the back of the postcard to complete the survey on a mobile device or they were directed to a website where the online survey was housed. Data collection commenced in June of 2018 and was concluded in August of 2019. In all, over 600 people responded to the survey. Those who did not finish it (or had incomplete responses for a majority of items) were dropped from the analyses. After data cleaning, the final sample size was 506 individuals (*M* age = 45.44 years, SD = 17.07 years, age range: 18–86 years).

Sociodemographic and flood status characteristics of the sample appear in Table 16.1.

Most respondents lived in East Baton Rouge, Livingston, and Ascension parishes at the time of the 2016 flood. The majority of the sample (52%) reported that the 2016 flood damaged their homes and/or properties, while the remainder (48%) had no flood-related damage in 2016. Nearly half of the sample reported that water had entered their homes (44%), while the rest of the sample did not have water in their homes (56%). Most respondents (whether they were flooded in 2016 or not) reported no previous flood damage due to prior flood(s) or hurricane(s). Thus, the 2016 flooding was the first hydrologic event for a vast majority (86.3%) of this sample.

Online Survey Features. The survey contained six modules. The first module assessed sociodemographic characteristics of the sample. The second and third modules assessed flood experience (including property loss, displacement, and prior severe weather exposures) and stressors (including moving out of flooded homes, filing for assistance with FEMA, dealing with insurance claims, the rebuilding process, cleaning and discarding possessions, locating people to do repairs, dealing with permits). The fourth, fifth, and sixth modules assessed mental health history, current mental health symptoms (post-traumatic stress, depression, anxiety), and self-rated health, respectively. In this chapter, we present select original data from four modules to estimate flood impact, stress and losses, and mental health symptoms (post-traumatic stress, depression, anxiety).

Table 16.1 Flood status and sociodemographic characteristics of the sample

	n (%)	*M* (SD) and Range
Current place of residence		
East Baton Rouge	297 (59%)	
Livingston	103 (20%)	
Ascension	74 (15%)	
Other	30 (6%)	
Did the flood damage the place you were living or any other personal property of yours?		
No	243 (48%)	
Yes	261 (52%)	
Did water enter your home in 2016?		
No	281 (56%)	
Yes	224 (44%)	
If yes, how deep was the water in your home (in feet)?		2.98 (1.76) 0.08–7.42
Other than the 2016 flood, have you ever had flood damage to your home in a prior storm or hurricane?		
No	435 (86%)	
Yes	69 (14%)	
Did you permanently relocate to the Baton Rouge area because of damage from the 2005 Hurricanes Katrina and Rita?		
No	477 (94.5%)	
Yes	28 (5.5%)	
Which of the following describes your relocation experience after Hurricanes Katrina or Rita?		
I relocated directly to the greater Baton Rouge area	16 (59%)	
I first relocated to somewhere outside of the greater Baton Rouge area	11 (41%)	
Gender		
Male	132 (26%)	
Female	372 (74%)	
Prefer to self-describe	1 (0.2%)	
Race		
African American or black	91 (18%)	
Asian	6 (1%)	
Caucasian or white	392 (77%)	
Biracial or multiracial	8 (2%)	
Other	6 (1%)	
Ethnicity		
Hispanic	11 (2%)	
Non-Hispanic	480 (98%)	

Notes. Numbers vary slightly across variables due to missing data, and percentages do not always add up to 100% due to rounding

To assess current mental health symptoms, the fifth module of the survey, we incorporated three subscales in the online assessment. These scales included the PCL-5, a 20-item self-report measure of DSM-5 PTSD symptoms (APA, 2013) to assess posttraumatic stress (Blevins, Weathers, Davis, Witte, & Domino, 2015). Participants rated how often they were bothered by each symptom in the past month using a 5-point scale (*not at all* to *extremely*). Psychometric qualities of the PCL-5 are well established (Blevins et al., 2015). Cronbach's alpha reliability coefficient for the PCL-5 in this sample was 0.96.

The 9-item Patient Health Questionnaire (PHQ-9) was used to measure symptoms of depression. Adapted from the PRIME-MD (Primary Care Evaluation of Mental Disorders) measurement (Spitzer, Kroenke, & Williams, 1999), the PHQ-9 is a self-report measure. Participants rated how often they were bothered by these problems over the last 2 weeks on a 4-point scale (*0 = not at all, 3 = nearly every day*). Cronbach's alpha reliability coefficient for the PHQ-9 in this sample was 0.93.

The GAD-7, also adapted from the PRIME-MD, consists of 7 symptoms of General Anxiety Disorder (Spitzer, Kroenke, Williams, & Löwe, 2006). Respondents rated how often they were bothered by these problems in the last 2 weeks on a 4-point scale (*not at all* to *nearly every day)*. For this sample, Cronbach's alpha reliability coefficient was 0.95.

For all mental health symptom measures, missing data were handled by the use of mean imputation on a case-by-case basis. If a respondent had at least 50% of the scale computed, then his or her responses on the remainder of the scale were assumed to be consistent with the mean of their observed responses.

Results

All statistical analyses were carried out using SPSS statistical software system (version 24). Frequencies and percentages (%) are reported for demographic characteristics, flood status, and flood-related stressors, including disruptions and losses (see Tables 16.1 and 16.2). Means and standard deviations are reported for symptoms of PTSD, depression, and anxiety (Table 16.3).

These data were analyzed according to the following plan. First, descriptive statistics were calculated for all categorical and continuous variables. Next, correlation analyses were run to examine associations among recovery stressors and mental health characteristics for the subset of respondents who reported flood damage to homes and/or properties. Lastly, regression analyses were carried out based on the full sample to identify predictors of mental health symptoms.

Table 16.3 presents mean symptom ratings for post-traumatic stress, depression, and anxiety by flood status group.

Elevations in symptoms for all three mental health indictors were apparent among those who reported property damage in the 2016 flood relative to those who did not have flood damage in 2016. To confirm these impressions statistically, an independent samples Mann-Whitney U-Test was carried out separately for each

Table 16.2 Flood stressors

	Frequency	%
Did you have flood insurance for the structure of your home?		
None	112	(50%)
Partial	24	(11%)
Full	86	(39%)
Did you have flood insurance for the contents of your home?		
None	164	(74%)
Partial	27	(12%)
Full	32	(14%)
Did you have to repair / rebuild your home as a result of the flood?		
No	24	(11%)
Yes	199	(89%)
Did you suffer the loss of:		
Household contents		
No	6	(3%)
Yes	213	(97%)
Sentimental possessions		
No	31	(14%)
Yes	187	(86%)
Automobiles, trucks		
No	80	(38%)
Yes	132	(62%)
Pets		
No	190	(96%)
Yes	8	(4%)
Crops, trees, garden		
No	99	(46.5%)
Yes	114	(53.5%)
Were you or are you still displaced from your home as a result of the flood?		
No	87	(39%)
Yes	137	(61%)
How long were you displaced from your home?		
Less than 1 month	8	(6%)
1–6 months	42	(31%)
6–12 months	45	(33%)
1–2 years	20	(15%)
Still displaced	22	(16%)

Notes. Entries reflect responses for participants who were flooded only (N = 224). Numbers vary slightly across variables due to missing data, and percentages do not always add up to 100% due to rounding

Table 16.3 Mental health outcomes

	Flooded in 2016?		
	No	Yes	*p*
Post-traumatic stress symptoms (PCL-5)	12.97 (15.39)	22.54 (19.09)	< 0.01
Depression symptoms (PHQ-9)	4.61 (5.95)	6.52 (6.41)	< 0.01
Anxiety symptoms (GAD -7)	4.79 (5.80)	6.37 (6.25)	< 0.01

Notes. Sample sizes differ slightly due to missing data (i.e., PCL-5: flooded n = 208, non-flooded = 249; PHQ-9: flooded n = 207, non-flooded n = 240; GAD-7: flooded n = 207; non-flooded n = 244)

symptom type. Results confirmed that the mean differences were statistically significant (all p's < 0.01).

Correlations Recovery stress was defined as the total sum of ratings of *didn't happen/not stressful*, *moderately stressful*, and *extremely stressful* for events during the recovery period. Results indicated that recovery stress was significantly correlated with symptoms of post-traumatic stress (PCL-5: $r = 0.42$; $n = 456$), anxiety (GAD-7: $r = 0.28$; $n = 449$), and depression (PHQ-9: $r = 0.31$, $n = 446$). To provide further insight into these relationships, we also calculated correlations among recovery events total, which was defined as the total sum of how many events respondents reported (happened vs. not happened, regardless of amount of stress experienced). Results yielded significant correlations, albeit smaller in size for symptoms of post-traumatic stress (PCL-5: $r = 0.33$; $n = 456$), anxiety (GAD-7: $r = 0.20$; $n = 449$), and depression (PHQ-9: $r = 0.23$, $n = 446$).

Regressions Separate regression analyses were carried out with age, gender, current flood status, and past flood experience as predictor variables for symptoms of PTSD, depression, and anxiety (see Table 16.4).

In each of these regressions, the first block examined the unique associations of current flood status and past flood experience after controlling for age and gender. Two-way interactions were entered in the second block of these analyses to determine whether age (or gender) and having flooded in the past modified the relationship between current flooding and mental health symptoms. Inspection of Table 16.4 indicates that none of the interaction blocks were significant.

With respect to post-traumatic stress symptoms, age, gender, and current flood status accounted for a significant 13.0% of the variance in PCL-5 scores, $F(4, 448) = 16.01$, $p < 0.01$. Past flood status was not a significant predictor. Analyses of symptoms of depression revealed that age, gender, and current flood status variables

Table 16.4 Multiple regressions with mental health scales as criterion variables

	Unstandardized coefficients		Standardized coefficients		
Symptom categories	B	SE B	β	P	R^2
Post-traumatic stress symptoms (PCL-5)					
Step 1 (demographics, flood status)					0.125
Age[a]	−0.138	0.048	−0.131	0.004	
Gender[b]	−6.681	1.832	−0.164	< 0.001	
Current flood[c]	10.523	1.611	0.293	< 0.001	
Past flood[d]	2.214	2.307	0.043	0.338	
Step 2 (interactions)					0.130
Age × current flood	−0.040	0.098	−0.030	0.612	
Age × past flood	−0.114	0.128	−0.058	0.261	
Gender × current flood	−3.653	3.709	−0.066	0.325	
Gender × past flood	−2.762	5.553	−0.029	0.619	
Depression symptoms (PHQ-9)					
Step 1 (demographics, flood status)					0.066
Age[a]	−0.061	0.017	−0.168	< 0.01	
Gender[b]	−1.366	0.665	−0.097	0.040	
Current flood[c]	2.256	0.586	0.180	< 0.01	
Past flood[d]	0.204	0.838	0.11	0.807	
Step 2 (interactions)					0.077
Age × current flood	0.005	0.035	0.009	0.883	
Age × past flood	−0.039	0.046	−0.046	0.398	
Gender × current flood	−2.481	1.344	−0.128	0.066	
Gender × past flood	1.787	2.012	0.053	0.375	
Anxiety symptoms (GAD -7)					
Step 1 (demographics, flood status)					0.094
Age[a]	−0.071	0.017	−0.202	< 0.001	
Gender[b]	−2.126	0.635	−0.154	0.001	
Current flood[c]	2.022	0.560	0.166	< 0.001	
Past flood[d]	0.884	0.800	0.050	0.270	
Step 2 (interactions)					0.105
Age × current flood	0.025	0.034	0.046	0.455	
Age × past flood	−0.058	0.044	−0.069	0.190	
Gender × current flood	−1.477	1.283	−0.078	0.250	
Gender × past flood	3.207	1.921	0.098	0.096	

Notes

[a]For all regression models, age was treated as a continuous variable
[b]Gender was coded with 0 = female
[c]Current flood status was coded as 0 = home did not flood
[d]Past flood status was coded as 0 = no prior flooded home experience

accounted for a significant 6.0% of the variance, $F(4, 439) = 7.81$, $p < 0.01$. Past flood status was not a significant predictor of PHQ-9 scores. Analyses of symptoms of anxiety indicated that age, gender, and current flood status accounted for a significant 9.0% of the variance, $F(4, 442) = 11.43$, $p < 0.01$. Past flood status was not a significant predictor of GAD-7 scores.

Conclusion and Implications

In this chapter, we examined the psychosocial impacts of a flood within a framework relating flood exposure, recovery stress, and prior flood experience to current mental health outcomes. Data from an online survey designed to assess post-flood wellbeing based on self-reported symptoms of post-traumatic stress, depression, and anxiety were collected 24–36 months after a flood that devastated south Louisiana in August of 2016. Three key findings emerged from the analyses. First, we found that respondents with flooded homes and properties showed elevated symptoms for all mental health indicators, relative to their non-flooded counterparts. Second, the recovery stressor variable was significantly correlated with the three mental health measures, confirming the contribution of post-disaster stressors to current mental health symptoms. Third, regression analyses confirmed the contribution of age, gender, and flood status to mental health symptoms, although prior flood experience was not a significant predictor of current symptoms. These findings and their implications for wellbeing and post-disaster mental and behavioral health are discussed in turn next.

Prior research documents the mental health impacts of disasters (Neria et al., 2009; Norris & Elrod, 2006). In response to challenges inherent in estimating community-wide wellbeing after a flood (Moles et al., Chap. 15, this volume), we created an online survey to assess damages and post-disaster distress using three traditional mental health indicators. Inspection of Table 16.1 reveals that a majority of the sample (52%) reported damage to homes and/or properties due to the flood. Nearly half of the sample (44%) reported that water entered their home during the 2016 flood. Estimates of how much water had entered homes ranged from inches to well over 7 feet. For those who responded "yes" to the question of whether the flood had damaged their homes and/or property, additional survey questions were queued to assess the damages (see Table 16.2). A striking finding was that half of those who were flooded (50.5%) had no flood insurance, yet a vast majority had to repair their homes (89.2%). Reported losses included household contents (89.8%), vehicles (58.2%), crops, trees, and gardens (53.2%). Less than a quarter of the sample with flooded homes reported losses of sentimental objects (22.3%) and pets (3.8%).

Analyses of post-traumatic stress, depression, and anxiety symptoms confirmed higher levels among those who reported that water had entered their home in the 2016 flooding compared to those who did not have water in their homes (see Table 16.3). This result was not surprising in that earlier reports have documented the mental health consequences of flood exposure (Bei et al., 2013; Phifer, 1990; Phifer, Kaniasty, & Norris, 1988; Phifer & Norris, 1989). Of greater interest was the finding that elevated symptoms were still observed 24–36 months post flood. This is a striking finding, which speaks to the longevity of disaster-related distress for those who were directly impacted by the August 2016 flooding in south Louisiana.

To provide greater insight into factors that may have been contributing to current disaster distress among respondents with flooded homes and/or properties, we examined associations among mental health symptoms and two indices of disaster disruption. To be precise, we calculated indices of recovery stress (defined as the total sum of ratings of *didn't happen/not stressful, moderately stressful, extremely stressful* for events during the recovery period) and total recovery events (defined as the total sum of how many events were experienced in the recovery stress section; i.e., just happened vs. not happened). Analyses indicated that recovery stress was positively correlated with symptoms of post-traumatic stress, depression, and anxiety. The same pattern of outcomes was also obtained for total recovery events. Together these results imply that community-wide support to manage the stresses of repairs to one's home and property may be needed for homeowners struggling with recovery after a flood. Another factor to consider is that the need for assistance with repairs at both the individual and community level may be longer in duration than local officials may expect (see Birch, Henry, & Nelson, Chap. 14, this volume).

To examine the role of age, gender, current flood status, and prior flood experience on mental health, follow-up regression analyses were conducted on the full sample. As can be seen in Table 16.4, female gender and younger age emerged as risk factors for elevations in symptoms. Greater symptom reporting for females is compatible with previous research in the disaster science literature (Galea, Tracy, Norris, & Coffey, 2008; Norris et al., 2002). The finding that older adults in the sample were less likely to report post-flood mental health symptoms implies resilience (Lowe, Sampson, Gruebner, & Galea, 2015), although interpretative caution is warranted as this finding may reflect a cohort effect in underreporting current mental health symptoms. As expected, current flood status predicted elevations in mental health symptoms as previously discussed.

Contrary to expectation, we found that those with prior flood experience were at no greater risk of distress than those for whom the 2016 flood was their first hydrologic event. A relatively small percentage of the sample (14%) reported previous flood damage to homes and/or properties due to a prior flood or hurricane. Even fewer of those with prior flood experience had relocated permanently due to lost homes and property after Hurricanes Katrina and Rita in 2005. It seems likely that the null effects of prior flood experience observed here might be due to the small number of people with prior flood exposures, which may also have been from long ago, although further research would be desirable before firm conclusions would be warranted (see Davis and Bellamy, Chap. 3, this volume for related discussion).

At least four methodological limitations of the present online assessment warrant brief mention. First, the sample size is relatively small and may not be representative, so broad generalizations based on these findings are not warranted. Second, the online survey necessarily involves self-report of flood-related experiences and mental health symptoms, which are subject to potential response bias. The present findings should be interpreted in light of this concern. Third, we did not assess income or educational attainment which may impact mental health outcomes (Cherry, Sampson, et al., 2015), so interpretative caution is warranted. Fourth, we did not have preflood data on respondents' mental health, so it is not possible to attribute the postflood elevations unequivocally. Furthermore, the cross-sectional design used here precludes causal inferences. Post-disaster mental health may vary over time as people adapt to new life circumstances in the months and years after a devastating flood. Future research that includes a longitudinal assessment is needed to examine how symptoms may covary over time. Future longitudinal research may also reveal individual difference factors associated with accelerated declines in mental health in the years after a disaster.

The present findings join others in the literature documenting mental health consequences for disaster survivors in the wake of disrupted social networks and infrastructure (Phillips, Chap. 10, this volume). The present results speak to the role that recovery stressors may play in mental health in the years after a flood. These results also have implications for person-centered interventions to lessen distress among survivors. SAMHSA (2015) has forwarded a slate of evidence-based interventions to address survivors' mental health needs for use in the immediate aftermath of a disaster and in the long-term recovery phase after a disaster (see Lucas & Brown, Chap. 11, this volume for review). Identifying individuals at risk for poor mental health outcomes and accelerated declines in health relative to their non-flooded counterparts over time is an important challenge, having implications for health care systems (Spurlock & Brown, Chap. 12, this volume) and policy (Kuriansky, Chap. 19, this volume). Additionally, individual difference factors (social support, religiosity) that may buffer health risks among disaster survivors is another important topic that awaits future research (Silva Brown et al., 2010).

In closing, severe flooding has long-term consequences for individuals and communities. The increased likelihood of future hurricanes, among other severe weather events, due to the warming of oceans over time is a worrisome prospect for social scientists and laypersons alike (Leaning & Guha-Sapir, 2013). The present findings add to a growing literature on the mental health impacts of a flood. Our findings imply that flood exposure and disaster recovery stressors are key variables that warrant continued attention and research in the years after a natural disaster. Future research in multiple disaster contexts, along with translational work to improve the lives of disaster survivors struggling with post-disaster challenges seems warranted.

Author Notes
We thank Kim Mosby, Marla Erwin, Quyen Nguyen, Bahareh Mohammadzadeh Moghanjooghi, and S. Zahra Fattahi for their help with the research effort.
Correspondence concerning this chapter should be addressed to Katie E. Cherry, Department of Psychology, Louisiana State University, Baton Rouge, LA 70803-5501 (telephone: 225-578-8745, fax: 225-578-4125; e-mail: pskatie@lsu.edu).
This research was supported by a grant from the Gulf Research Program of the National Academies of Sciences, Engineering, and Medicine and the Robert Wood Johnson Foundation Award Number 2000008299. This support is gratefully acknowledged.

References

American Psychiatric Association (APA). (2013). *Diagnostic and statistical manual of mental disorders* (5th ed.). Arlington, VA: American Psychiatric Publishing.

Bei, B., Bryant, C., Gilson, K.-M., Koh, J., Gibson, P., Komiti, A., et al. (2013). A prospective study of the impact of floods on the mental and physical health of older adults. *Aging & Mental Health, 17*(8), 992–1002.

Benight, C. C. (2004). Collective efficacy following a series of natural disasters. *Anxiety, Stress & Coping, 17*(4), 401–420.

Benight, C. C., & McFarlane, A. C. (2007). Challenges for disaster research: Recommendations for planning and implementing disaster mental health studies. *Journal of Loss and Trauma, 12*(5), 419–434.

Blevins, C. A., Weathers, F. W., Davis, M. T., Witte, T. K., & Domino, J. L. (2015). The posttraumatic stress disorder checklist for DSM-5 (PCL-5): Development and initial psychometric evaluation. *Journal of Traumatic Stress, 28*(6), 489–498.

Bukvic, A., Gohlke, J., Borate, A., & Suggs, J. (2018). Aging in flood-prone coastal areas: Discerning the health and well-being risk for older residents. *International Journal of Environmental Research and Public Health, 15*, 2900. https://doi.org/10.3390/ijerph15122900

Cherry, K. E. (Ed.). (2009). *Lifespan perspectives on natural disasters: Coping with Katrina, Rita and other storms*. New York: Springer.

Cherry, K. E. (Ed.). (2015). *Traumatic stress and long-term recovery: Coping with disasters and other negative life events*. New York: Springer.

Cherry, K. E. (2020). *The other side of suffering: Finding a path to peace after tragedy*. New York: Oxford University Press.

Cherry, K. E., Lyon, B. A., Marks, L. D., Nezat, P. F., Adamek, R., Fitzgerald, K. B., et al. (2015). After the BP Deepwater Horizon Oil Spill: Financial and health concerns among coastal residents and commercial fishers. *Current Psychology, 34*, 576–586.

Cherry, K. E., Marks, L. D., Adamek, R., & Lyon, B. A. (2015). Younger and older coastal fishers face catastrophic loss after Hurricane Katrina. In K. E. Cherry (Ed.), *Traumatic stress and long-term recovery: Coping with disasters and other negative life events* (pp. 327–348). New York: Springer.

Cherry, K. E., Sampson, L., Nezat, P. F., Cacamo, A., Marks, L. D., & Galea, S. (2015). Long-term psychological outcomes in older adults after disaster: Relationships to religiosity and social support. *Aging & Mental Health, 19*, 430–443.

Ferreira, R., Buttell, F., & Ferreira, S. (2015). Ethical considerations for conducting disaster research with vulnerable populations. *Journal of Social Work Values and Ethics, 12*(1), 29–40.

Galea, S., Maxwell, A. R., & Norris, F. (2008). Sampling and design challenges in studying the mental health consequences of disasters. *International Journal of Methods in Psychiatric Research, 17*(S2), S21–S28.

Galea, S., Tracy, M., Norris, F., & Coffey, S. (2008). Financial and social circumstances, the incidence and course of PTSD in Mississippi during the first two years after Hurricane Katrina. *Journal of Traumatic Stress, 21*, 357–368.

Goldmann, E., & Galea, S. (2014). Mental health consequences of disasters. *Annual Review of Public Health, 35*(1), 169–183. https://doi.org/10.1146/annurev-publhealth-032013-182435

Leaning, J., & Guha-Sapir, D. (2013). Natural disasters, armed conflict, and public health. *New England Journal of Medicine, 369*(19), 1836–1842. https://doi.org/10.1056/NEJMra1109877

Lowe, S. R., Sampson, L., Gruebner, O., & Galea, S. (2015). Psychological resilience after Hurricane Sandy: The influence of individual- and community-level factors on mental health after a large-scale natural disaster. *PLoS One, 10*(5), e0125761. https://doi.org/10.1371/journal.pone.0125761

Neria, Y., Galea, S., & Norris, F. H. (Eds.). (2009). *Mental health and disasters*. New York: Cambridge University Press.

Norris, F. H., & Elrod, C. L. (2006). Psychosocial consequences of disaster: A review of past research. In F. H. Norris, S. Galea, M. J. Friedman, & P. J. Watson (Eds.), *Methods for disaster mental health research* (pp. 20–42). New York: The Guilford Press.

Norris, F. H., Friedman, M. J., Watson, P. J., Byrne, C. M., Diaz, E., & Kaniasty, K. (2002). 60,000 disaster victims speak: Part I. An empirical review of the empirical literature, 1981-2001. *Psychiatry, 65*, 207–239.

Osofsky, H. J., & Osofsky, J. D. (2013). Hurricane Katrina and the Gulf Oil Spill: Lessons learned. *The Psychiatric Clinics of North America, 36*, 371–383.

Phifer, J. F., Kaniasty, K. Z., & Norris, F. H. (1988). The impact of natural disaster on the health of older adults: A multiwave prospective study. *Journal of Health and Social Behavior, 29*, 65–78.

Phifer, J. F., & Norris, F. N. (1989). Psychological symptoms in older adults following a natural disaster: Nature, timing, duration, and course. *Journal of Gerontology: Social Sciences, 44*, S207–S217.

Phifer, J. M. (1990). Psychological distress and somatic symptoms after natural disaster: Differential vulnerability among older adults. *Psychology and Aging, 5*, 412–420.

Silva Brown, J., Cherry, K. E., Marks, L. D., Jackson, E. M., Volaufova, J., Lefante, C., et al. (2010). After Hurricanes Katrina and Rita: Gender differences in health and religiosity in middle-aged and older adults. *Health Care for Women International, 31*, 997–1012.

Spitzer, R. L., Kroenke, K., Williams, J. B., & Löwe, B. (2006). A brief measure for assessing generalized anxiety disorder: The GAD-7. *Archives of Internal Medicine, 166*, 1092–1097.

Spitzer, R. L., Kroenke, K., & Williams, J. B. W. (1999). Patient Health Questionnaire Primary Care Study Group. Validation and utility of a self-report version of PRIME-MD: The PHQ Primary Care Study. *JAMA, 292*(18), 1737–1744.

Substance Abuse and Mental Health Services Administration (SAMHSA). (2015, May). *Disaster technical assistance center supplemental research bulletin: Disaster behavioral health interventions inventory*. Retrieved from https://www.samhsa.gov/sites/default/files/dtac/supplemental-research-bulletin-may-2015-disaster-behavioral-health-interventions.pdf

Chapter 17
Behavioral Health After a Natural Disaster

Aaron F. Waters and Amy L. Copeland

Introduction

Natural disasters happen all over the world and lead to a number of poor health outcomes. Initially following a disaster, focus shifts towards treating and managing acute pathology that has the potential to become life threatening. Less acutely severe pathology, such as behavioral health problems, are generally not addressed directly following a disaster as the effects of poor behavioral health are not as severe short-term. However, research has highlighted that this area warrants much more attention as untreated behavioral health problems can lead to a number of poor short and long-term physical and psychological health outcomes.

The purpose of this chapter is to explore how natural disasters affect behavioral health. The definition of behavioral health is the emotional and behavioral problems related to an individual. This chapter will examine the effects of natural disasters on different behavioral-health-related problems. Additionally, it will discuss how natural disasters differentially affect those across the developmental trajectory and treatment considerations.

Behavioral health encompasses a wide range of behaviors and psychological conditions including, but not limited, to diet/obesity, physical activity, cigarette use, substance use, mental health disorders, sleep, and other health-related behaviors (Buysse, 2014; Lipari, Hedden, & Hughes, 2014; Raghupathi & Raghupathi, 2018; Spring, Moller, & Coons, 2012). Behavioral health can encompass behaviors directly related to physical health conditions, including medication or medical treatment adherence, which have been found to be a major problem for those who experience a natural disaster (Arrieta, Foreman, Crook, & Icenogle, 2009). Behavioral-health-related problems such as smoking, poor diet, and low physical activity are modifiable risk factors significantly related to the probability of

A. F. Waters (✉) · A. L. Copeland
Department of Psychology, Louisiana State University, Baton Rouge, LA, USA
e-mail: awater7@lsu.edu

© Springer Nature Switzerland AG 2021

K. E. Cherry, A. Gibson (eds.), *The Intersection of Trauma and Disaster Behavioral Health*, https://doi.org/10.1007/978-3-030-51525-6_17

physiological conditions, such as stroke occurrence (Feigin et al., 2016). Additionally, among older adults, depression has been found to be significant risk factor for the development of several chronic conditions, including diabetes, heart problems, and arthritis (Karakus & Patton, 2011). Mortality among those with mental health conditions reveals a similar picture. Those with mental health conditions tend to die earlier due to natural causes, including heart disease, cancer, and cerebrovascular, respiratory, and lung diseases (Colton & Manderscheid, 2006). The 2010 Global Burden of Disease study determined that mental and substance use disorders were the leading causes of years lived with disability (Whiteford et al., 2013). Additionally, the 2010 Global Burden of Disease Study concluded that mental health disorders accounted for the greatest proportion of disability-adjusted life years, the proportion of years lived with disability and years lost due to premature death (Whiteford, Ferrari, Degenhardt, Feigin, & Vos, 2015). The burden of behavioral health on long-term outcomes, such as years lived with disability and years lost to premature death, highlight the importance of making sure behavioral health is a consideration following natural disasters.

To begin, this chapter will provide a brief overview of behavioral health and natural disasters. Following the introduction, there is an overview of literature pertaining to specific behavioral health concerns (e.g., chronic illness, diet, sleep, substance use disorders, physical activity) and natural disasters. Lastly, this chapter will focus on how natural disasters differentially affect those across the developmental trajectory and treatment recommendations for addressing behavioral health concerns within the context of natural disasters.

Overview of Behavioral Health and Natural Disaster

Natural disasters pose numerous threats to the physiological and psychological well-being of those affected. Understanding barriers in how chronic physiological illnesses are affected post-natural disaster is important in order to treat disorders where acute increases in symptomology can become fatal (Mensah et al., 2005). However, understanding the consequences of untreated behavioral health conditions and problems following natural disasters, and how these untreated conditions can additionally exacerbate physiological conditions, is arguably of equal importance. Research has shown that behavioral health conditions, including mental health, tend to experience the greatest reductions in care following a natural disaster compared to other conditions (Kessler, 2007). Natural disasters have been linked with the development of numerous mental health disorders and behavioral-health-related problems, including anxiety disorders and mood disorders (Hussain, Weisaeth, & Heir, 2011), obesity (Hikichi, Aida, Kondo, Tsuboya, & Kawachi, 2019), sleep problems (Bavafa, Khazaie, Khaledi-Paveh, & Rezaie, 2019; Li et al., 2018), substance use (Flory, Hankin, Kloos, Cheely, & Turecki, 2009; Rohrbach, Grana, Vernberg, Sussman, & Sun, 2009), and access and adherence to healthcare (Kessler, 2007). While behavioral and mental health problems may seem like secondary problems to address following intervention of more acute conditions, there is a definite need to implement preventive measures as well as treatment for these conditions and problems.

Chronic Illness and Medical Adherence

Natural disasters can pose many problems related to behavioral health among survivors. Several studies have demonstrated across a range of different disasters the potential consequences on behavioral health outcomes. Kessler (2007) found that following Hurricane Katrina, 73.9% of survivors in affected areas who were 18 years of age or older reported some chronic illness (e.g., cardiovascular, digestive, musculoskeletal, respiratory, psychiatric, and other) in the 12 months prior to Katrina. Of these survivors with chronic illness, 20.6% reported experiencing disruption in treatment for at least one chronic illness. The greatest probability of treatment cutback among chronic conditions following Hurricane Katrina were seen among "other mental disorders" in comparison to chronic physiological conditions (Kessler, 2007). This is not surprising given the focus of treatment following a natural disaster is aimed towards treating chronic illnesses that can become fatal if briefly left untreated. These chronic illnesses include respiratory, cardiovascular, and musculoskeletal conditions. Additionally, Linkage with providers was the number one reason for treatment disruptions, followed by lack of access to medications, and finally financial/insurance problems. It is interesting to note that the fewest reductions in treatment care occurred for conditions potentially life threatening if not treated, while conditions that were not life threatening (e.g., mental health disorders) had the greatest reductions in treatment care. Other research has found the disruption in medication continuity following a natural disaster as the main barrier for the treatment of chronic disease (Arrieta et al., 2009). While it is critical to prioritize the treatment of chronic illnesses that present immediate symptomology if untreated, the potential damage of delaying treatment for those with mental health disorders can have a significant effect on individual and community recovery. Bethel, Foreman, and Burke (2011) determined that vulnerable populations (e.g., those with poor health, disabilities, and chronic illness) were more likely to have a 3-day supply of medication, but less likely to have preparedness items (i.e., food, water, flashlight, radio). Health-related behaviors such as medication management and ability for connection with appropriate providers following a natural disaster is important as delays can lead to a number of adverse outcomes on chronic illnesses and overall health.

Diet Obesity worldwide has risen significantly over the years and is associated with numerous health conditions, including diabetes, hypertension, and cancer, with rates of obesity being highest in the United States (Arroyo-Johnson & Mincey, 2016). Additionally, dietary risks, including increased consumption of sodium and trans-fatty acids as well as low consumption of omega-3 fatty acids, have been linked to increased probaility of mortality (Danaei et al., 2009). Research has oriented towards the necessity of adequate food availability and dietary needs following natural disasters as food accessibility is significantly affected by these events (Mundorf, Willits-Smith, & Rose, 2015). One potential explanation for increased risk of obesity following exposurer to a natural disaster is forced relocation, which creates a change in availability of food (Hikichi et al., 2019). Specifically, Hikichi and colleagues found that following the 2011 Japan earthquake and tsunami, increased eating out and alcohol consumption were found to be associated with increased body mass index (2019). An explanation for increased body mass index following a natural disaster is that relocation and shortened distance to restaurants and bars leads to increased eating out, in turn leading to increased body mass index (BMI) (Hikichi et al., 2019). Research has also identified that those of lower socioeconomic status tend to experience more worry about food following a natural disaster when compared to their higher socioeconomic status counterparts (Subaiya, Moussavi, Velasquez, & Stillman, 2014). The relationship between socioeconomic status and food can be accounted for by the lack of resources (e.g., damaged vehicle) to travel longer distances for food (Subaiya et al., 2014). Access to clean water and dehydration are other concerns following natural disasters. Swerdel, Rhoads, Cosgrove, and Kostis (2016) found a significant increase in rates of hospitalizations due to dehydration following Hurricane Sandy. Poor diet following a natural disaster has the potential to lead to a number of long-term health consequences, as poor diet is associated with mortality and morbidity.

Sleep The effect of natural disasters on sleep can be complicated to interpret as sleep is highly comorbid with other psychiatric conditions (Morin & Ware, 1996). Bavafa et al. (2019) found a significant positive correlation between sleep quality and severity of stress, depression, and anxiety following the occurrence of a natural disaster (earthquake). Specifically, it was determined that stress was the unique variable that predicted sleep quality (Bavafa et al., 2019). Research has also identified that sleep disturbances occurring prior to natural disaster exposure can increase the risk of psychiatric morbidity (Mellman, David, Kulick-Bell, Hebding, & Nolan, 1995). Three months following the evacuation of the 2016 Fort McMurray wildfires, the most prevalent diagnosis identified was insomnia (Belleville, Ouellet, & Morin, 2019). Additionally, among those who met criteria for post-traumatic stress disorder (PTSD) following the Fort McMurray wildfires almost half also met criteria for insomnia (Belleville et al., 2019). There has also been research to identify what aspects of a natural disaster are predictive of sleep disturbances. In a study examining sleep disturbances among older adults 2.5 years after the 2011 Japan earthquake and tsunami, it was determined that financial hardship significantly predicted the majority of sleep disturbances and was associated with less sleep, not feeling well rested in the morning, poorer sleep quality, and insomnia symptoms (Li et al., 2018). Following the 2011 Japan earthquake and tsunami, the loss of a close relative or friend was not associated with sleep disturbances among older adults, which is surprising (Li et al., 2018). It could be hypothesized that following a natural disaster attention is more focused on reestablishing resources such as shelter and food (Li et al., 2018). As a result of having to provide for basic needs such as shelter and food, natural disaster survivors may not be in a position to grieve for the loss of close ones. While sleep disturbances are common symptoms of a number of psychiatric conditions (Morin & Ware, 1996), sleep has been identified as a behavioral-health-related condition experienced by many exposed to a natural disaster (Belleville et al., 2019). Furthermore, sleep problems prior to a natural disaster have been found to be associated with increased psychiatric morbidity following the event (Mellman et al., 1995).

Substance Use Disorders Within the literature, there has been extensive documentation of the effect of natural disasters on substance use among those who are exposed. Rohrbach et al. (2009) found increases in substance use (e.g., cigarettes, marijuana, and alcohol) at 7-month and 19-month follow-up for high school students who were affected by Hurricane Rita. Additionally, negative life events (e.g., arguments with parents, illness in family, and money problems) following Rita predicted increases in substance use. This is troubling given the long-term consequences of substance use among adolescents. North, Ringwalt, Downs, Derzon, and Galvin (2011) conducted an analysis examining the prevalence of alcohol use disorders prior and post occurrence of 10 natural and nonnatural disasters (4 mass murder episodes, floods, a tornado, an earthquake, a firestorm, a plane crash into a hotel, and the Oklahoma City terrorism, and bombing). Interestingly, it was determined that a majority of alcohol use disorder cases identified post-disaster were not new cases, but instead either an ongoing disorder or a relapse into a past disorder. Identifying individuals with substance abuse problems post-natural disaster may be beneficial for the prevention of relapse into a past alcohol use disorder as well as to help those with ongoing problems. Flory et al. (2009) found significant increases in cigarette use, alcohol use, and alcohol-related problems following Hurricane Katrina compared to the prevalence of use nationally and locally pre-disaster. Psychosocial risk factors (e.g., stressors, PTSD symptoms, negative affect) were found to be related to increased cigarette and alcohol use as well as misuse. Psychosocial protective factors (i.e., positive affect, hope, social support) were not as important in understanding this relationship (Flory et al., 2009). Findings related to increased substance use among survivors of other types of natural disasters are consistent. Within a sample of Swiss survivors of the tsunami in the Indian Ocean basin, increases in substance use were observed and substance use was also associated with the severity of PTSD symptoms (Vetter, Rossegger, Rossler, Bisson, & Endrass, 2008). Among young adults exposed to major bushfires in Australia, higher number of fire-related experiences (e.g., trauma experiences/exposure) were found to be associated with increased tobacco use (Parslow & Jorm, 2006). Interestingly, Parslow and Jorm (2006) found that exposure to fire-related experiences predicted tobacco use even when controlling for PTSD.

Physical Activity Research supports a strong inverse relationship between physical activity and mental health/psychological distress such that those who engage in regular physical activity tend to experience less poor mental health days (Fluetsch, Levy, & Tallon, 2019; Hamer, Stamatakis, & Steptoe, 2009). There is limited research highlighting the importance of physical activity following a natural disaster. Within the limited research, there has been a focus on incorporating physical activity programs for children exposed to natural disasters. Following Hurricane Katrina, the Lifetime Exercise and Physical Activity Service-Learning Program was developed to help children cope during the aftermath of Katrina (Carson, 2008). Through the Lifetime Exercise and Physical Activity Service-Learning Program, children were able to engage in physical activity through organized sports, movement games, and interactive play (Carson, 2008). While mental health outcomes for the Lifetime Exercise and Physical Activity Service-Learning Program are not currently available, the program did grow substantially and was an effective model in getting children engaged in physical activity. Wahl-Alexander and Sinelnikov (2013) developed a similar after-school program for children in the southeastern United States affected by tornadoes in 2011. Within their program, children were engaged in physical activity along with group sharing and art therapy with the goal of promoting emotional recovery following exposure to tornadoes (Wahl-Alexander & Sinelnikov, 2013). While no quantitative outcome data is available for either of these physical activity programs implemented, the general consensus from these researchers is that the inclusions of these programs was beneficial for mental health following a natural disaster.

Development and Behavioral Health Following a Natural Disaster Exposure to natural disasters can affect behavioral health differently for those across developmental stages. Identifying ways in which natural disasters affect behavioral health across different developmental stages can help to understand ways to provide effective interventions for these different groups. Dancause et al. (2015) contacted and followed women who were pregnant during a flood and found that greater objective hardship and greater subjective maternal distress was associated with total adiposity of children at 2.5 and 4 years of age. Additionally, Dancause et al. (2015) found that exposure to a natural disaster during the gestation period significantly predicted increased adiposity at 2.5 years of age and was a significant predictor of (BMI) at 2.5 and 4 years of age. Dancause et al. (2015) findings suggest that gestation period stress, due to a natural disaster, can have long-term consequences in regards to an infant's weight years after birth. Research looking at depressive, anxious, and stress symptoms endorsed by mothers' during pregnancy have been found to predict behavioral problems and depressive symptoms in offspring during early adulthood (Betts, Williams, Najman, & Alati, 2015). Providing immediate intervention in reducing stress among pregnant mothers is crucial as a preventative measure to shield fetuses from potential harm following a natural disaster.

Young children are also at a greater risk for developing behavioral health problems post-natural disaster. Preschool children exposed to a hurricane have been found to have greater anxiety, withdrawal, and behavioral problems when compared to preschool children not exposed to such an event (Swenson et al., 1996). When looking at the effect of a natural disaster on children and long-term outcomes, similar trends are found. Specifically, for children 5 years of age or younger, exposure to a natural disaster significantly increases the likelihood of experiencing an anxiety disorder during adulthood (Maclean, Popovici, & French, 2016). Among adolescents between 12–17 years of age exposed to tornadoes, increased comorbidities for PTSD, major depressive episodes, and substance use disorders have been documented (Adams et al., 2015). While there is an intuitive increased likelihood of developing post-traumatic stress disorder following a natural disaster, the increased probability of developing comorbid conditions such as depression or a substance use disorder is concerning.

Among adults, the effects of natural disasters can lead to negative long-term behavioral health outcomes. Within a sample of Norwegian tourists exposed to a tsunami in Pang Nga, nearly 2/3 of those contacted reported psychiatric conditions including anxiety and depressive disorders at 2.5 years after exposure (Hussain et al., 2011). However, it was determined that Norwegian survivors experienced lower rates of morbidity compared to natives, which was accounted for by Norwegian survivors being able to return to a stable society (Hussain et al., 2011). Adults exposed to a natural disaster, who are able to return to a stable environment following the natural disaster, may have a protective factor to decrease chronic behavioral health problems (Hussain et al., 2011).

While the detrimental effects of natural disasters on behavioral health among children and adults are quite clear, the research on behavioral health outcomes among elderly adults is mixed. Some literature supports that elderly populations are vulnerable for increased psychiatric morbidity, while other studies have found older age to be a protective factor against the development of psychiatric conditions (Jia et al., 2010; Rafiey et al., 2016). Older adults have been found to be more likely to develop PTSD and other psychiatric morbidity in comparison to younger individuals exposed to natural disaster (Jia et al., 2010). However, other research has found elderly survivors of natural disaster to report higher emotional, social, and psychological well-being compared to their younger counterparts (Rafiey et al., 2016). A few hypotheses Rafiey et al. (2016) posed in regards to why elderly adults may experience more positive mental health outcomes following a natural disaster include the inoculation and maturation theories. The inoculation theory posits that exposure to natural disasters earlier in life provides protection from reactivity to natural disasters later in life, while the maturation theory proposes that the elderly are able to focus on more positive thoughts and emotions (Rafiey et al., 2016).

Treatment of Behavioral Health Following Natural Disaster

Research has identified the increased prevalence of behavioral-health-related problems following a natural disaster as well as the potential barriers following a natural disaster for those experiencing behavioral-health-related problems pre-disaster. However, little research has identified the exact methods of creating and executing a network for preventative measures as well as treatment for behavioral-health-related problems in relation to natural disasters. Theoretical models of mental health treatment in the months following a natural disaster have suggested treatment focusing on educational campaigns to educate those effected on mental health risk factors, emotional reactions, persistence of emotions across time, sources of social support, and available resources for recovery (Allen, 1993). Additionally, in the following months after a disaster, acute informal stress management resources have been proposed as sufficing for most exposed (Allen, 1993). The use of a multipronged approach has been conceptualized as a way to treat behavioral health following exposure to a natural disaster (Manzoor & Ali, 2018). Specifically, utilizing all available resources following a disaster including family, community, social workers, and mental health professionals (Manzoor & Ali, 2018). Physical activity programs have been implemented as a means for emotional recovery among children exposed to natural disasters, which observationally have been found to be effective (Carson, 2008; Wahl-Alexander & Sinelnikov, 2013). Social resources have been proposed as a mechanism for increasing resilience to promote positive mental health following a disaster, whether natural, manmade, or technological (Abramson et al., 2015). The premise of how social resources promote positive mental health following a disaster is that individuals are inherently resilient and that providing access to social support, spiritual guidance, health and human services can elicit individual resiliency towards exposure to a disaster (Abramson et al., 2015). In the aftermath of Hurricanes Katrina and Rita, the REACH NOLA Mental Health Infrastructure and Training Project was developed to address and increase the reach of mental health services to underserved communities (Springgate et al., 2011). The Mental Health Infrastructure and Training Project's focus was to provide evidence-based treatment for depression and trauma and to integrate these resources into a vast number of community primary care and social services (Springgate et al., 2011). Due to the Mental Health Infrastructure and Training Project's development as a service and not a research study, outcome data for efficacy is not available. However, the implementation of such a program not only assists in the recovery of those exposed to a prior natural disaster, but also acts as a resource for natural disasters in the future. It is important to identify and provide preventative support to combat behavioral health problems following exposure to a natural disaster. More research should further evaluate the efficacy and effectiveness of the proposed models as well as services that have been put into application.

Conclusion

Natural disasters pose a risk to the development of psychopathology and exacerbation of prior behavioral health problems. The literature to date has identified increases in different behavioral health problems and mental health disorders following exposure to different types of natural disasters. Chronic physiological illnesses where increased symptomology is potentially fatal are vital in catching and treating as early as possible following a natural disaster. However, it is becoming increasingly notable that untreated behavioral health conditions exacerbate chronic illnesses and lead to overall poorer long-term health outcomes.

Additionally, natural disasters differentially effect behavioral health for those across the developmental spectrum. There have been several proposed theoretical models for responding to increases in behavioral health problems, but more research is needed to understand the efficacy and effectiveness of these proposed models when employed. Future research should focus on understanding how to best address and treat behavioral health concerns following a natural disaster.

References

Abramson, D. M., Grattan, L. M., Mayer, B., Colten, C. E., Arosemena, F. A., Bedimo-Rung, A., et al. (2015). The resilience activation framework: A conceptual model of how access to social resources promotes adaptation and rapid recovery in post-disaster settings. *Journal of Behavioral Health Services & Research, 42*(1), 42–57. https://doi.org/10.1007/s11414-014-9410-2

Adams, Z. W., Danielson, C. K., Sumner, J. A., McCauley, J. L., Cohen, J. R., & Ruggiero, K. J. (2015). Comorbidity of PTSD, major depression, and substance use disorder among adolescent victims of the spring 2011 tornadoes in Alabama and Joplin, Missouri. *Psychiatry, 78*(2), 170–185. https://doi.org/10.1080/00332747.2015.1051448

Allen, R. D. (1993). *Handbook of post-disaster interventions* (Vol. 8, No. 5). Corte Madera, CA: Select Press.

Arrieta, M. I., Foreman, R. D., Crook, E. D., & Icenogle, M. L. (2009). Providing continuity of care for chronic diseases in the aftermath of Katrina: From field experience to policy recommendations. *Disaster Medicine and Public Health Preparedness, 3*(3), 174–182. https://doi.org/10.1097/DMP.0b013e3181b66ae4

Arroyo-Johnson, C., & Mincey, K. D. (2016). Obesity epidemiology worldwide. *Gastroenterology Clinics of North America, 45*(4), 571–579. https://doi.org/10.1016/j.gtc.2016.07.012

Bavafa, A., Khazaie, H., Khaledi-Paveh, B., & Rezaie, L. (2019). The relationship of severity of symptoms of depression, anxiety, and stress with sleep quality in earthquake survivors in Kermanshah. *Journal of Injury and Violence Research, 11*(2), 225–232. https://doi.org/10.5249/jivr.v11i2.1203

Belleville, G., Ouellet, M. C., & Morin, C. M. (2019). Post-traumatic stress among evacuees from the 2016 Fort McMurray Wildfires: Exploration of psychological and sleep symptoms three months after the evacuation. *International Journal of Environmental Research and Public Health, 16*(9). https://doi.org/10.3390/ijerph16091604

Bethel, J. W., Foreman, A. N., & Burke, S. C. (2011). Disaster preparedness among medically vulnerable populations. *American Journal of Preventive Medicine, 40*(2), 139–143. https://doi.org/10.1016/j.amepre.2010.10.020

Betts, K. S., Williams, G. M., Najman, J. M., & Alati, R. (2015). The relationship between maternal depressive, anxious, and stress symptoms during pregnancy and adult offspring behavioral and emotional problems. *Depression and Anxiety, 32*(2), 82–90.

Buysse, D. J. (2014). Sleep health: Can we define it? Does it matter? *Sleep, 37*(1), 9–17. https://doi.org/10.5665/sleep.3298

Carson, R. L. (2008). Introducing the lifetime exercise and physical activity service-learning (LE PAS) program. *Journal of Physical Education, Recreation & Dance, 79*(1), 18–35.

Colton, C. W., & Manderscheid, R. W. (2006). Congruencies in increased mortality rates, years of potential life lost, and causes of death among public mental health clients in eight states. *Preventing Chronic Disease, 3*(2), A42.

Danaei, G., Ding, E. L., Mozaffarian, D., Taylor, B., Rehm, J., Murray, C. J., et al. (2009). The preventable causes of death in the United States: Comparative risk assessment of dietary, lifestyle, and metabolic risk factors. *PLoS Medicine, 6*(4), e1000058. https://doi.org/10.1371/journal.pmed.1000058

Dancause, K. N., Laplante, D. P., Hart, K. J., O'Hara, M. W., Elgbeili, G., Brunet, A., et al. (2015). Prenatal stress due to a natural disaster predicts adiposity in childhood: The Iowa flood study. *Journal of Obesity, 2015*, 570541. https://doi.org/10.1155/2015/570541

Feigin, V. L., Roth, G. A., Naghavi, M., Parmar, P., Krishnamurthi, R., Chugh, S., et al. (2016). Global burden of stroke and risk factors in 188 countries, during 1990-2013: A systematic analysis for the Global Burden of Disease Study 2013. *Lancet Neurology, 15*(9), 913–924. https://doi.org/10.1016/S1474-4422(16)30073-4

Flory, K., Hankin, B. L., Kloos, B., Cheely, C., & Turecki, G. (2009). Alcohol and cigarette use and misuse among Hurricane Katrina survivors: Psychosocial risk and protective factors. *Substance Use & Misuse, 44*(12), 1711–1724. https://doi.org/10.3109/10826080902962128

Fluetsch, N., Levy, C., & Tallon, L. (2019). The relationship of physical activity to mental health: A 2015 behavioral risk factor surveillance system data analysis. *Journal of Affective Disorders, 253*, 96–101. https://doi.org/10.1016/j.jad.2019.04.086

Hamer, M., Stamatakis, E., & Steptoe, A. (2009). Dose-response relationship between physical activity and mental health: The Scottish Health urvey. *British Journal of Sports Medicine, 43*(14), 1111–1114. https://doi.org/10.1136/bjsm.2008.046243

Hikichi, H., Aida, J., Kondo, K., Tsuboya, T., & Kawachi, I. (2019). Residential relocation and obesity after a natural disaster: A natural experiment from the 2011 Japan Earthquake and Tsunami. *Scientific Reports, 9*(1), 374. https://doi.org/10.1038/s41598-018-36906-y

Hussain, A., Weisaeth, L., & Heir, T. (2011). Psychiatric disorders and functional impairment among disaster victims after exposure to a natural disaster: A population based study. *Journal of Affective Disorders, 128*(1–2), 135–141.

Jia, Z., Tian, W., Liu, W., Cao, Y., Yan, J., & Shun, Z. (2010). Are the elderly more vulnerable to psychological impact of natural disaster? A population-based survey of adult survivors of the 2008 Sichuan earthquake. *BMC Public Health, 10*, 172. https://doi.org/10.1186/1471-2458-10-172

Karakus, M. C., & Patton, L. C. (2011). Depression and the onset of chronic illness in older adults: A 12-year prospective study. *Journal of Behavioral Health Services & Research, 38*(3), 373–382. https://doi.org/10.1007/s11414-011-9234-2

Kessler, R. C., & Hurricane Katrina Community Advisory, Group. (2007). Hurricane Katrina's impact on the care of survivors with chronic medical conditions. *Journal of General Internal Medicine, 22*(9), 1225–1230. https://doi.org/10.1007/s11606-007-0294-1

Li, X., Buxton, O. M., Hikichi, H., Haneuse, S., Aida, J., Kondo, K., et al. (2018). Predictors of persistent sleep problems among older disaster survivors: A natural experiment from the 2011 Great East Japan earthquake and tsunami. *Sleep, 41*(7). https://doi.org/10.1093/sleep/zsy084

Lipari, R. N., Hedden, S. L., & Hughes, A. (2014). Substance use and mental health estimates from the 2013 National Survey on Drug Use and Health: Overview of findings *The CBHSQ Report*: Substance Abuse and Mental Health Services Administration (US).

Maclean, J. C., Popovici, I., & French, M. T. (2016). Are natural disasters in early childhood associated with mental health and substance use disorders as an adult? *Social Science & Medicine, 151*, 78–91. https://doi.org/10.1016/j.socscimed.2016.01.006

Manzoor, S., & Ali, M. (2018). Disaster and mental health: A need for multipronged approach. *Indian Journal of Health & Wellbeing, 9*, 3.

Mellman, T. A., David, D., Kulick-Bell, R., Hebding, J., & Nolan, B. (1995). Sleep disturbance and its relationship to psychiatric morbidity after Hurricane Andrew. *The American Journal of Psychiatry, 152*(11), 1659–1663. https://doi.org/10.1176/ajp.152.11.1659

Mensah, G. A., Mokdad, A. H., Posner, S. F., Reed, E., Simoes, E. J., Engelgau, M. M., & Chronic Diseases and Vulnerable Populations in Natural Disasters Working Group (2005). When chronic conditions become acute: Prevention and control of chronic diseases and adverse health outcomes during natural disasters. *Preventing chronic disease, 2*(Spec No A04).

Morin, C. M., & Ware, J. C. (1996). Sleep and psychopathology. *Applied and Preventive Psychology, 5*(4), 211–224.

Mundorf, A. R., Willits-Smith, A., & Rose, D. (2015). 10 years later: Changes in food access disparities in New Orleans since Hurricane Katrina. *Journal of Urban Health, 92*(4), 605–610. https://doi.org/10.1007/s11524-015-9969-9

North, C. S., Ringwalt, C. L., Downs, D., Derzon, J., & Galvin, D. (2011). Postdisaster course of alcohol use disorders in systematically studied survivors of 10 disasters. *Archives of General Psychiatry, 68*(2), 173–180. https://doi.org/10.1001/archgenpsychiatry.2010.131

Parslow, R. A., & Jorm, A. F. (2006). Tobacco use after experiencing a major natural disaster: Analysis of a longitudinal study of 2063 young adults. *Addiction, 101*(7), 1044–1050.

Rafiey, H., Momtaz, Y. A., Alipour, F., Khankeh, H., Ahmadi, S., Sabzi Khoshnami, M., et al. (2016). Are older people more vulnerable to long-term impacts of disasters? *Clinical Interventions in Aging, 11*, 1791–1795. https://doi.org/10.2147/CIA.S122122

Raghupathi, W., & Raghupathi, V. (2018). An empirical study of chronic diseases in the United States: A visual analytics approach. *International Journal of Environmental Research and Public Health, 15*, 3. https://doi.org/10.3390/ijerph15030431

Rohrbach, L. A., Grana, R., Vernberg, E., Sussman, S., & Sun, P. (2009). Impact of Hurricane Rita on adolescent substance use. *Psychiatry: Interpersonal and Biological Processes, 72*(3), 222–237.

Spring, B., Moller, A. C., & Coons, M. J. (2012). Multiple health behaviours: Overview and implications. *Journal of Public Health, 34*(Suppl 1), i3–i10. https://doi.org/10.1093/pubmed/fdr111

Springgate, B. F., Wennerstrom, A., Meyers, D., Allen 3rd, C. E., Vannoy, S. D., Bentham, W., et al. (2011). Building community resilience through mental health infrastructure and training in post-Katrina New Orleans. *Ethnicity & Disease, 21*(3 Suppl 1), S1–20-29.

Subaiya, S., Moussavi, C., Velasquez, A., & Stillman, J. (2014). A rapid needs assessment of the Rockaway Peninsula in New York City after Hurricane Sandy and the relationship of socioeconomic status to recovery. *American Journal of Public Health, 104*(4), 632–638. https://doi.org/10.2105/AJPH.2013.301668

Swenson, C. C., Saylor, C. F., Powell, M. P., Stokes, S. J., Foster, K. Y., & Belter, R. W. (1996). Impact of a natural disaster on preschool children: Adjustment 14 months after a hurricane. *American Journal of Orthopsychiatry, 66*(1), 122–130. https://doi.org/10.1037/h0080162

Swerdel, J. N., Rhoads, G. G., Cosgrove, N. M., & Kostis, J. B. (2016). Rates of hospitalization for dehydration following hurricane sandy in New Jersey. *Disaster Medicine and Public Health Preparedness, 10*(2), 188–192. https://doi.org/10.1017/dmp.2015.169

Vetter, S., Rossegger, A., Rossler, W., Bisson, J. I., & Endrass, J. (2008). Exposure to the tsunami disaster, PTSD symptoms and increased substance use - an internet based survey of male and female residents of Switzerland. *BMC Public Health, 8*, 92. https://doi.org/10.1186/1471-2458-8-92

Wahl-Alexander, Z., & Sinelnikov, O. A. (2013). Using physical activity for emotional recovery after a natural disaster. *Journal of Physical Education, Recreation & Dance, 84*(4), 23–28.

Whiteford, H. A., Degenhardt, L., Rehm, J., Baxter, A. J., Ferrari, A. J., Erskine, H. E., et al. (2013). Global burden of disease attributable to mental and substance use disorders: Findings from the Global Burden of Disease Study 2010. *Lancet, 382*(9904), 1575–1586. https://doi.org/10.1016/S0140-6736(13)61611-6

Whiteford, H. A., Ferrari, A. J., Degenhardt, L., Feigin, V., & Vos, T. (2015). The global burden of mental, neurological and substance use disorders: An analysis from the Global Burden of Disease Study 2010. *PLoS One, 10*(2), e0116820.

Chapter 18
The 3 Es of Psychosocial Recovery After Disaster

Darlyne G. Nemeth, Judy Kuriansky, and Yasuo Onishi

Introduction

The importance of assisting people in addressing their emotional suffering post disaster was proposed by World Council for Psychotherapy (WCP) representatives to the United Nations (UN) on September 7, 2005. At that time, WCP delegates José Thomé, M.D., and Guillermo Garrido, M.D., first suggested a ninth Millennium Development Goal (MDG), "Freedom From Emotional Suffering." This proposal, co-authored by Alfred Pritz, Ph.D. (Austria), Judy Kuriansky, Ph.D. (USA), Darlyne G. Nemeth, Ph.D. (USA), Gloria Alvernaz Mulcahy, Ph.D. (Canada), Neil Walsh (USA), and Sylvester Ntomchukwu Madu, Ph.D. (South Africa), was designed to bring to the attention of this astute world body the importance of psychotherapeutic intervention for survivors of trauma (Pritz et al., 2005).

As defined in the Substance Abuse and Mental Health Services Administration's (SAMHSA) 2014 position paper, "trauma does not occur in a vacuum" (p. 17). It can occur "as a result of violence, abuse, neglect, loss, disaster, war, and other emotionally harmful experiences… trauma has no boundaries" (SAMHSA, 2014, p. 2). Regardless of its etiology, as stated in WCP's UN proposal, "When a person's emotional suffering is addressed, the personal, interpersonal, social and financial costs lessen. [Thus] by restoring an individual's human dignity, and personal and cultural identity through psychotherapy, people can regain emotional resilience and achieve

The authors wish to acknowledge Kayla Mckenzie Chustz for her research assistance in the preparation of this chapter.

D. G. Nemeth (✉)
The Neuropsychology Center of Louisiana, Baton Rouge, LA, USA

J. Kuriansky
Teachers College, Columbia University, New York, NY, USA

Y. Onishi
Higashi Nippon Internatonal University, Iwaki, Fukushima, Japan

© Springer Nature Switzerland AG 2021

K. E. Cherry, A. Gibson (eds.), *The Intersection of Trauma and Disaster Behavioral Health*, https://doi.org/10.1007/978-3-030-51525-6_18

sustainable mental health to overcome fear and deal effectively with personal suffering and trauma in the world" (Pritz et al., 2005, p. 2).

Since 2005, WCP's concept of psychotherapy has evolved to include not only individual intervention, but also group intervention (e.g., Wellness Workshops) (Nemeth et al., 2011). As concluded in the 2014 SAMHSA report, "Unaddressed trauma significantly increases the risk of mental and substance use disorders and chronic physical diseases" (p. 2). Yet, most disaster interventions seldom address emotional sequelae. As Garrido noted in his 2007 presentation at the United Nations, "the severity of a great natural disaster lies in its collective tragedy" (see Nemeth & Whittington, 2012, for a related discussion).

Two important concepts must now be defined – trauma and disaster. SAMHSA stated that "Individual trauma results from an event, series of events, or set of circumstances that is experienced by an individual as physically or emotionally harmful or life threatening and that has lasting adverse effects on the individual's functioning and mental, social, emotional, or spiritual well-being" (2014, p. 7). Drawing on a considerable number of interventions worldwide providing psychosocial support post disaster, psychologist Dr. Judy Kuriansky cited Tierney's (1989) definition of disaster as being "in the context of community to signify stressful situations occurring relatively suddenly or unexpectedly in a geographic area that result in loss and interfere with the ongoing social life of the collective." She then noted that, "when disaster strikes, the resulting trauma affects not only individuals but their social context, disrupting the quality of life of the community and the fabric of society" (Kuriansky, 2012, p. 141).

Symptoms of Trauma

Harvey (1990) concluded that "trauma survivors have symptoms instead of memories." According to Fisher (2007), these symptoms may include, but not be limited to, intrusive memories, nightmares/flashbacks, decreased concentration, startle response, chronic pain, headaches, self-destructive behaviors, depression, irritability, dissociation, and insomnia.

As most survivors of environmental disaster have been traumatized, many have experienced the aforecited symptoms. When such symptoms occur on or around anniversaries of these disasters, they are commonly referred to as Anniversary Reactions (Kuriansky & Nemeth, 2013). These involve reactions to an emotionally charged episode that holds tremendous significance for an individual or group. When the initial event is experienced as traumatic, individuals may tend to become sensitized to re-experience the symptoms under reminiscent circumstances (Bornstein & Clayton, 1972). As discovered in the Katrina 1 Wellness Workshops implemented by Nemeth and Kuriansky, these symptoms often include, but are not limited to, the following: panic, grief, conflict, intrusive memories, emotional numbness, reactivity, reminders of the event, sleep disturbance, irritability, depression, anxiety, and substance abuse (Nemeth et al., 2010).

The 3 Es of Trauma

According to the 2014 SAMHSA report, the three important components of trauma exposure are: (1) the event(s), (2) the experience of the event(s), and (3) the effect of the event(s) (p. 8) (see Kusmaul, this volume). If the event wasn't significant, or deeply felt, the effect would be minimal. In major environmental disasters, however, the effects have usually been very significant. The SAMHSA report defines these 3 Es of trauma as follows:

1. "*Events* and circumstances may include the actual or extreme threat or physical or psychological harm (i.e., natural disasters, violence, etc.). One such example is the 1982 Livingston Train Derailment (Lemoine, Rostow, Nemeth, & Ellis, 1984).
2. "The individual's *experience* of these events or circumstances helps to determine whether it is a traumatic event … how the individual labels, assigns meaning to, and is disrupted physically and psychologically by an event will contribute to whether or not it is experienced as traumatic" (p. 8). The outcome research from the 1982 Train Derailment in Livingston, Louisiana, concluded that not all individuals experienced the event as equally traumatic. Lemoine et al. (1984) found that the severity of residents' symptoms was directly related to their physical distance from the cite of the derailment explosions.
3. "The long-lasting adverse *effects* of the event are a critical component of trauma … the duration of these effects can be short to long term" (p. 8). Thus, the effects of an environmental trauma may be quite variable and may reappear at anniversaries of these events.

Post-Trauma Recovery

Psychosocial recovery begins after a traumatic *event*. An individual's or community's *experience* of that event and the *short- or long-term effect* or outcome of that trauma on the individual and/or community must be addressed. Therefore, SAMHSA's 3 Es of trauma occur before the process of psychosocial recovery begins. Simply stated, trauma pre-dates recovery.

As Garrido stated (2007), order must be created first during the post-disaster period. In the example of Hurricane Katrina, it was only when Lieutenant General Russell Honoré came to New Orleans after Hurricane Katrina hit that order was restored (Nemeth, 2015b, p. 5; Brinkley, 2006, p. 616–617). Before, chaos and lawlessness prevailed (Honoré, 2012, p. 27; Nemeth, 2015b, p. 5). Honoré, with his commanding presence and African-American Creole heritage, was able to reduce the chaos, organize the available resources, and manage the crisis. As Honoré stated, "people are looking for leaders who are clearly aligning strategy, tactics, and logistics … leaders have to get those three things right. That's their mission" (Honoré, 2012, p. 91). As Honoré stated, "That's what leading is – getting people to willingly do what must be done to complete the mission" (Honoré, 2012, p. 92). This was the beginning of the psychosocial recovery of New Orleans in the wake of the 2005 Hurricane Katrina (see Mosby, this volume).

New Orleans' post-disaster recovery involved a tripartite process. Instead of the 3 Es of trauma defined by SAMHSA, the authors propose the 3 Es of Psychosocial Recovery. These are Economic Rehabilitation, Environmental Restoration, and Emotional Resilience. Once conditions are stabilized, Economic Rehabilitation must begin. Without money, psychosocial recovery is not possible. Governments, organizations, and people themselves cannot recover without an influx of capital. Economic Rehabilitation post environmental trauma is essential, as is Environmental Restoration. Yet, both are not enough. Oftentimes, governments, leaders, and traumatized people do not fully understand or appreciate the importance of emotional factors in recovery. They may believe that money alone will solve the problem, but it does not. If people are afraid to return to their homes, they will not return. Money helps, but it is only one component in the journey to recovery. It takes all three components working together to rebuild after a disaster – Emotional Resilience, Environmental Restoration, and Economic Rehabilitation. According to nuclear scientist Yasuo Onishi (2012, p. 25), "An accident causes sudden dramatic changes to the environment." Environmental historian Sarah B. Pritchard noted that, in Fukushima, Japan, "a magnitude 9.0 earthquake struck the region on March 11, 2011, which resulted in a fourteen meter tsunami, and the meltdown of half of six nuclear reactors at the Fukushima Daiichi nuclear power plant" (Ermus, 2018, p. 2), which Pritchard labeled, by this chapter's authors, as "a triple disaster." Likewise, it takes a triple rehabilitation process, now labeled as the 3 Es of Psychosocial Recovery, to bring about resolution.

It has been almost 9 years since that nuclear accident, yet, many people have not returned to Fukushima, out of "imagined as well as real radiation fears" (Onishi, 2012, p. 29). As economic and environmental incentives have not been enough, efforts to facilitate the emotional recovery of those affected are crucial (see also Maeda & Oe, 2015). True resolution is not possible without addressing peoples' feelings (Bar-Levav, 1988). Wellness workshops conducted after disasters have been shown to be helpful in allowing people to address and resolve their feelings both immediately and long after a crisis (Kuriansky, Nemeth, Ray, & Songy, 2014). For example, missions were conducted in Sendai, Japan, after the 2011 earthquake, tsunami (Kuriansky, 2019). Lectures on resilience were subsequently given by the authors (via Skype) to community leaders and students at the Higashi Nippon International University and National Institute of Technology, Fukushima College in Iwaki, Fukushima, Japan. As Gibson and Ermus stated, "disasters are generally measured in three ways: economic costs, environmental devastation, and loss of human life …" but "the stakes are always much higher when human lives are at risk" (2018, p. 119).

The Experience of Hurricane Katrina

In the aftermath of Hurricane Katrina (hereafter referred to as "Katrina"), flooding occurred in 85 percent of the greater New Orleans area and contributed to over 1400 deaths (Gotham, 2018, p. 161). Keim and Muller called this New Orleans disaster "one of the greatest American natural disasters waiting to happen" (2009). Katrina was caused by a combination of lack of leadership, lack of holistic planning, and lack of community involvement (see Mosby, this volume). Governmental leadership was too self-oriented, the holistic plans developed by Charles Ellet over a century earlier were never fully implemented, and the community became complacent over time (Kelman, 2007). A self-oriented, risk-denying, risk-avoiding mentality prevailed (Scruton, 2012, p. 120). The results of such diminished awareness were catastrophic. In the Lower Ninth Ward area of New Orleans, flooding exceeded 13 feet (Barrios, 2018, p. 132). Sociocultural anthropologist Barrios explained that "the practices that shaped Hurricane Katrina's destruction were the extensive levee construction, river and port engineering, wetland degradation, and urbanization" (p. 133).

Learning from this Katrina tragedy must become a priority. Given that "43 percent of the U.S. population lives in counties [parishes] with levees," sociologist Kevin Fox Gotham concluded that "Katrina is Coming to Your City" (2018, p. 161), because as we live on a dynamic planet, we must be ready for change. Geologist, Donald F. Nemeth (2018), offered the following formula: Knowledge + Behavioral Strategies = Survival. As "90% of the people in the world have experienced some sort of trauma in their lives" (Rehm, 2015), people must be prepared to cope with disaster. Nemeth points out that people must head the warnings; be prepared to cope; arrange safety for family, friends, and pets; and have supplies including medicine, food, and important documents available (Nemeth et al., 2018).

As Honoré concluded, "It's a whole lot more expensive to fix a problem than prevent it (2012, p. 96). In fact, according to Onishi, who lives in Richland, Washington, "The U.S. government spends $3.5 billion every year on our local area and will continue to do so for the next 50 years for the environmental clean-up" (Onishi, personal communication, 2019) of the Hanford Site, also referred to as "America's Chernobyl" (Farrow & McHugh, 2016). In spite of the scientific warnings, half of the world's population lives within 36 miles of the coast and 75% of the world's largest cities are located on the coast (Nemeth et al., 2018). According to Onishi and colleagues (2007), a holistic process of managing climate change is essential. Therefore, they included the concepts offered by Ury and Patton (Fisher, Ury, & Patton, 1991) in their 2007 UN presentation. These involved four steps: (1) Identify the problem and history, (2) Aftermath analysis, (3) Effective approaches to resolution, and (4) Ideas and actions. These holistic concepts, which were then divided into five segments: water, earth, wildlife, people, and biosphere, included, but were not limited to, the following recommendations (Garrido et al., 2007):

1. *Water*
    - Develop a global program for sustainable use (Onishi et al., 2007).
2. *Earth*
    - Environmental management plans must be based on geologic principles (Onishi et al., 2007).
3. *Wildlife*
    - Anticipate and prepare for further habitat changes (Hamilton, 2007).
4. *People*
    - Develop a sense of cultural appreciation of nature and ecosystems (Garrido, 2007).
5. *Biosphere*
    - Determine the consequences on the biosphere of increasing human numbers (Onishi et al., 2007).

Nemeth then outlined some of the immediate and far-reaching consequences of climate change. He noted that the Arctic and Antarctic are defrosting and releasing melt water into the world's rising oceans. A worldwide rise in sea level is occurring. More than 100 million people live within three feet of mean sea level. Changes occurring along today's retreating coastal areas offer an indication of what can be expected with future rise in sea level (Garrido et al., 2007) (see also Birch, Henry, & Nelson, this volume).

As Honoré concluded, it would be easier and less expensive if people "would take a little responsibility for themselves, [and] become more resilient" (2012, p. 96). This is especially important, because people will not move away from their homes, their culture, their oikophilia (their love of home) (Scruton, 2013). They will only do so when forced. Like the Eskimos in northern Alaska, with the significant melting of the ice and permafrost, (Garrido et al., 2007; Eurich, 2018; Waldholz, 2018), they only did so when their entire village was relocated, at an estimated cost of over $100 million (Waldholz, 2018). Now they must resiliently adapt to new ways of providing for themselves and their families.

This will soon happen to other "at risk" areas (e.g., New Orleans, Miami), as the glaciers are melting and the waters are rising. How will these cities and their residents adapt to this climate change? Are there any holistic long range plans? The 2007 WCP UN Workshop suggested Strategies to Facilitate Biosphere Management and Lifestyle Changes: Measures to Protect the Environment and Prevent Disaster Sequelae of Current and Future Climate Changes (Garrido et al., 2007). During this workshop, noted scientists offered solutions. But, will the people who make the policies listen?

Perception, Insight, and Oikophilia

Historically, while having been warned to evacuate, most New Orleanians did not. They did not perceive the need to do so. Perception is often based on past experiences (Nemeth & Olivier, 2015). For many of those who did not head the Katrina warnings and evacuate, their past experiences, their visual and auditory histories, did not support the need to do so (Myers, 2007). Perhaps their insight, "the faculty involved in grasping the inner character or underlying truth" (Wolman, 1989, p. 179), was lacking. Or, perhaps, remaining in denial was far easier.

According to McCright and Dunlap (2011, p. 147), concerted efforts are being made to promote this denial. Dunlap and McCright have characterized these efforts as the "Climate Change Denial Machine," which has allegedly been an organized effort of the fossil fuels industries, corporate America, and conservative foundations to undermine the future of our planet and our people (Nemeth & Olivier, 2015, p. 193). Thus, the development and marketing of perceptions has become a major player in politics today. Or, perhaps, people just want to live in the present, avoid the past, and not think about the future? Perhaps, for some people who believe the future is not likely to include them, climate change does not really matter to them. Or, perhaps, they simple perceive climate change issues as "fake news."

For example, on that calm, sunny, pre-Katrina day in August 2005, people did not perceive or understand the need for change. Yet, during hurricane season (June 1 – November 30), with all of its possibilities for danger and distress, they preferred homeostatic perceptions (Norgaard, 2011), rather than insightful awareness. New Orleanians were warned, but many did not act on those warnings. As during past hurricanes, many avoided the warnings that caused distress. What prevailed was the desire to seek comfort and remain in the safety of their home, their "oikophilia" (Scruton, 2012, p. 3). This unconscious desire tends to prevail over insightful awareness and/or conscious perception. Even today, New Orleanians know that they should not live in a place that is 5 to 15 feet below sea level, but they do. The illusion of safety remains.

Facts Versus Illusions

Giving up the illusions of safety and accepting the fact of uncertainty can be very frightening. Although we live in an ever-changing world, we tend to perceive the world as static. Yet, it evolves daily. People must understand and embrace the three principles of change – people must *respond* appropriately to the warnings of hurricanes, tornados, floods, and other disasters. This requires robust leadership, such as that provided by Lieutenant General Russell Honoré post Katrina (Honoré, 2012). In this regard, people must develop *resilience*, which is defined as the ability to bounce back from a crisis, to engage in positive coping actions, to request and accept support, and to look forward toward a better future (Walker & Heffner, 2010;

see also Wilks, Choate, Brown, Du, & Curry, this volume). Rodin outlines the five main characteristics of resilience as "Aware, adaptive, diverse, integrated, and self-regulating." As Mulcahy stated, "This requires positive relationships, individual responsibility, mutual reciprocity, and overall *respect* for the environment and the ever-changing world (2013, p. 194). Thus, people must summon their resilience to address the forthcoming danger, and they must prepare to deal with the changes and uncertainties that lie ahead (Nemeth & Olivier, 2015, p. 207).

In order for change to take place in how people respond, their perceptions of their environment must evolve. Individuals must move away from complacency and move toward insightful awareness. As Tucker stated, they must embrace the concept of being stewards, not owners, of their environment (Garrison Institute, 2012). Native Americans, who were caretakers of the land, understood and continue to understand this concept, but the Europeans who came after to conquer and plunder did not and still do not. Gloria Mulcahy, a psychologist and Native American, pointed out that all who live on planet Earth must learn to foster respect, reciprocity, responsibility, and relationships for and with "Mother Earth" (2013, pp. 193–194). Mulcahy concluded that this planet is not ours; we are not owners; we are merely visitors and caretakers. Similarly, Onishi pointed out that "Water is a part of a delicate, interactive, balancing biosphere system. It is the lifeblood of people and all living things. Respect for our environment must include *respect* for our living waters" (Onishi, 2012, p. 37). Onishi concluded that, through our painful experiences, we now know that it is more economical and beneficial to take care of the environment from the beginning. Onishi proposed the need to address our water resource conflicts (agriculture, industries, and nations) and coordinate and collaborate "to conserve water and energy, to improve water usage efficiency, to improve energy sources, to promote more renewable energy, and to recycle materials to promote sustainable lifestyles" (p. 38).

Planet Earth must outlive us, not vice versa. Will New Orleans and the Louisiana Coast outlive us? With the coastline's loss of "approximately a football field an hour" due to mostly manmade intrusions (Baurick, 2017), is this even possible? Onishi notes, "Because we realize that human actions affect both people and our environment, the concept of environmental sustainability is taking hold" (Onishi, 2012, p. 25). But, is sustainability enough? According to Zolli, "where sustainability aims to put the world back into balance, resilience looks for ways to manage an unbalanced world" (2012).

Resilience Leads to Resolution

Rodin concludes that, "building resilience is one of our most urgent social and economic issues because we live in a world that is defined by disruption (2014, p. 6–7). Starting over after environmental trauma requires resilience, which is a biopsychosocial phenomenon. All five characteristics of individual and community resilience are involved in moving forward post environmental trauma. Thus, in order to move toward the resolution required to manage our unbalanced world, perception and insight must be integrated. Resolution is an evolving, not static, process that requires a reconciliation of the mind (perception) and the heart (insight). Perception requires insight. It is not either-or; both must be seamlessly blended for responsible action. This responsible action, according to Lefebvre, is a product of "three distinct but interrelated moments, the lived, the conceived, and the perceived" (Barrios, 2018, p. 134). The *lived* is conceptualized as the ways in which space is organized and structured (e.g., a neighborhood). The *conceived* is defined as "the ways politically influential groups conceptualized space in a given social context or influenced the conceptualization of space in urban planning and architectural design." The *perceived* is cited as the ways people experience and remember a space.

In order to blend perception and insight, it is necessary to understand space as a social product, which involves "three distinct 'moments': the lived, the conceived, and the perceived." In this regard, the perceived space "include[s] memories, rituals and symbolic practices through which people make space meaningful in everyday practice" (Barrios, 2018, p. 134). According to Mulcahy, the implementation of these concepts is well understood by Native Americans (2013, p. 193–194). Scruton refers to this concept of perceived space in which people live and conceptualize their world as their "oikophilia" (2013, p. 3). Lefebvre notes that all three of these moments in time will produce "an understanding of the connections between the ways people think about, experience, and do things with/in a space in a particular social context" (Barrios, 2018, p. 134). In other words, this requires the blending of perception and insight.

Ecodynamics and Ecokinetics

In order for this blending to occur, people must understand two processes – *Ecodynamics*, which refers to how people affect the environment, and *Ecokinetics*, which refers to how the environment affects people. These two basic principles must work together to achieve a meaningful outcome to protect people's ever-changing world. They are the yin and yang of resolution. In this way, "people and environment are inseparably intertwined" (Barrios, 2018, p. 154).

It is not just about people, however, it is also about the animals and plants that live on this earth. Hamilton addressed the importance of habitat movement; specifically, many pipelines are being built through habitat routes causing migratory animals, such as the caribou, severe difficulties (Onishi et al., 2007). The mining and smelting of copper in Southeast Tennessee destroyed a 60-mile radius of vegetation, thus creating the Ducktown Desert (Muller, 2012, p. 52). Deforestation is being experienced in the Brazilian rainforest where for-profit individuals are setting parts of the rainforest afire to produce more land available for agricultural development (Gibbens, 2019). This is being done even though there is only a small layer of topsoil in which to grow crops. Therefore, when balancing the concepts of Ecodynamics and Ecokinetics, ecologist Charles Elton noted that three questions must be asked,

"What plants and animals live in it, what beauty and interest may be lost, and what extra risk changing it will add to the accumulating instability of communities" (Gibson & Ermus, 2018, pp. 121–122), for risk implies uncertainty (Gotham, 2018, p. 163).

Bottom-Up Problem-Solving to Manage Uncertainty

A meaningful resolution of one disaster does not prevent another. Rather, it can help a community prepare for what is to come. It can instill hope for a better outcome – for it is not *if* another environmental disaster occurs, but *when*. Since many people will not leave their "oikophilia," they must adapt and prepare. They must change the way they choose to rebuild, to protect their environment, and to work together as a community. As a resident-organizer from the Holy Cross area of New Orleans, stated,

> "Once we rid ourselves of traditional thinking, we can move on to creating the future. I say a lot of times, if you hold on to the past, you can't move on to a brighter future. Not to say you don't look back to the past to get some wisdom" (Barrios, 2018, p. 146).

Wisdom comes from "Bottom-Up" community problem-solving, in which solutions to community issues are identified and implemented by individuals within the community itself (Netabay, 2007), whereas "Top-Down" approaches are typically developed by governmental or social agencies and imposed on local people. According to Onishi, "In Fukushima's case, the Japanese government has spent so much money for environmental clean-up and economic development. But Fukushima recovery is hardly achieved, because all these efforts have been top down" (Onishi, personal communication, 2019). These "Top-Down" problem-solving strategies often do not take into account the cultural wisdom of the local people. This occurred in the aftermath of Katrina, where agencies such as the Red Cross would not allow local churches to cook culturally appropriate food for those afflicted (Cramer, personal communication, 2005). At the same time, the Red Cross did not separate diabetics with special needs from the other victims and make appropriate foods available to them (Hayes, personal communication, 2005). Instead of fruit appropriate for their condition, diabetics were being served honey buns. Local wisdom, when allowed, typically prevents such errors. Rather, a wise plan of action developed by the community, for the community, is usually more effective. People must be encouraged to become a part of the process of discovering new ways of doing things. This is achieved via community communication and connection. As Onishi affirmed, "The local Fukushima people truly need to strengthen their emotional resilience, so they can see their future and plan to achieve it" (Onishi, personal communication, 2019). People, not governmental or social agencies, are the ultimate agents of change, but this requires effective leadership.

Three Types of Leaders

According to Heifetz and Laurie (2001), "getting people to do adaptive work is the mark of leadership." At times of great distress, three types of leaders tend to emerge – the robust leader, who commands authority via competence; the psychoid leader, who demands authority via bullying; and the altruistic leader, who invites authority via example. The robust leader displays ethical behaviors. As Campbell stated, "There are no wrong ways to do what is right and there are no right ways to do what is wrong" (Engster & Campbell, 2014). Robust leaders display decision-superiority capabilities (Honoré, 2012) and know how to lead. They are respectful, prosocial individuals who command authority via their solid presence.

Then there are the charismatic psychoid leaders, who offer appealing slogans and half-truths. They demand loyalty while ignoring the facts. They bully others in order to maintain power and control. These psychoid leaders, who are "crazy like a fox," so to speak, must be avoided. They promote "groupthink," which is defined as "a pattern of thought characterized by self-deception, forced manufacture of consent, and conformity to group values and ethics" (The Merriam-Webster.com Dictionary, n.d.). They offer simple solutions to complex problems, solutions that only work in the short term, and often spell disaster in the long term (Nemeth, 2015a, pp. 51–52). These rather narcissistic individuals do not understand or care about holistic thinking. They are merely present-oriented.

Altruistic leaders, on the other hand, are humble individuals. They lead by example, not by fear or intimidation. Such leaders seem to be motivated by a calling; they are the caretakers of the earth and its people (Nemeth, Hamilton, & Kuriansky, 2015, p. 415). These moral leaders do what they believe they were called upon to do. One example was Mother Theresa, another was Mahatma Gandhi. Altruistic leaders encourage people to build or rebuild together. With such encouragement, people respond cooperatively. Thus, when a trauma happens either to a group or community, altruistic leaders invite people to address it together (Nemeth & Songy, 2015, p. 102).

For example, at the governmental level, two such altruistic leaders, Speier and Prats (2015, p. 15-34), were in charge of the Louisiana Department of Health and Hospitals' Public Health Response to Hurricane Katrina (see Speier, Osofsky, & Osofsky, 2009). This involved addressing behavioral health initiatives, resource development, building a long-term response strategy, partnership development, and respecting cultural diversity. In 2015, 9 years post Katrina, Speier and Prats concluded that "the practices for emergency preparedness planning and response are more nimble than they were nine years ago. Communication and clarity of roles and available resources enhances the likelihood of building successful immediate and long-term response to major disaster incidents" (2015, p. 32).

When individuals, communities, and governmental agencies work together, hope for recovery is restored. But, as D.F. Nemeth concluded, minimizing the consequences of natural disasters is a three-fold process. We must: (1) Be aware of the past, (2) Monitor the present, and (3) Head the warnings. This requires the attention of all stakeholders.

The Six Stages of Psychosocial Recovery

In the end, however, as Charles Cameron stated, "tragedy or crisis is never purely economic, political, or military. It is always pre-eminently psychological" (Nemeth, Cameron, Creveling, Draeger, & Schexnayder, 2000). Trauma is so emotionally laden that it interferes with people's ability to think logically and problem-solve effectively (Gay, Nemeth, Kuriansky, Olivier, & Songy, 2015, p. 53). Physical stability and emotional stability must be given equal priority in recovery. Six stages are involved in this universal process. These stages have been observed around the world post hurricanes, earthquakes, nuclear accidents, oil spills, etc. They include Shock, Survival Mode, Assessment of Basic Needs, Awareness of Loss, Susceptibility to Spin and Fraud, and Resolution (Nemeth & Whittington, 2012, pp. 130–131).

First, *Shock* or disbelief abounds. How could this happen to us? The second stage is *Survival Mode*. People do whatever it takes to continue to live. In Katrina's flooding waters, for example, some swam to safety; while others were rescued, often by the Cajun Navy; some climbed into their attics or on top of their roofs to be rescued by helicopters. Eventually, most sought shelter by walking to the safety of public buildings (e.g., the Superdome). Nevertheless, they weren't safe there; despite the illusion of safety, crimes abounded.

The third stage, *Assessment of Basic Needs* post Katrina included the needs for blankets, medicine, food – all of which were difficult to obtain/provide. Those with medical needs (e.g., diabetes) were especially "At Risk." Many died from lack of health care services. New Orleans was not prepared for a direct hit from a major hurricane. The levees failed. The pumps failed. The police failed. The politicians failed. It was a dire time for New Orleanians. People felt like refugees in their own home city – their precious "oikophilia." They either slowly or abruptly entered the fourth stage, *Awareness of Loss*. Many lost their homes, or their loved ones, or their neighborhoods, or their friends, but mostly, their illusions were shattered. In this regard, the most vulnerable were the senior citizens, who often did not survive the drama of the trauma (Kuriansky, 2012, p. 148). Brunkard and colleagues reported that "forty nine percent of victims were people 75 years and older" (2008). When respected authority figures, such as Lt. General Russell Honoré, came to soothe their souls and restore order, hope began to re-emerge (Nemeth, 2015b, p. 5). Although people expected financial help from the Federal government, in order to return to their precious "oikophilia" and begin anew, government officials instead cited the Federal Control Act of 1928 as a justification for not helping (Verchick, 2010, p. 180).

During Katrina, rebuilding for many quickly turned into a nightmare. In light of stories about *Susceptibility to Spin and Fraud*, people were worried about who they could trust. Toxic Chinese drywall was used to repair flooded homes, such as the home of Sean Payton, Coach of the New Orleans' Saints NFL football team, and his family (Armstrong, 2009). Workers built by day, and stole by night. Copper wiring was especially at risk (Hunter, 2011). Contractors promised results, but often left homeowners adrift after their first down payment was received and cashed. The trauma was bad enough, but the drama that followed was even worse. Many New Orleanians cried, "Can anyone anywhere be trusted?"

At times of great distress, the comforting power of local groups can prevail and can lead to a supportive outcome. With the restoration of hope and the trust of local communities post Katrina, people were able to move toward the final stage of psychosocial recovery, *Resolution*. Community groups were typically formed in religious centers. People of many different religious orientations began helping one another and, by doing so, they eventually helped themselves. They moved from fixed to flexible, from rigid to adaptive. During their journey to recovery, however, people were exposed to both acute and chronic psychosocial stress. What they learned in the Hurricane Wellness Group Workshops helped them to cope (Kuriansky et al., 2014).

Healing Through Wellness Workshops' Group Process

When a community has been deeply traumatized, outcome research shows that group process is an essential component of psychosocial recovery (Bloom, 1998; Haen, 2005; Phillips, 2005; Jordan, 2003). In reference to the Lower Ninth Ward, Ward "Mack" McClendon commented that, in the 8 years post Katrina, "We got less than one third of my community back" (Barrios, 2018, p. 155). In fact, "the Lower Ninth Ward became iconic of the [Katrina] disaster" (Barrios, 2018, p. 131). Therefore, the workshop team, led by first and second chapter authors, was invited into the Lower Ninth Ward during the tenth year Anniversary of Katrina to conduct a wellness workshop at the Lutheran Community Center. It should be noted that "the Lower Ninth Ward sustained catastrophic damage and flooding exceeding fourteen feet of water" (Barrios, 2018, p. 132) due to "multiple drainage/levee failures" (Barrios, 2018, p.139). People in the Lower Ninth Ward were encouraged by robust and/or altruistic leaders to come together to address their pain and suffering from loss. Without addressing their psychological distress, creative planning for recovery would likely be ineffective. The value of community wellness workshops has been demonstrated in Haiti by the chapter's second author, in New Orleans post Katrina (Nemeth et al., 2012), and after the 2016 Great Flood in Louisiana (Nemeth & Olivier, 2017, pp. 59–76) (see also Cherry, Calamia, & Elliott, this volume). The 2016 Great Flood Wellness Workshops were conducted to prepare newly homeless greater Baton Rouge families for the holidays. In retrospect, some families took years to rebuild (Schexnayder, personal communication, February 25, 2017).

When communities are suffering environmental trauma, rebuilding emotionally in the collective safety of a group is an effective approach. These group experiences, which focus on rebuilding resilience, are typically held in churches where community groups are familiar with gathering and feeling safe. In the aftermath of Katrina, such workshops were held in Catholic, Baptist, and Lutheran settings. In the aftermath of the 2016 Great Flood, a Baptist church coordinator offered her facility for these wellness workshops.

Helping One Another in Post-Disaster Wellness Workshops

In these workshops, facilitated by psychologists with the blessings of religious leaders, community people were guided in activities designed to foster resilience and form connections. They told their own stories, shared their feelings of loss and discovery, and bonded together. This process was a very healing experience. People shared songs, prayers, food, drawings, exercises, and altruistic activities. For example, in the 2016 Great Flood Wellness Workshops, during an exercise facilitated by Dr. Judy Kuriansky, children received small pillows constructed by their counterparts in Haiti, during the Hurricane Matthew Wellness Workshops, and created similar pillows (based on the psychological principle of promoting contact comfort) to be brought to children in other disaster situations (Kuriansky, 2019; Nemeth & Olivier, 2017, p. 73, Fig. 4.7 – Global Kids Project Pillows from Haiti and Louisiana).

Connecting with others is a healing and cathartic experience. In the 2016 Great Flood Wellness Workshop, children located Baton Rouge on a world map, and then connected a string to Haiti, and then another string to Japan. In this way, they learned that such tragedies are worldwide. They saw the connection with other children, supporting healing based on the psychological principle that, when faced with a trauma, it is comforting to know that "you are not alone." This perspective became a part of their healing journey. Parents shared in this experience and, like their children, found it heartening.

For the adults who attended the 2016 Great Flood Wellness Workshops, another creative arts activity allowed them to draw their individual experiences. This turned out to be emotionally releasing as the pain of the trauma was shared, and dissipated. Then, the groups came together to create a composite drawing. This process allowed group members to begin problem-solving, moving from the rigidity of their individual emotional pain and static memories to the flexibility of creation and wellbeing. As they began drawing and talking, a sense of resilience emerged.

As defined earlier in this chapter, resilience is about being firmly grounded in today, while learning from yesterday, so that people can see themselves in tomorrow. As was evident during Hurricane Katrina, some of those people, especially the elderly, whose circumstances would not allow them to see themselves in tomorrow, did not make it to tomorrow. Typically, these were seniors who did not have the money or physical strength to endure. Frequently, they, such as the French Quarter artist Harold "Napolean" King, had medical problems that could not be addressed (Nemeth & Whittington, 2012, p. 120). Due to the non-availability of accessible dialysis, Napolean died.

The 2016 Great Flood Wellness Workshops

The primary goal of the 2016 Great Flood Wellness Workshops, especially for seniors, was to help them make it to tomorrow. The focus was for them to survive and thrive. This was a significant lesson learned from the earlier experiences of the tragedy of Katrina. In fact, in the slow recovery of the Lower Ninth Ward after Katrina, "sixty-five percent of the property was owned by elderly," who were less than 5 percent back to their original community (Barrios, 2018, p. 152). As Ronald W. Lewis stated, "Without the people, we are nothing" (Barrios, 2018, p. 154). In the later tragedy of the 2016 Great Flood in Louisiana, only 13 people died. The Cajun Navy was credited for many effective rescues. The way people came together to help one another was remarkable. These facts further represent a bottom-up approach (by the people) to problem-solving (Nemeth & Olivier, 2017, pp. 59–75).

Wellness Workshops Promote Emotional Recovery

Community re-connection has been shown to be vital for recovery (Kuriansky, 2012). Psychologists are in a unique position to foster community rebuilding. Wellness Workshops are effective in helping people address their emotional pain so that they can open the doors to their tomorrows.

Nuclear scientists Onishi, Voitsekhovich, and Zheleznyak (2007) have written extensively about the effect of the Chernobyl Nuclear Disaster and indicated that one of the most long-lasting effects of this environmental trauma has been psychological. A similar conclusion was presented by Cameron at the American Psychological Association's 108th Convention in 2000 (Outcome of Millennium, 2000). Yet, as in the case of the Fukushima power plant accident in Japan in 2011, people's emotional pain and memories were seldom addressed (Maeda & Oe, 2015). Without group process intervention via wellness workshops, people remain stuck in the past. When they are firmly grounded in the present, however, they can move forward into the future. Even robust leadership to promote Economic Recovery and Environmental Restoration is not enough; the third "E," Emotional Resilience, must be addressed. Facing trauma, people's visual and auditory histories, their memories, and their unresolved feelings overpower their ability to think logically and to overcome their fears. Fear is a very important emotion. Only when people's fears are addressed and resolved can they move forward. Wellness workshops demonstrate how addressing people's emotions and fears can be resolved in a group setting (Kuriansky & Nemeth, 2013). Community sharing is a healing experience, oftentimes precluding the need for individual intervention. Sharing one's stories with others who have had similar experiences soothes the soul, frees the spirit, dissipates fears, and allows hope to return.

Renewed hope generates movement and allows people to see themselves in the future. They are no longer victims of the past. They are re-empowered to think, feel, and act, to use healthy problem-solving skills, and to plan for and implement their future.

Thus, the integration of the "3 Es" of Psychosocial Recovery is an essential aspect of a comprehensive disaster program. Usually, the emotional sequelae post disaster is under-recognized; oftentimes it manifests in physical symptoms. These manifestations, which may include headaches, panic, grief, emotional numbness, intrusive memories, reactivity, reminders of the event, sleep disturbance, irritability, depression, anxiety, and substance abuse, must be addressed and resolved (Nemeth et al., 2012) (see also Waters & Copeland, this volume). Wellness workshops have been shown by the authors to play a valuable role in assisting people to navigate their own journey through the six stages of psychosocial recovery.

Conclusion

The 3 Es of Psychosocial Recovery are intricately woven together. Disaster causes emotional trauma and economic and environmental destruction. Conversely, rebuilding the environment and the economy fortifies, but does not fully achieve, emotional resilience. By participating in the efforts noted in this chapter, mental health professionals and psychologists (the first two authors), working together with scientists (the third author), governmental experts and other stakeholders, can contribute to the 3 Es of Emotional, Environmental, and Economic psychosocial recovery from disasters. Although disasters cause trauma symptoms and subsequent Anniversary Reactions, appropriate intervention via Wellness Workshops can reduce their effects on affect, behavior and cognition (Gay et al., 2015, p. 65). Without thoughtful problem-solving and ongoing advocacy, the need for restoring emotional resilience post trauma fades from awareness, leaving people unprepared to face the future. The overarching goal of this chapter is to highlight the need for people, governmental agencies, and other stakeholders to include programs to promote psychosocial resilience in all planning considerations and interventions related to local, national, and/or global disasters.

References

Armstrong, J. (2009). A behind-the-scenes look at a day in the life of New Orleans Saints Coach Sean Payton. *The times-picayune*. Retrieved from https://www.nola.com/saints/2019/05/saints-punter-thomas-morstead-to-give-away-snowballs-in-old-metairie.html.

Bar-Levav, R. (1988). *Thinking in the shadow of feelings: A new understanding of the hidden forces that shape individuals and societies*. New York: Simon & Schuster.

Barrios, R. E. (2018). Political-ecological emergence of space and vulnerability in the lower ninth ward, New Orleans. In C. Ermus (Ed.), *Environmental disaster in the gulf south: Two centuries of catastrophe, risk, and resilience* (pp. 131–160). Baton Rouge: Louisiana State University Press.

Baurick, T. (2017). Is Louisiana really losing a football field of land per hour? *The times-picayune*. Retrieved from https://www.nola.com/environment/2017/05/is louisiana_losing_a_football.html.

Bloom, S. L. (1998). By the crowd they have been broken, by the crowd they shall be healed: The social transformation of trauma. In R. G. Tedeschi, C. L. Park, & L. G. Calhoun (Eds.), *Posttraumatic growth: Positive changes in the aftermath of crisis* (pp. 179–213). Mahwah, NJ: Lawrence Erlbaum.

Bornstein, P. E., & Clayton, P. J. (1972). The anniversary reaction. *Diseases of the Nervous System, 33*(7), 470–472.

Brinkley, D. (2006). *The great deluge: Hurricane Katrina, New Orleans, and the Mississippi gulf coast*. New York, NY: HarperCollins.

Brunkard, J., Namulanda, G., & Ratard, R. (2008). Hurricane Katrina deaths, Louisiana, 2005. *Disaster Medicine and Public Health Preparedness*. Retrieved from https://www.cambridge.org/core/services/aop-cambridge-core/content/view/8A4BA6D478C4EB4C3308D7DD48DEB9AB/S1935789300001142a.pdf/hurricane_katrina_deaths_louisiana_2005.pdf.

Engster J. & Campbell, F. (2014). Public service commissioner Foster Campbell [Radio broadcast]. In K. Gallagher (Producer), *The Jim Engster show*. Baton Rouge, LA: WRKF 89.3.

Ermus, C. (2018). Introduction. In C. Ermus (Ed.), *Environmental disaster in the gulf south: Two centuries of catastrophe, risk, and resilience* (pp. 1–13). Baton Rouge: Louisiana State University Press.

Eurich, J. (2018). Newtok is on the move. Retrieved from https://www.alaskapublic.org/2018/12/28/newtok-is-on-the-move/.

Farrow, R. & McHugh, R. (2016). Welcome to 'the most toxic place in America.' NBC News. Retrieved from https://www.nbcnews.com/news/us-news/welcome-most-toxic-place-america-n689141.

Fisher, R., Ury, W., & Patton, B. (1991). *Getting to yes: Negotiating agreement without giving in* (2nd ed.). New York: Penguin Books.

Fisher, J. (2007). Psychoeducational flipchart. Retrieved from https://janinafisher.com/flip-chart

Garrido, G., Onishi, Y., Hamilton, R., Nemeth, D.F., Albuquerque, J., & Nemeth, D.G. (2007). *Strategies to facilitate biosphere movement and lifestyle change: Measures to protect the environment and prevent drastic sequelae of current and future climate changes*. Symposium presented at the 60th Annual United Nation DPI/NGO Conference Midday Workshops, New York, NY.

Garrido, G. (2007). *Trauma resolution and lifestyle change*. Paper presented at the 60th Annual United Nations DPI/NGO Conference Midday Workshops, New York, NY.

Garrison Institute. (2012). *Comprehensive report on the 2012 climate, mind, and behavior initiative*. Retrieved from https://www.garrisoninstitute.org.

Gay, R. K., Nemeth, D. G., Kuriansky, J., Olivier, T. W., & Songy, C. (2015). The effects of environmental trauma on our thinking: Research on cognition and implications for public policy. In D. G. Nemeth & J. Kuriansky (Eds.), *Ecopsychology: Advances from the intersection of psychology and environmental protection* (Vol. 2, pp. 53–76). Santa Barbara, CA: Praeger.

Gibson, A. H., & Ermus, C. (2018). Swamp things: Invasive species as environmental disaster in the gulf south. In C. Ermus (Ed.), *Environmental disaster in the gulf south: Two centuries of catastrophe, risk, and resilience* (pp. 103–130). Baton Rouge: Louisiana State University Press.

Gibbens, S. (2019). The Amazon is burning at record rates – and deforestation is to blame. *National Geographic*. Retrieved from https://www.nationalgeographic.com/environment/2019/08/wildfires-in-amazon-caused-by-deforestation/#close.

Gotham, K. F. (2018). Katrina is coming to your city: Storm- and flood-defense infrastructure in risk society. In C. Ermus (Ed.), *Environmental disaster in the gulf south: Two centuries of catastrophe, risk, and resilience* (pp. 161–183). Baton Rouge: Louisiana State University Press.

Groupthink. (n.d.). *The Merriam-Webster.com Dictionary*, Merriam-Webster Inc., https://www.merriam-webster.com/dictionary/groupthink.

Haen, C. (2005). Rebuilding security: Group therapy with children affected by September 11. *International Journal of Group Psychotherapy, 55*, 391–414.

Harvey, M. R. (1990). An ecological view of psychological trauma and trauma recovery. *Journal of Traumatic Stress, 9*(1), 3–23.

Heifetz, R. A., & Laurie, D. L. (2001). The work of leadership. *Harvard Business Review, 79*(11), 130–141.

Honoré, R. L. (2012). *Leadership in the new normal: A short course*. Lafayette, LA: Acadian House Publishing.

Hunter, M. (2011). Air conditioners stolen whole for the copper wire inside. *The Times-Picayune*. Retrieved from https://www.nola.com/crime/2011/07/thieves_dragging_away_entire_a.html.
Jordan, K. (2003). What we learned from the 9/11 first anniversary. *The Family Journal: Counseling and Therapy for Couples and Families, 11*, 110–116.
Keim, B. D., & Muller, R. A. (2009). *Hurricanes of the gulf of Mexico*. Baton Rouge, LA: LSU Press.
Kelman, A. (2007). Boundary issues: Clarifying New Orleans's murky edges. *Journal of American History, 94*, 695–703. https://doi.org/10.2307/25095130
Kuriansky, J. (2019). Expressive arts for helping children heal in crisis and disaster. In I. A. Serlin, S. Krippner, & K. Rockefeller (Eds.), *Integrated health care for the traumatized: A whole person approach*. Lathan, NY: Rowman & Littlefield.
Kuriansky, J. (2012). Our communities: Healing after environmental disasters. In D. G. Nemeth, R. B. Hamilton, & J. Kuriansky (Eds.), *Living in an environmentally traumatized world* (pp. 141–167). Santa Barbara, CA: ABC-Clio (Praeger).
Kuriansky, J., & Nemeth, D. G. (2013). A model for post-environmental disaster wellness workshops: Preparing individuals and communities for hurricane anniversary reactions. *Ecopsychology, 5*(S1), S-38–S-45. https://doi.org/10.1089/eco.2013.0006
Kuriansky, J., Nemeth, D. G., Ray, K. P., and Songy, C. M. (2014). Intervention for environmental trauma: Resiliency, responsiveness, & recovery approaches. Symposium presented at the American Psychological Association's 122nd Annual Convention, Washington, DC.
Lemoine, R. L., Rostow, C. D., Nemeth, D. G., & Ellis, J. S. (1984). *The Livingston train derailment accident: An evaluation of the psychological impact on community residents*. An Evaluation Report Prepared for the Plaintiff's Steering Committee by Baton Rouge Psychological Associates.
Maeda, M., & Oe, M. (2015). The great East Japan earthquake: Tsunami and nuclear disaster. In K. E. Cherry (Ed.), *Traumatic stress and long-term recovery: Coping with disasters and other negative life events* (pp. 71–90). New York: Springer.
McCright, A. M., & Dunlap, R. E. (2011). The politicization of climate change and polarization in the American public's views of global warming, 2001–2010. *The Sociological Quarterly, 52*(2), 155–194. https://doi.org/10.1111/j.1533-8525.2011.01198.x
Mulcahy, G. A. (2013). Our indigenous people: Healing ourselves and mother earth. In D. G. Nemeth, R. B. Hamilton, & J. Kuriansky (Eds.), *Living in an environmentally traumatized world: Healing ourselves and our planet* (pp. 179–197). Santa Barbara, CA: Praeger.
Muller, R. A. (2012). Our atmosphere: Friend or foe? In D. G. Nemeth, R. B. Hamilton, & J. Kuriansky (Eds.), *Living in an environmentally traumatized world: Healing ourselves and our planet* (pp. 41–55). Santa Barbara, CA: Praeger.
Myers, D. (2007). *Psychology* (8th ed.). New York: Worth Publishers.
Nemeth, D. F. (2015a). Understanding our environment. In D. G. Nemeth & R. B. Hamilton (Eds.), *Ecopsychology: Advances from the intersection of psychology and environmental protection* (Vol. 1, pp. 37–64). Santa Barbara, CA: Praeger.
Nemeth, D. F., Nemeth, D. G., & Kuriansky, J. (2018). *Resilience and recovery post natural and human-caused disaster.* Symposium presented at the II Congress on Mental Health in Moscow, Russia.
Nemeth, D. G. (2015b). From chaos to community: The federal response – An account of Lieutenant General Russel Honoré's leadership during Hurricane Katrina. In D. G. Nemeth & J. Kuriansky (Eds.), *Ecopsychology: Advances from the intersection of psychological and environmental protection* (Vol. 2, pp. 5–13). Santa Barbara, CA: Praeger.
Nemeth, D. G., Hamilton, R. B., & Kuriansky, J. (2015). Reflections and recommendations: The need for leadership, holistic thinking, and community involvement. In D. G. Nemeth & J. Kuriansky (Eds.), *Ecopsychology: Advances from the intersection of psychological and environmental protection* (Vol. 2, pp. 413–428). Santa Barbara, CA: Praeger.

Nemeth, D., Kuriansky, J., Olivier, T., Whittington, L., May, N., Hamilton, J., et al. (2010). *Group Interventions for Disaster/Trauma Anniversary Reactions*. Presented at Returning to Katrina: Bringing Hurricane Katrina Research Back to the Community, Long Beach, MS

Nemeth, D. G., Kuriansky, J., Olivier, T. W., Whittington, L. T., May, N., Hamilton, J., et al. (2011). Group interventions for disaster: Trauma anniversary reactions. *Global Horizons: The Center for Policy and Resilience, 4*(1), 51–64.

Nemeth, D. G., Kuriansky, J., Reeder, K. P., Lewis, A., Marceaux, K., Whittington, T., et al. (2012). Addressing anniversary reactions of trauma through group process: The Hurricane Katrina anniversary wellness workshops. *International Journal of Group Psychotherapy, 62*(1), 129–142.

Nemeth, D. G., & Olivier, T. W. (2015). Perceptions of the environment. In D. G. Nemeth & R. B. Hamilton (Eds.), *Ecopsychology: Advances from the intersection of psychology and environmental protection* (Vol. 1, pp. 193–217). Santa Barbara, CA: Praeger.

Nemeth, D. G., & Olivier, T. W. (2017). *Innovative approaches to individual and community resilience: From theory to practice*. Cambridge, MA: Academic Press.

Nemeth, D. G., & Songy, C. (2015). Robust leadership and problem-solving in the face of environmental trauma. In D. G. Nemeth & J. Kuriansky (Eds.), *Ecopsychology: Advances from the intersection of psychology and environmental protection* (Vol. 2, pp. 89–104). Santa Barbara, CA: Praeger.

Nemeth, D. G., & Whittington, T. (2012). Our robust people: Resilience in the face of environmental trauma. In D. G. Nemeth, R. B. Hamilton, & J. Kuriansky (Eds.), *Living in an environmentally traumatized world* (pp. 113–140).

Netabay, N. (2007). Bottom-up approach: A viable strategy in solving the Somali conflict. Retrieved from https://www.beyondintractability.org/casestudy/netabay-bottom.

Norgaard, K. M. (2011). Climate denial: Emotion, psychology, culture, and political economy. In J. S. Dryzek, R. B. Norgaard, & D. Schlosberg (Eds.), *The Oxford handbook of climate change and society* (pp. 399–413). New York: Oxford University Press.

Onishi, Y. (2012). Our living waters: Polluting or cleansing? In D. G. Nemeth, R. B. Hamilton, & J. Kuriansky (Eds.), *Living in an environmentally traumatized world: Healing ourselves and our planet* (pp. 23–39). Santa Barbara, CA: ABC-CLIO/Praeger.

Onishi, Y., Garrido, G., Hamilton, R., Nemeth, D.F., Albuquerque, J., and Nemeth, D.G. (2007). Strategies to facilitate biosphere management and lifestyle change: Measures to protect the environment and prevent drastic sequelae of current and future climate changes. Presented at the 60th Annual UN/DPI/NGO Conference, New York, NY, Thursday, September 6, 2007.

Onishi, Y., Voitsekhovich, O. V., & Zheleznyak, M. J. (Eds.). (2007). *Chernobyl – What have we learned?: The successes and failures to mitigate water contamination over 20 years* (Environmental Pollution) (Vol. 12). Dordrecht: Springer.

Outcome of Millennium. (2000). Historical, technical, psychological, and research-based perspectives. Presented by D.G. Nemeth, C. Cameron, C.C. Creveling, R.M. Dreger, & M. Schexnayder at the 108th Annual Convention of The American Psychological Association in Washington, DC, 8/4–8/2000.

Phillips, S. B. (2005). The role of the bereavement group in the face of 9/11: A self-psychology perspective. *International Journal of Group Psychotherapy, 55*(4), 507–525.

Pritz, A., Kuriansky, J., Nemeth, D. G., Mulcahy, G. A., Walsh, N., & Madu, S. N. (2005). *Introducing a holistic and practical vision of human rights from different perspectives: The psychotherapy perspective*. Position paper presented at the United Nations annual conference, New York.

Rehm, D. (2015). The science of resilience and how it can be learned [Audio blog post]. Retrieved February 13. 2017 from http://dianerehm.org/show/2015-08-24/the-science-of-resilience-and-how-it-can-be-learned.

Rodin, J. (2014). *The resilience dividend: Being strong in a world where things go wrong*. New York: Public Affairs.

Scruton, R. (2013). Conservatism and the Environment. Retrieved from https://www.roger-scruton.com/articles/281-conservatism-and-the-environment.

Scruton, R. (2012). *How to think seriously about the planet*. New York: Oxford University Press.

Speier, A. H., Osofsky, J. D., & Osofsky, H. J. (2009). Building a disaster mental health response: Louisiana and hurricane Katrina. In K. E. Cherry (Ed.), *Lifespan perspectives on natural disasters: Coping with Katrina, Rita and other storms* (pp. 241–260). New York: Springer.

Speier, A. H., & Prats, R. (2015). Post hurricane Katrina and the preparedness-response-recovery cycle: Integrating behavioral health into a state's disaster-response capabilities. In D. G. Nemeth & J. Kuriansky (Eds.), *Ecopsychology: Advances from the intersection of psychology and environmental protection* (Vol. 2, pp. 89–104). Santa Barbara, CA: Praeger.

Substance Abuse and Mental Health Services Administration. (2014). *SAMHSA's concept of trauma and guidance for a trauma-informed approach*. Prepared by SAMHSA's Trauma and Justice Strategic Initiative (pp. 1–20).

Tierney, K. J. (1989). The social and community contexts of disaster. In R. M. Gist & B. Lubin (Eds.), *Psychosocial aspects of disaster* (pp. 11–39). New York: John Wily and Sons.

Verchick, R. R. M. (2010). *Facing catastrophe: Environmental action for a post-Katrina world*. Cambridge, MA: Harvard University Press.

Waldholz, R. (2018). Newtok to Congress: Thank you for serving our village. Retrieved from https://www.alaskapublic.org/2018/03/27/newtok-to-congress-thank-you-for-saving-our-village/.

Walker, J. & Heffner, F. (2010). Resilience as a Critical Factor in the Workplace. *The New Worker*.

Wolman, B. (1989). *Dictionary of behavioral science* (2nd ed.). San Diego, CA: Academic Press.

Zolli, A. (2012). Learning to bounce back. *The New York Times*. Retrieved from http://www.nytimes.com.

Chapter 19
Psychosocial Recovery after Natural Disaster: International Advocacy, Policy, and Recommendations

Judy Kuriansky

Introduction

Natural disasters throughout the world have claimed vast numbers of lives, decimated livelihoods, and subverted economic and growth and social development of communities, societies, and countries, especially in already fragile states with weak infrastructures and low human capacity to provide services. These events leave citizens and governments in desperate straits to rebuild on several levels. These levels have been identified as the 3 Es, postulated as Economic Recovery, Environmental Reconstruction, and Emotional Resilience (Nemeth, Kuriansky & Onishi, Chap. 18, this volume). While all three operate in a nexus, emotional resilience of the people is fundamental to the restoration of a functional social system. For emotional resilience to be achieved, it is important to ensure that policies on local, national, and international levels support the development of programs, as well as the training of personnel, allocation of resources, and building of health systems infrastructure to support recovery. These policies must be comprehensive, allow access and affordability of care, and be sustainable. This means that they must have the human capacity and financial resources to continue, and to be scaled up to serve many people.

This chapter has three purposes. First, I describe important advocacy on the part of psychologists regarding psychosocial resilience in disaster risk reduction and recovery. Second, I review historical and ground-breaking government agreements and policies that currently support such efforts which have all been formulated in recent times and that are important for all psychologists and stakeholders in the field of disaster risk reduction and recovery. Third, I describe the importance of these agreements as a foundation for interventions. In the conclusion, I offer suggestions for future advocacy directives and research efforts to advance emotional and psychological resilience in the wake of disasters. These are supported by my extensive

J. Kuriansky (✉)
Teachers College, Columbia University, New York, NY, USA

© Springer Nature Switzerland AG 2021

K. E. Cherry, A. Gibson (eds.), *The Intersection of Trauma and Disaster Behavioral Health*, https://doi.org/10.1007/978-3-030-51525-6_19

experience in both advocacy and innovative program implementations in this field (Kuriansky, 2018, 2019a; Masangkay, 2015).

Background About the Importance of Policy in Implementations

The literature includes many reports on the importance of community and social support after a disaster (e.g., Kuriansky, 2012; Phillips, 2015) and coping strategies to increase resilient outcomes (e.g., Silva Brown, 2015). Other evidence indicates the need for psychological interventions to build resilience post natural disaster (Jean-Charles, 2011; Nemeth & Olivier, 2017; Norris, Friedman, & Watson, 2002; Norris et al., 2002; Koyama et al., 2014; Rich & Sirikantraporn, 2018). Post-disaster psychosocial support interventions include trainings and workshops. For example, after Hurricane Katrina and in Baton Rouge, Louisiana USA, clinicians conducted "anniversary wellness workshops" on the anniversary of the traumatic event (Nemeth, et al. Chap. 18, this volume), given that anniversary dates trigger memories and re-experiences of the disaster (Kuriansky & Nemeth, 2013; Nemeth, Kuriansky, Oliver et al., 2011; Nemeth et al., 2012). Many other such trainings and workshops have been conducted by the first author and teams worldwide, including in Haiti, China, and Iran after earthquakes there; in Japan and Sri Lanka after the tsunami-earthquakes; in Sint Maarten after Hurricane Irma; and in the New York and New Jersey areas after Superstorm Sandy; as well as in the case of deadly epidemics like HIV/AIDS, in Hong Kong during SARS, and in Sierra Leone during the outbreak of the Ebola virus (Art Saves Lives Foundation, 2019; Kuriansky, 2013a, 2013b, 2016, 2017, 2018, 2019a; Kuriansky, Margevich, Jean-Charles, & Daisey, 2018; Kuriansky, Zinsou, Arunagiri, et al., 2015; Kuriansky, Wu, Bao, et al., 2015; Luce, 2010). These workshops have been based on guidelines and activities from my toolbox of interventions applied in varied cultural contexts that includes activities of expressive arts, as well as activities that facilitate feelings of safety, personal empowerment, and community re-connection (Art Saves Lives Foundation, 2019; Kuriansky, 2008, 2019b; Kuriansky & Jean-Charles, 2012).

Those interventions have requirements on several important and necessary levels. These requirements include that they build capacity, which means ensuring adequate resources of personnel, especially in low resource settings and rural areas in developing countries, where there are few or no trained psychologists, social workers, or other healthcare workers, and where the number of providers can be enhanced by task-shifting, i.e., training para-professionals and volunteers. Also, interventions must be sustainable, which means meeting the needs of the people in the present and also for the future without requiring the presence of original experts or trainers, by empowering the community to provide for its own needs. In my work, I achieve these requirements by applying a tiered train-the-trainer model, whereby master trainers train local health providers, community leaders and volunteers who then train other cohorts in the community, that has been shown to be effective (Kuriansky, 2019d; Kuriansky et al., 2018; Kuriansky, Zinsou, Arunagiri et al., 2015).

These projects by myself and my team focus on three pillars: building psychosocial resilience, empowerment, and community re-connection. These pillars form the basis of my advocacy about the importance of psychosocial resilience in disaster recovery and risk reduction, and recommendations for policy. The overarching goal outlined in this chapter is to highlight the need for governments and other stakeholders to recognize and include contributions of psychology and the concept of psychosocial resilience in all planning considerations and interventions related to national or global disasters. The ultimate intention is to ensure the mental health and well-being of all peoples before, during, and even long after any such tragic event (for a review of injury and mortality, health systems and infrastructure, mental health, infectious disease, chronic disease, and health behavior after disaster, see Johnson & Galea, 2009).

The Nature of Policy

Policy is necessary and important to support disaster interventions and to implement recommendations. Policy is defined as "a definite course or method of action related to a specific social problem or issue in light of given conditions to guide and determine present and future decisions" (Merriam-Webster, n.d.). Such a plan is designed and then implemented on the highest level possible that outlines the general goals and acceptable procedures. Policy is generally formed by a governmental body and meant to guide their actions.

Policy can be drafted by a small group of people in an inner circle, or it may be a vast and extensive process over a long period of time. Drafting policy takes into account the opinions and needs of a vast number of constituents, especially those who will be the beneficiaries of such decisions. International agreements, compacts or resolutions at the highest level related to disaster recovery have generally followed a process taking many years and taking into account the views and input of many experts and stakeholders, with the ultimate adoption of these documents ultimately resting on voting by the governmental delegates on the international stage, i.e., at the United Nations or other convenings.

The method of enforcing such policy is as important as its drafting. Legal systems are usually set in place in policy asking for government action. However, agreements made on the international level at the United Nations are not legally binding. Even though the governments at the United Nations, made up of 193 Member States, work towards consensus on such agreements, these are no means of enforcement. Additionally, policy phrases are qualified by the context of the sovereignty of individual states (UNISDR, 2015; United Nations, 2018). Despite these limitations, the body strives for unanimous agreement.

The Nature of Advocacy

Advocacy is the process of making a point that supports a desired action, to influence another party to accept and act upon that point in interest. A key aspect of advocacy is presenting your argument in a statement. The advocacy statement should be concise, presenting evidence and support through a list of relevant agreements or research projects that support the point, much like a legal brief. This statement should make very straightforward easily understood recommendations, which ideally fit the wording and process of the current situation or negotiations.

I have conducted this advocacy in my role as a representative of psychological nongovernmental organizations (NGOs) that are accredited to the Economic and Social Council of the United Nations and affiliated with the Department of Global Communication/NGO Section. These accreditations, which require applications to meet criteria, allow NGO representatives to participate in meetings at the United Nations and to have access to interact with government representatives who are appointed to represent their state at the United Nations in negotiations on international topics and agreements. These offices are called Missions, headed by Permanent Representatives, called Ambassadors, along with Deputies and other staff who handle specific topic areas.

Disaster Recovery in the International System: Relevant Offices, Agreements, and Conferences

Given the vast numbers of disasters throughout the world, and their impact on both national stability as well as international relations, the United Nations has maintained an office since 1999 that coordinates international efforts related to disaster risk reduction. As of May 2019, this office is called the United Nations Office for Disaster Risk Reduction (UNDRR), renamed from the International Strategy for Disaster Risk Reduction (ISDRR), to better reflect its activities. The office reports on progress of a major agreement about governments' commitments regarding disaster, called the Sendai Framework, and convenes biennial conferences to discuss disaster issues, called the Global Platform on Disaster Risk Reduction, as discussed more fully in the sections that follow.

The UNDRR is led by a United Nations Special Representative of the Secretary-General for Disaster Risk Reduction and has over 100 staff located in headquarters in Geneva, Switzerland, with five regional offices (Nairobi, Africa; Panama City, Panama, for the Americas; Cairo for the Arab State; Bangkok for the Asia-Pacific region; and Brussels for Europe), with other field presences in Addis Ababa, Ethiopia, Almaty, Kazakhstan; Bonn, Germany; Incheon, South Korea; Kobe, Japan; Rio de Janeiro, Brazil and Suva, Fiji, and New York City at UN Headquarters.

Important international frameworks and conferences relevant to disaster are described next.

The Sendai Framework

A major job of the UNDRR is to report on the implementation and review of the Sendai Framework for Disaster Risk Reduction (UNISDR, 2015), so named because the governmental negotiations took place in Sendai, Japan, following the practice that many UN agreements are named for the city in which the meetings take place. This voluntary, nonbinding, and people-centered framework outlines a 15-year agreement (from 2015–2030) by governments of the world regarding the responsibilities, targets, and priorities for reducing and preventing global disaster risk, and for efficient and effective disaster recovery. Adopted on March 18, 2015, the agreement built on the previous agreement called the Hyogo Framework, agreed upon at a conference in Hyogo, Japan, that covered the years 2005–2015.

The framework recognizes that, while the State has the primary role to reduce disaster risk, responsibility should be shared with local government, the private sector, and other stakeholders, including civil society. The directives aim to reduce losses in lives, livelihoods and health, for the benefit of the economic, physical, social, cultural, and environmental welfare of people, businesses, communities, and countries.

A single sentence in the Framework supports psychosocial resilience, and hence has guided my activities and advocacy. Paragraph 33 (o) says that to achieve Priority 4: Enhancing disaster preparedness for effective response and to "Build Back Better" in recovery, rehabilitation, and reconstruction, on national and local levels, it is important "To enhance recovery schemes to provide psychosocial support and mental health services for all people in need."

This sentence has provided the foundation for my extensive advocacy highlighting "psychosocial resilience" as distinct from infrastructural resilience, in order to underscore the importance of one of the 3 Es, namely, Emotional Resilience (Nemeth et al., Chap. 18, this volume). As such, I have presented at all the meetings of the UN disaster office, dating back to 2007, and the critical conferences evaluating the progress in achieving the agreements of the Sendai Framework, in Sendai in 2015, in Cancun in 2017, and in Geneva in 2019 (Kuriansky, 2019a; Masangkay, 2015).

My advocacy efforts to promote psychosocial resilience are supported by the fact that the framework is intended to be "people-centered" and also by extensive research indicating the extent of emotional sequelae of disaster as not only immediate but also long-term. While I have observed that many statements at United Nations conferences about disaster – such as the 2017 Hurricanes Harvey, Irma and Maria that devastated the Caribbean islands as well as Texas, Florida, Puerto Rico and the U.S. Virgin Islands – mention the word "resilience," these generally refer to infrastructure like alarm systems and building codes, as opposed to the emotional needs of the people affected. At many of these meetings, I have distributed written statements with recommendations, and also made public statements whenever possible, about the importance of psychosocial resilience and the focus on people (Kuriansky, 2015).

Thus, the goal of my advocacy has been to underscore and amplify the importance of this sentence in the Sendai Framework and to encourage States and other stakeholders to increase efforts to realize this call to "enhance recovery schemes to provide psychosocial support and mental health services for all people in need." My further intention is to lay the groundwork for elaborating on this call for action in future iterations of the Framework, i.e., during negotiations that will start a few years before the framework expiration date of 2030.

It is interesting to note that the precursor to the Sendai Framework was the "Hyogo Framework for Action 2005–2015: Building the Resilience of Nations and Communities to Disasters" (agreed upon in Kobe, Hyogo, Japan, at the World Conference on Disaster Reduction, in January 2005, International Strategy for Disaster Reduction, 2005). In that framework, the words "psycho-social" and "psychological" appeared, to read, "Enhance recovery schemes including psycho-social training programmes in order to mitigate the psychological damage of vulnerable populations, particularly children, in the aftermath of disasters."

Each agreement provides an opportunity to further advocate about the importance and inclusion of psychological issues in disaster planning. This makes efforts by psychologists, including myself, continually important and relevant, and potentially impactful. As the former head of UNISDR, Margareta Wahlström once told me, "Judy, never stop reminding us of the importance of psychology and psychosocial resilience."

The United Nations Agenda 2030 for Sustainable Development

The historical inclusion of the call for the "promotion of mental health and well-being" in target 3.4 of the United Nations 2030 Agenda for Sustainable Development, provides a foundation, in a major international document, for advocacy about post-disaster psychological recovery. This target was ensured by a highly successful campaign during the intergovernmental negotiations led by my good friend, then-Ambassador of the Republic of Palau to the United Nations, public health physician Dr. Caleb Otto, representing the Member States (governments) of the UN, in partnership with me, as a representative of psychologists and civil society (Forman, 2014; Kuriansky & Zinsou, 2019). This effort was an excellent example of combined and committed efforts by government and civil society that leads to such accomplishments. The effort involved several years of drafting and distributing policy papers and holding meetings with United Nations delegates involved in the negotiations held over 2 years at the UN New York headquarters that led to the ultimate adoption of the agenda in 2015. Those five words – "promote mental health and well-being" – can now form the basis for huge steps forward in disaster recovery as well as every aspect of social, economic, and emotional development (i.e., the 3 Es).

The Global Compact for Safe, Orderly, and Regular Migration

Advocacy regarding psychosocial resilience was also conducted during the negotiations for the Global Compact for Safe, Orderly and Regular Migration, adopted in December 2018 in Marrakesh, Morocco, in which I participated actively as part of civil society and as a member of the UN NGO Committee on Migration (Kuriansky, 2019c; for the compact, see United Nations, Global Compact for Safe, Orderly and Regular Migration - Intergovernmentally Negotiated and Agreed Outcome, 13 June 2018). The importance of emotional resilience is importantly mentioned in this Global Compact, which includes a section about "Natural disasters, the adverse effects of climate change, and environmental degradation" with actions "to address the vulnerabilities of persons affected by sudden-onset and slow-onset natural disasters, by ensuring they have access to humanitarian assistance that meets their essential needs with full respect for their rights wherever they are, and by promoting sustainable outcomes that increase resilience and self-reliance, taking into account the capacities of all countries involved." Notably, several references in the document are made to provision of psychosocial services, using the words "psychosocial," "mental health," and "counseling."

The importance of psychosocial resilience for migrants, as well as for refugees, is underscored by slow-onset climate disaster occurring in countries threatened by climate change and rising sea levels, leading to what is being referred to as "climate refugees." Such slow-onset disaster has been recognized in the Global Compact. Populations on islands such as in the Small Island Pacific States, e.g., the Marshall Islands, are in grave danger of losing their homeland as the land sinks into the sea, causing intense trauma. Highlighting the plight of these "climate refugees" is a major focus of my ongoing advocacy (Kuriansky, 2019a).

The Political Declaration on Health Coverage

Advocacy about the importance of mental health and of psychosocial resilience in disaster recovery was conducted by myself and colleagues for the important "Political Declaration on Universal Health Coverage, Moving Together to Build A Healthier World," adopted in October 2019 by the government Member States of the United Nations (United Nations General Assembly, 2019). This highly significant declaration represents their commitment to provide access to health coverage to all people. In this regard, I drafted and submitted a statement on behalf of the Psychology Coalition of NGOs accredited at the United Nations (PCUN) to the co-facilitators at the Missions of Georgia and of Thailand to the United Nations, in time for this reference to be considered and successfully included (as acknowledged by return email). The first paragraph of the declaration affirms "the right of every human being, without distinction of any kind, to the enjoyment of the highest attainable standard of physical and mental health" and later, recognizes "the consequence of the adverse impact of climate change, natural disasters, extreme weather events as well as other environmental determinants of health… underlining that resilient and people-centered health systems are necessary to protect the health of all people, in particular those who are vulnerable or in vulnerable situations." As Weems and Banks (2015) have pointed out, an ecological needs-based perspective can be valuable for addressing post-disaster mental health issues particularly among adolescents as a vulnerable population (see also Weems & Overstreet, 2009).

The Astana Declaration on Primary Health Care

Another document important to note in this chapter, as it is another recent international agreement that includes mental health and consequently affords a foundation for advocacy about psychosocial resilience, is the Astana Declaration on Primary Health Care, adopted at a conference in Kazakhstan in October 2018. As in previous work, I co-drafted a major statement to support this advocacy, which was presented in person by a colleague from the UN Major Group for Children and Youth at the conference. Participation in these two steps, drafting a statement and presence to present the points at a conference, was again evident, as in the cases described earlier, and as described in more detail in the example in the next section.

Advocacy at United Nations Conferences on Disaster Risk Reduction: A Case Study of Process and Progress

Advocacy regarding the importance of mental health and psychosocial resilience on the international stage dates back to 2007, at the disaster risk reduction conference held in Geneva, Switzerland, headquarters of the International Strategy for Disaster Risk Reduction (so-named at that time), which brings together representatives from governments, UN agencies, financial and academic institutions, NGOs, and other members of civil society. Future meetings followed similar formats. The 3-day meeting consisted of high-level plenaries, sector workshops, and side events, mainly focused on government negotiations but with opportunities for input and meetings with civil society (NGOs). The advocacy on behalf of the United Nations NGO Committee on Mental Health was conducted by executive committee board members, myself, and clinical psychologist Inka Weissbecker, then Research Scientist at the Center for Hazards Research and Policy Development at the University of Louisville (now at the World Health Organization).

The advocacy was on many levels, that serves as an example for how efforts are done, and how success can be achieved to include awareness of psychology and mental health in disaster planning. One method is to speak up at meetings. At a thematic workshop on "Disaster Risk Reduction in the Health Sector," which emphasized the importance of what WHO means by "Safe Hospitals" in terms of structural, nonstructural, and functional aspects (e.g., the poor planning of a hospital built on a waterbank), in the discussion period, I emphasized the importance of including mental health in all languaging, planning, and summary statements from the health sector. Due to my collaboration with the civil society group, mental health was included in the group's finals statement submitted to the co-chairs. Another important effort, as noted above in this chapter, is to distribute a statement, prepared before the meeting, about the importance of psychosocial issues in disaster risk reduction, and recommended actions. Statements should be endorsed by as many organizations as possible, as an impressive showing of the number of stakeholders who support the points. Of course these should include psychology-related organizations but should also include organizations from related fields. This statement was endorsed by eleven organizations.[1] We also distributed an announcement about an upcoming Department of Public Information NGO conference in September about "Climate Change: How It Impacts Us All."

On behalf of the Committee on Mental Health (CMH), we also presented a "side event" on the topic of "The Integration of Mental Health and Psychosocial Issues into Disaster Risk Reduction and the Hyogo Framework," constituting a major opportunity to highlight the issue in such a major international forum. Exceptionally significant – and groundbreaking – was that very high-level persons agreed to be on the panel, including then-Assistant Secretary-General for Humanitarian Affairs and Deputy Emergency Relief Coordinator of the Office for the Coordination of Humanitarian Affairs Margareta Wahlström (mentioned previously in this chapter), who supported the importance of integrating psychosocial and mental health interventions into the Framework, as part of a comprehensive program of disaster preparedness, response, and recovery. Her support emboldened our advocacy. Of note, Robert Glasser, who took over the head position of the UNDRR, made similar encouraging remarks to me at the meeting in Cancun, encouraging me to advocate about psychosocial resilience.

The panel was also groundbreaking in that Mark van Ommeren from the World Health Organization's Department of Mental Health and Substance Abuse announced the launch of the new "Guidelines on Mental Health and Psychosocial Support in Emergency Settings" (Kuriansky, 2011). These guidelines were drafted by the Inter-Agency Standing Committee (IASC) with endorsement by 27 agencies representing 10 UN agencies (including OCHA, UNFPA, UNHCR, UNICEF, and WHO), the Red Cross/Red Crescent movement, 16 leading NGOs, and the World Bank, and remain to this day as the gold standard in interventions in disaster. Ommeren asked the attendees to reflect on their positive and negative experiences

[1] the World Council for Psychotherapy, International Association of Applied Psychology, International Association of Schools of Social Work, International Council of Psychologists, International Psychoanalytical Association, International Union of Psychological Science, MindFreedom Support Coalition International, Soroptimist International, American Psychological Association, World Association for Psychosocial Rehabilitation and World Federation for Mental Health.

related to mental health and psychosocial support in emergencies, and encouraged integrated preparation and response, inter-agency consensus on mental health and psychosocial support, and self-care of responders. Considerations, that also remain to this day, include cultural sensitivity, assessments beyond PTSD, and ensuring the welfare of healthcare staff.

As I do in many events to this day, serving as moderator and panelist, I reviewed research and experience that disasters impact the mental health and psychosocial well-being of individuals, families, and communities immediately and long term, resulting in threats to overall health, human rights, and social development (Maeda & Oe, 2015). I further emphasized that psychosocial/mental health interventions should be an integral part of any comprehensive disaster program given that adults and especially children suffer from losses and worry about recurrences. Also, since a locally trained psychological staff is scarce in developing countries, community resources need to be enhanced and local people trained. And importantly, such services need to be integrated into primary health care.

Weissbecker described a unique multi-stakeholder model being applied in a developing country vulnerable to climate change, namely, Belize, involving government, school and community agencies and her university departments, that also has low resources (i.e., only one psychiatrist and one clinical psychologist). The program, following the IASC guidelines, developed handouts for the general public regarding disaster preparedness, response, and coping, especially for children. It also included plans to train parents to foster more resilient families. Another panelist explained how applied psychology can support the framework with assessments and by rebuilding social networks and resettling survivors, that are essential steps in disaster recovery.

All such meetings provide valuable opportunities for connecting with government delegates and also to connect with other NGOs to further the campaign of including psychosocial issues in disaster risk reduction planning.

It is always important to ensure that the advocacy points are included in final summaries. At the closing plenary of this meeting, because health was not mentioned in the short summary even though the Health sector representative submitted recommendations, Weissbecker and myself hurriedly and successfully submitted a request to the Secretariat to be on the list of speakers. Luckily, I was included, as the last to speak, and able to make a verbal statement about the importance of including attention to psychosocial issues in a holistic risk-reduction program to address community needs.

Further, for the final report, we drafted an extensive summary about the importance of mental health and psychological well-being, submitted to the UN Under-Secretary General for Humanitarian Affairs and to Director of the Secretariat of the International Strategy for Disaster Reduction.

The statement read:

> Health and mental health/well-being are integral to humanitarian efforts and social development, negatively impacted by disaster, and essential to build resilient communities for disaster risk reduction. Therefore, it is recommended that physical and mental health issues and experts be integrated into all policies, plans and programs in ISDR and in the

implementation of the Framework for Action. This includes assessing and monitoring physical and mental health needs, providing resources for appropriate interventions, and developing education and training programs about physical and mental health preparedness in community and institutional settings.

Further Advocacy at Global Conferences

To influence international agreements, it is important to continually be involved and to take advantage of every opportunity to advocate about an issue. Thus, I followed up the meeting in Geneva by co-organizing and moderating a panel later that year (2007) in September, at the United Nations Department of Public Information/NGO conference, held at UN headquarters in New York. The panel was on "Partnerships to Mobilize Community Health and Mental Health Resources for Recovery, Resilience and Risk Reduction of Climate-Related Disasters: What Multi-Stakeholders and NGOs Can Do."

Continuing to participate in meetings at psychology conferences with colleagues is also critical, to consolidate support with civil society groups. The next year, in October, 2008, I presented a symposium on "Climate Change, Social Trauma, and Lifestyle Modification" at the World Congress of the World Council for Psychotherapy (WCP), in Beijing, China, during with colleagues from WCP as well as scientists who are expert on climate change. This was an extension of the WCP World Congress in Buenos Aires, Argentina, in 2005, when a position paper was developed calling for freedom from emotional suffering, including the need for post-disaster recovery.

This advocacy became very intensive at the Global Platforms on Disaster Risk Reduction held in Sendai in 2015, in Cancun in 2017, and in Geneva in 2019 referred to earlier in this chapter. For example, at the Third UN World Conference on Disaster Risk Reduction, in Sendai in 2015, I and youth representatives (Joel Zinsou and Quyen Nguyen) from the International Association of Applied Psychology (IAAP), an NGO which I represent at the UN, presented a report about my team's mission to Japan after the tsunami-earthquake. This presentation was made at a session called the "Ignite Stage," meant to promote innovative approaches to disaster risk reduction and recovery. Following my principle to take advantage of multiple opportunities to make presentations in different formats at any one meeting, I also participated on panels at the pre-conference hosted by my colleagues at the United Nations Major Group for Children and Youth (UNGCY), where I included a fellow IAAP representative, Father Wismick Jean-Charles, an educational psychologist and Haitian Catholic priest with whom I had conducted many disaster interventions during the earthquakes and hurricanes in Haiti (Kuriansky & Jean-Charles, 2012; United Nations Office for Disaster Risk Reduction, 2015). Also, I lobbied, and was selected, to make a statement at the major government session about the importance of psychosocial resilience given the ongoing suffering of the Japanese people since the tsunami (Kuriansky, 2015, April 20).

At the fifth session of the Global Platform for Disaster Risk Reduction held in Cancun, Mexico, in 2017, I again presented at the "Ignite Stage," this time about my trainings and workshops for students and community leaders to help children and community members cope after Hurricane Matthew devastated Haiti in 2016. I also demonstrated these methods at the UNMGCY exhibit booth. As I usually do, I brought a young professional along to assist, but mainly to learn and gain experience in co-presenting, this time, my teaching assistant, Alexandra Margevich, from my class on "Psychology and the United Nations" at Columbia University Teachers College.

Two years later, in Geneva in 2019, when the UNDRR meeting focused on the topic of "Resilience Dividend: Towards Sustainable and Inclusive Societies," I made use of the word "resilience" as highly relevant for advocacy about psychosocial resilience. At the conference, I organized and presented at a 3-hour session on "Innovations in Resilience for Disaster Risk Reduction: Artificial Intelligence for Planning, Policy and Psychosocial Support." I partnered in this project with another IAAP UN representative, Russell Daisey, who is a member of my team on many disaster missions, and with Dr. Shariha Khalid Erichsen, my colleague on the "Health in Your Hands" initiative described later in this chapter, as well as a representative of an international technology company. This session was timely, given three recent natural disasters: (1) the recent Cyclone Fani in India and Bangladesh, (2) an earthquake in southern Iran the previous month, and (3) cyclones in Mozambique. Notably, deaths from Cyclone Fani were kept at a minimum due to evacuation, but the emotional trauma nonetheless deserved major consideration. At the session, a video was shown of my interventions around the world, to build emotional resilience for children, including expressive arts of music and art constructions (Kuriansky, 2019b). The use of technology tools like artificial intelligence and robotics was described to facilitate communication as well as deployment of first responders and delivery of needed supplies and coordination of all efforts. A panel addressing "Technologies and Innovations For the Last Mile in Disaster Risk Prevention, Reduction and Recovery – What Is The Reality On The Ground?" included the following speakers: (1) Dr. Siyana Mahroof-Shaffi, a physician and Chair & Founder of Kitrinos Health that provides mobile and static health care to refugees in Greece; (2) Dr. Kenneth Carswell, a clinical psychologist and consultant for the World Health Organization Mental Health Innovation Network, who is developing technology tools for mental health, including a Self-Help+, a guided multimedia psychosocial self-help package delivered by facilitators with minimal training, and "Step-by-Step," an e-mental health brief and scalable psychological intervention delivered through the internet, being applied in Lebanon; and (3) Pradip Khatiwada, a member of the UN Major Group for Children and Youth and Executive Director of the Youth Innovation Lab, a civic tech company that recently developed a platform called BIPAD (Building Information Platform Against Disaster), an integrated Disaster Information Management Platform for Nepal which includes Open Data and the prospects of AI and Drone technology, for the Ministry of Home Affairs' National Emergency Operation Centre.

The role of the private sector in disaster risk reduction and recovery is increasingly being recognized. Accordingly, another panel addressed "Resilience dividends and returns on investment – Innovative Approaches to Financing Disaster Risk Reduction & Recovery using an innovation, climate and gender lens," focusing on the private sector contributions to financing such innovations. Fostering emotional resilience in disaster was described by Daisey, who distributed soft globes called "Hugg-A-Planet" (an original design by environmentalist Robert Firenza) to attendees, accompanied by the original anthem co-written by Daisey and me called "Happy People, Happy Planet" that had been debuted at a major event at the United Nations on the International Day of Happiness. Daisey and I described the interlinkage of Emotional Resilience and Environmental Rebuilding in disaster recovery, two of the three Es in the model of the "3 Es" of disaster recovery referred to earlier in this chapter (Nemeth et al., Chap. 18, this volume). Complementary copies were also distributed of the book I co-edited, *Living in an Environmentally Traumatized World: Healing Ourselves and Our Planet* (Nemeth, Hamilton & Kuriansky, 2012). The session was well attended and attendees reported significant benefits and learning, and welcomed this approach at the conference.

Again making use of every opportunity to publicly refer to the importance of psychological resilience in disaster, I also made a statement during one of the major sessions about disaster. In this comment, I raised the need for attention to the long-term effects of disaster on survivors and the importance of distinguishing psychosocial resilience from structural resilience, emphasizing the importance of a focus on people, not just buildings.

Outcome of the 2019 Geneva ISDRR Conference

A strong focus at the 2019 conference was on the "climate–security nexus," meaning the stresses when a natural disaster strikes a country already struggling with civil conflict, heightening the humanitarian crisis. Noting the vulnerability of women in disaster, another focus was on gender-sensitive approaches to risk reduction; regarding gender parity, women represented half of the panelists and 40% of the participants. Also, more than 120 persons with disabilities attended.

The co-chairs' summary stated that "Achieving the 'Resilience Dividend' will require countries, communities, enterprises, and individuals to increase capacities to participate in preventing and reducing risks." The summary further noted that nature- and ecosystem-based approaches should be promoted to achieve the objectives of resilience dividend and integrated in disaster risk reduction strategies at all levels, and urged increased budgetary allotment, and more attention to boosting resilience in least developed countries, landlocked developing countries, and small island developing States. The overarching message was that resilience pays off, resulting in a strong call to leaders at all levels to ensure resilience dividends for all. The summary made the social, environmental, and economic case for disaster risk reduction; noted that mortality is on a downward trend and that the bulk of the human cost occurs in low- and middle-income countries; and mentioned "health" twice (applauding new and innovative guidance, tools, and instruments in that area, and urged enhancing investments in resilient health facilities. Yet, wording about mental health and psychosocial resilience did not appear.

The Global Platform conference was meant to contribute to the 2019 High-Level Political Forum on Sustainable Development that was held in July 2019, and the Climate Action Summit on 23 September 2019 held at UN headquarters in New York City. The conclusions called for accelerated action in achieving the seven targets of the Sendai Framework and highlighted the importance of disaster risk reduction to achieve the 2030 Agenda for Sustainable Development, and the contribution of the Sendai Framework towards the Paris Agreement, the Agenda for Humanity, the New Urban Agenda, and the SAMOA Pathway.

At the closing of the Geneva conference, UN disaster prevention chief Mami Mizutori of Japan, appointed on January 31, 2018, by the UN Secretary-General as Assistant Secretary-General and Special Representative of the Secretary-General for Disaster Risk Reduction, noted disappointment that only 91 out of 197 UN member states (governments) had developed national and local strategies to ward off disasters ahead of a 2030 deadline.

The UN Summit Climate Change in September 2019

Leaders from government, business, and civil society announced steps to confront climate change at the United Nations Secretary-General's one-day historic Climate Action Summit at UN headquarters in New York held on September 23, 2019, during the high-level United Nations General Assembly. Sixty-five countries, and even the US state of California, committed to cut greenhouse gas emissions to net zero by 2050, 70 countries announced they will boost their national action plans by 2020, and over 100 business leaders committed concrete actions to align with the Paris Agreement targets and speed up the transition from the gray to green economy. In closing the Summit, The UN Secretary-General said, "You have delivered a boost in momentum, cooperation and ambition. But we have a long way to go."

Notably, an entire day on the previous Sunday was devoted to youth involvement, with youth from around the world invited and selected to attend. This is consistent with the international emphasis on involving youth in this issue and the awarding of the Nobel Peace Prize to 15-year-old activist Malala Yousafzai.

The UN Summit on Universal Health Coverage in September 2019

On the same day as the major Climate Change Summit, 23 September 2019, a major Summit took place on Universal Health Coverage, held at UN headquarters in New York during the UN General Assembly. Impressively, the Director General of the World Health Organization, Dr. Tedros Adhanom Ghebreyesus, announced that health is at the center of the UN Agenda 2030 for Sustainable Development. Governments made commitments to the Political Declaration, some of which specifically mentioned the importance of mental health and psychosocial support (MHPSS). These included Finland, Argentina, and the Netherlands who also represented Nepal, Tunisia, New Zealand, Australia, Uruguay, Denmark, El Salvador, and others, and whose statement was consistent with the theme of the meeting they hosted the next month, described in a following section of this chapter. Civil society organizations also submitted statements on the issue.

The International Conference on Mental Health and Psychosocial Support in Crisis Situations, October 2019

To continue advocacy and form further collaborations with colleagues from civil society and government, I attended the international conference on Mental Health and Psychosocial Support in Crisis Situations that was held in the Netherlands in October 2019. The 2-day meeting, hosted by the government of the Netherlands and specifically the Ministry of Foreign Trade and Development Cooperation, brought together national Ministers of Health, Public Health, and other related Ministries of various countries with civil society advocates to discuss disaster recovery policy and programs. An important theme, in response to inadequate funding of this issue was, "Do not ask whether you can afford MHPSS, ask whether you can afford not to." Important commitments were made by government representatives about psychosocial support in crisis, including Uganda, India, and Italy. Regarding the consistent theme to establish best practices (with proven outcomes) in the field, three examples were presented. These were: (1) efforts in Sri Lanka to offer comprehensive mental health and PSS after decades of suffering from the aftereffects of the tsunami and the added trauma of conflict in the region; (2) accomplishments in the Philippines, which had been devastated by Typhoon Haiyan, to now offer trained personnel with social care assistants to manage mental health conditions; and (3) more widely available MHPSS in Syria, despite challenges of ongoing conflict, through community and women's centers and school-based programs rather than previously through mental hospitals and private-practices.

Key recommendations included to: (1) support a continuum of care, including MHPSS; (2) implement long-term programs; (3) offer MHPSS to displaced and host communities; (4) address underlying factors, like poverty, unemployment, and exposure to violence. (5) strengthen the voice of persons with disabilities and other excluded, vulnerable, and at-risk groups; (6) integrate MHPSS into health and community services; (7) enhance research on evidence-based approaches; (8) utilize effective delivery platforms and technology; (9) support further research; (10) establish competencies and provide training for MPHSS workplace; (11) enact relevant policy; (12) encourage donors for needed funding. Civil society workshops focused on sharing practices for disaster recovery and methodologies in emergency settings, including those from organizations like WarChild.

The WHO Mental Health Forum, October 2019

Also in September, 2019, civil society organizations gathered with governments at the WHO Mental Health Forum at WHO headquarters in Geneva, which I also attended. Relevant to the discussions and outcome regarding mental health interventions in humanitarian settings, governments made commitments and case studies were presented of what is possible to achieve in challenging settings, including Peru, Lebanon, and the Ukraine. The mhGAP Community Toolkit was launched, aimed at increasing and facilitating country action and implementation of mental health. Also, WHO hosted a panel and discussion group on interventions in humanitarian and emergency settings. Practices were shared by the NGO group IsraAID delivering MHPSS after disasters worldwide, with whom I had collaborated in Sierra Leone during the Ebola outbreak (Kuriansky, Polizer, & Zinsou, 2016).

Advocacy Statements at UN Disaster Conferences

Statements are a critical part of advocacy to influence policy. At all opportunities at disaster conferences, statements are drafted about the importance of psychological resilience. The more organizations supporting such statements, the more impact they can have. My advocacy efforts about psychosocial resilience are supported by the Psychology Coalition of NGOs accredited at the United Nations (PCUN), consisting of psychology-related NGOs authorized by successful application to the Economic and Social Council (ECOSOC) to advocate within the United Nations system to promote the contribution of psychological science and practice to issues on the global agenda ranging from poverty eradication and women's empowerment to combating climate change and ensuring peace, health, and well-being.

On behalf of PCUN, as well as the World Council for Psychotherapy (WCP) and the International Association of Applied Psychology (IAAP), both of which I represent at the UN, I drafted and distributed a statement at the Global Platforms for Disaster Risk Reduction held in Sendai in 2015, in Cancun in 2017, and in Geneva in 2019. An example of the statement in 2017 about the Importance of Psychosocial Resilience and Mental Health and Well-being in the 2030 Agenda:

> *The Global Platform for Disaster Risk Reduction, From Commitment to Action Cancun, Mexico, May 2017*
>
> The importance of psychosocial resilience is key to disaster risk reduction and recovery. Target 3.4 of the UN Agenda 2030 for Sustainable Development urges governments to "promote mental health and wellbeing." Also, 33 (o) of the Sendai Framework indicates: "To enhance recovery schemes to provide psychosocial support and mental health services for all people in need." Now, there is growing support to go a step further, to recommend the following:

1. To support programs that promote and implement action to enhance recovery schemes to provide psychosocial support and mental health services for all people
2. To develop national strategies for the prevention and treatment of psychosocial and mental health conditions

3. To allocate necessary funds to carry out the above
4. To integrate psychosocial services into all DRR plans and policies
5. To work with multi-stakeholders in partnerships to achieve the above
6. With regard to the Framework current wording, to recognize that where 'resilience' is mentioned, it refers not only to infrastructural but also psychosocial resilience

The statement from the 2019 meeting in Geneva, that builds on the previous statements of 2015 and 2017, includes references to supportive documents, as follows:

> We assert the importance of psychosocial resilience as key to disaster risk reduction and recovery. In an historic act, Target 3.4 in the UN Agenda 2030 for Sustainable Development urges governments to 'promote mental health and wellbeing.' Also, paragraph 33 (o) of the Sendai Framework indicates: 'To enhance recovery schemes to provide psychosocial support and mental health services for all people in need'.

Psychosocial Support in Epidemics

While this chapter focuses on natural disasters, it is important to point out that epidemics have similarly affected the psychological and social structures of many countries. The similarities of these naturally occurring crises with those from other sources, e.g., that are human-induced, has been pointed out (Kuriansky, 2013c; Kuriansky, 2016; Kuriansky et al., 2018).

The "Health in Your Hands" Initiative

The crucial need for policy and programs regarding psychological resilience is urgent for low-resource settings where population are "at the last mile," meaning that they have little access to services. To address the global intention that "no one is left behind," as called for in the UN 2030 Agenda for Sustainable Development, an initiative called "Health in Your Hands" was developed to make health services accessible to those "at the last mile." The platform was formulated at the convening of Global Councils related to the Sustainable Development Goals by the Prime Minister's Office of the United Arab Emirates during the World Government Summit in February 2017, in their initiative called "SDGs in Action." This project, by Global Council 3 (referring to the UN Agenda 2030's SDG 3 on Health and Wellbeing for All at All Ages), which I chair, was awarded as the most innovative in the meeting's competition, and then presented at the United Nations during the Science, Technology and Innovation Forum in May 2017. Other experts in the GC3 include from the World Health Organization, the World Bank, the World Federation of Public Health Associations, academics, and others. The "Health in Your Hands" project – co-founded by myself and Dr. Shariha Khalid Erichsen, mentioned earlier, who is a former surgeon, now a private sector impact advisor and investor as Managing Partner of Mission & Co. – was then launched at the World Health Assembly in Geneva in May 2019, during an event sponsored by UNAIDS about their innovation marketplace and presented at two events during the United Nations General Assembly meetings in September 2019. One of these was sponsored by the United Arab Emirates, on "Collaboration toward achieving our shared goals" and the other, that I organized and moderated, was on "Achieving SDG 3 and Universal Health Coverage: Innovative Projects, Policies and Financing Models to Reach Those Left Furthest Behind." Innovations are featured on the platform that have proven impact, management, and financial sustainability, and that reach vulnerable, marginalized, and at-risk groups, including women, children, persons with disabilities, and those living in poverty. Provision and scaling up of such innovations are particularly relevant and important for disaster-prone communities and those that have suffered tragic disasters.

The Importance of Partnerships

Increasingly, stakeholders are recognizing the critical role that partnerships play in disaster recovery. This is consistent with Goal 17 of the UN 2030 Agenda for Sustainable Development that identifies working together by various parties as essential for a coordinated and effective effort. For example, The Global Facility for Disaster Reduction and Recovery (GFDRR) is a global partnership working with over 400 subnational, national, regional, and international partners that "helps developing countries better understand and reduce their vulnerabilities to natural hazards and adapt to climate change" (Global Facility for Disaster Reduction and Recovery, 2019). Managed by the World Bank and supported by 37 countries and 11 international organizations, the GFDRR provides grant financing, technical assistance, training, and knowledge sharing activities to mainstream disaster and climate risk management in policies and strategies. The GFDRR seeks strategic partnerships across the United Nations system, including the United Nations Office for Disaster Risk Reduction, nongovernmental organizations, national and subnational governments, regional organizations, academic institutions, and others.

To advance its mission, GFDRR seeks partnership with the UNDRR, NGOs (led by Save the Children), the private sector, warning systems, city authorities, the International Federation of Red Cross and Red Crescent Societies, the Global Network of Civil Society Organizations for Disaster Reduction, the World Bank, European Union, United Nations Development Programme and other groups, to enhance global response to disasters, coordinate actions, and develop tools for reconstruction.

Historic National Policy Efforts

A major and historical step forward on the national stage that supports the importance of mental health, well-being, and psychosocial support in disaster recovery is the adoption of a policy on "Addressing the Impacts of Climate Change on Mental

Health and Well-being" by the American Public Health Association (APHA, 2019). This policy was co-drafted by myself with two other APHA members on behalf of the Mental Health section. APHA is a membership organization dedicated to improve the health of the public, achieve equity in health status, and influence federal policy to improve the public's health. The policy includes the importance of mental health and well-being in recovery from disasters caused by climate change, a review of interventions, and recommended actions by varied stakeholders.

Importantly, APHA considers climate change a priority, including its effect on physical as well as mental health, evidenced in the fact that climate change was the theme for the APHA 2017 annual meeting on "Creating the Healthiest Nation: Climate Changes Health," marking a Year of Climate Change and Health, asserting that,

> Climate Change is happening today and it undeniably poses many risks to our health, making it the greatest public health challenge. Public health professionals need to work towards safeguarding all communities against climate risks by protecting people's health, wellbeing, and quality of life from climate change impacts.

Also, the new APHA *Center for Climate, Health and Equity* intends to inspire action on climate and health, advances equitable climate policies, and galvanizes the public health field to address climate change.

Climate Change Inter-Linkage with Emotional Resilience from Disasters

Disaster recovery is intricately linked to climate change, as reflected in the policy statement reported in a previous section of this chapter. In terms of the 17 Sustainable Development Goals (SDGs) outlined in the UN 2030 Agenda, the goals and targets of "Combatting Climate Change" is the subject of SDG #13, the goals and targets of preserving "Life Below Water" are outlined in SDG #14, and SDG #15 addresses the goals and targets of preserving "Life On Land." Advocacy about these issues is integrally tied to the promotion of disaster recovery. Thus, the author and colleagues attend conferences related to the environment and also to the oceans, and to climate change in general, which present opportunities to highlight the importance of emotional resilience and influence policy.

For example, in light of the extreme devastation and destruction caused by the 2017 major hurricanes in the Caribbean and the USA (Harvey, Irma, and Maria), I was invited to moderate a panel convened by the NGO Committee on Sustainable Development-NY to discuss the aftermath and steps that can be taken. The event, held on Friday, 29 September 2017 in the UN Church Center, highlighted the need for ongoing help from individuals and the international community. Speakers included three UN Ambassadors from the affected region: Ambassador Dr. Walton Alfonso Webson, of the Permanent Mission of Antigua and Barbuda; Ambassador Loreen Ruth Bannis-Roberts of the Permanent Mission of Dominica; and Ambassador Pennelope Beckles of the Permanent Mission of Trinidad and Tobago.

Ambassador Webson was very candid about the reaction of several nations that did not adhere to the basic vows made at the UN in reference to the SDGs and who refused to extend help to Barbuda and other small island states. Three speakers who had been on their home islands during the hurricanes described their harrowing experiences. I gave the psychological perspective concerning emotional needs and urgency to provide psychosocial support.

Importantly, this topic was also addressed at an event on "Building Support for Our Ocean" held on 31 July 2019 at UN headquarters, organized by the NGO Committee on Sustainable Development-NY, and sponsored by Ambassador Pennelope Beckles of the Permanent Mission of Trinidad and Tobago to the United Nations. Other missions whose ambassadors spoke were from Antigua and Barbuda (Ambassador Dr. Walton Alfonso Webson, mentioned above) and from Vanuatu (Ambassador Odo Tevi). The Ambassador of Kenya to the UN, H.E. Lazarus O. Amayo, also came, and talked about an upcoming Oceans Summit hosted by his country with Portugal to take place in Nairobi in June 2020. Ambassador Beckles talked passionately about the turtles and other sea life being strangled by discarded plastics. When I mentioned the importance of these countries paying attention to emotional long-term aftereffects of the severe Hurricane Irma in their region – given, for example, that Barbuda had been 90% destroyed – they all agreed and admitted sadly that not enough attention is paid to this issue. Other speakers, including academicians, private sector actors, and climate activists all explained the dangers to the oceans and marine life, critical for food supply and economies, especially from dumping plastic waste. All agreed that urgent action is necessary.

Discussions about climate change related to disasters, and the deleterious impact on people and the planet, have been ongoing at the United Nations, by both governments and civil society and also by youth groups. Notably, at a Conference in the Dominican Republic, students at a model UN – an event where youth enact roles of the real governments – produced, under my direction of the International Student Journalism Program, a video appealing to governments to "Seal the Deal," meaning come to agreement about climate change (Bateson, Chireno, & Kuriansky, 2009). Also, intergovernmental meetings occur annually at what's called the Conference of the Parties (COP) to The United Nations Framework Convention on Climate Change (UNFCCC).

Advocacy for Policy About the Environment

Advocacy and research focused on the environment is exceptionally important in efforts concerning disaster recovery and psychosocial resilience, since preservation of the environment is critical to adaptation and mitigation of climate change and strategies regarding disaster risk, prevention, and recovery. Such efforts relate to the field of environmental psychology and a subset of this field in applied psychology called Ecopsychology that focuses on the relationship with nature and ecosystems. In a two-volume anthology about this field, topics explore examples of implementation, e.g., transformational growth workshops in the biodiverse environment of Belize as well as policy, e.g., paradigm shifts in UN policies and programs about the environment and climate change related to nature and well-being (Kuriansky, LeMay, & Kumar, 2015: Nemeth & Kuriansky, 2015). In this regard, prosocial behavior is important. Studies on causes of prosocial or pro-environment behavior (PEB) reveal several points. These include that energy-saving behavior was more encouraged by following others than by providing information; that visual messages influence PEB more than text; and that walkable neighborhoods positively affect mental health. Concepts like "ecological resilience" and "cosmopolitan identity" cultivate socially responsible values and behavior (e.g., recycling, purchasing biodegradable products). Significantly, research shows that volunteering helps the helper (Nemeth, Hamilton, & Kuriansky, 2015, pp.419–424; Silva, Marks, & Cherry, 2009). Research in areas like these, and projects presented in this volume, are greatly promising given the inclusion of well-being in the UN global agenda. I have made statements at the UN about preservation of the environment and the respect for nature (Kuriansky, 2014; Kuriansky, LeMay, & Kumar, 2015).

Involving Media

Involving media in advocacy efforts is critical, and often overlooked (Kuriansky, 2005, 2006, 2009). At a UN meeting on disaster, a three-day training on covering climate change issues was held for invited journalists, organized by an independent Geneva-based company, Media 21, and initiated by the press agency, InfoSud, a nonprofit journalism organization. At one session, "Health: New or Aggravated Diseases?" about climate change and disease spread, as a panelist, I stressed the importance of considering psychosocial issues in addressing problems related to climate change. Many journalists requested more information and asked for advice developing programs for their country (e.g., Sri Lanka and the Philippines) regarding psychosocial recovery and risk reduction.

Conclusions

To ensure Emotional Resilience, one of the triad of the 3 Es for disaster resilience – with Economic Recovery and Environmental Restoration – it is crucial to ensure that policy is put in place to support programs that assist people in restoring their sense of safety and empowerment to rebuild. Psychological representatives at the United Nations have engaged in extensive advocacy in order to encourage governments to support the emotional recovery of populations affected by natural disaster, specifically focused on "psychosocial resilience." Such efforts involve ongoing awareness-raising in the form of presentations at international conferences, not just for professionals but for policymakers on all levels, and in individual meetings with high-level delegates. More psychologists need to be involved in these efforts both on the international stage as well as on their national level. Further, documentation of the process of advocacy and the outcome of such efforts needs to be done, in order to offer training to develop competency and to engage more psychologists. With such advocacy, restoring emotional resilience post trauma and allowing opportunities for post-traumatic growth is possible for vast numbers of populations affected by disaster now and in the future.

Equally important is developing a body of evidence-based disaster interventions focused on mental health and well-being and psychosocial resilience, in order to support, and also apply, these policies. With colleagues, I am committed to ongoing efforts to include psychosocial resilience as a language in inter-governmental agreements, as well as developing and implementing interventions for psychosocial disaster recovery. The recommendations in the new policy adopted by the American Public Health Association for addressing the impact of climate change, and thereby natural disasters, on mental health provide an excellent path for the way forward.

References

APHA, American Public Health Association (2019). Addressing the Impacts of Climate Change on Mental Health and Well-being. Policy Number: 20196. Retrieved from https://www.apha.org/policies-and-advocacy/public-health-policy-statements/policy-database/2020/01/13/addressing-the-impacts-of-climate-change-on-mentalhealth-and-well-being.

Art Saves Lives Foundation. (2019). We are resilient people. Mission report: Resilience, empowerment, and connection building training. Access at https://drive.google.com/open?id=17Gf6uOVZMD3N-Ak4igihhTRewTC5BoE1.

Bateson, T., Chireno, E. & Kuriansky, J. (Co-Executive Producers) (2009). *Seal the deal at CILA 2009* (2009, Nov 6). [Video]. Retrieved April 6, 2015, from https://www.youtube.com/watch?v=vP-yjpdXO8M.

Forman, A. (2014). Five words that can change the world. *Jewish Journal, 39*(4), 1–9.

Global Facility for Disaster Reduction and Recovery. (2019). Access at https://www.gfdrr.org/en.

International Strategy for Disaster Reduction (2005). Hyogo framework for action 2005–2015: Building the resilience of nations and communities to disasters. Retrieved from https://www.unisdr.org/2005/wcdr/intergover/official-doc/L-docs/Hyogo-framework-for-action-english.pdf.

Jean-Charles, W. (2011). Rebati: After the earthquake, the IAAP UN team continues to remember Haiti. *IAAP Bulletin of the International Association of Applied Psychology, 23*, 32–38. Retrieved from https://iaapsy.org/site/assets/documents/apnl_v23_i1-2.pdf.

Johnson, J., & Galea, S. (2009). Disasters and population health. In K. E. Cherry (Ed.), *Lifespan perspectives on natural disasters: Coping with Katrina, Rita and other storms* (pp. 281–326). New York: Springer.

Koyama, S., Aida, J., Kawachi, I., Kondo, N., Subramanian, S. V., Ito, K., et al. (2014). Social support improves mental health among the victims relocated to temporary housing following the Great East Japan Earthquake and Tsunami. *The Tohoku Journal of Experimental Medicine., 234*(3), 241–247.

Kuriansky, J. (2005). Working effectively with the mass media in disaster mental health. In G. Reyes & G. A. Jacobs (Eds.), *Handbook of international disaster psychology* (Vol. 1, pp. 127–146). Westport, CT: Praeger Press.

Kuriansky, J. (2006). Getting it to the hill. Your voice counts: Public policy advocacy for psychologists. *Amplifier, Media Psychology. Submitted to* division 46 media psychology of the American Psychological Association.

Kuriansky, J. (2008). A clinical toolbox for cross-cultural counseling and training. In U. P. Gielen, J. G. Draguns, & J. M. Fish (Eds.), *Principles of multicultural counseling and therapy* (pp. 295–330). New York: Taylor and Francis/Routledge.

Kuriansky, J. (2009). Communication and media in mass trauma: How mental health professionals can help. In J. T. Thome, M. Benyakar, & I. H. Taralli (Eds.), *Intervention in destabilizing situations: Crises and traumas* (pp. 195–232). Rio de Janeiro: Associação Brasileira de Psiquiatria.

Kuriansky, J. (2011). Guidelines for mental health and psychosocial support in response to emergencies: Experience and encouragement for advocacy. *The IAAP Bulletin. The International Association of Applied Psychology Covering the World of Applied Psychology, 23*, 30–32.

Kuriansky, J. (2012). Our communities: Healing after environmental disasters. In D. G. Nemeth, R. B. Hamilton, & J. Kuriansky (Eds.), *Living in an environmentally traumatized world: Healing ourselves and our planet* (pp. 141–167). Santa Barbara, CA: ABC-CLIO/Praeger.

Kuriansky, J. (2014). Statement as Invited Discussant. United Nations General Assembly Interactive Dialogue on Harmony with Nature. The promotion of a balanced integration of the economic, social and environmental dimensions of sustainable development through harmony with nature. New York: United Nations Trusteeship Council Chamber. June 22.

Kuriansky, J. (2013a). *Japan Mission: Healing workshops by Dr Judy Kuriansky & team on tsunami/earthquake anniversary* [Video file]. Retrieved from https://bit.ly/2LXh8CY.

Kuriansky, J. (2013b). Superstorm sandy 2012: A psychologist first responder's personal account and lessons learned about the impact on emotions and ecology. *Ecospsychology, 5*(S1), S-30–S-37. https://doi.org/10.1089/eco.2013.0010

Kuriansky, J. (2013c). Thoughts on Katrina vs. Sandy: Essays on nature-induced, human-induced, and nature + human-induced environmental trauma. *Ecopsychology, 5*(S1), S-20–S-26. https://doi.org/10.1089/eco.2013.0039

Kuriansky, J. (2015). *DrJudy Kuriansky: Statement at World Conference on Disaster Risk Reduction, Sendai Japan 2015, April 20* [Video file]. Retrieved from https://bit.ly/2JS6Ao2.

Kuriansky, J. (Ed.). (2016). *The psychosocial aspects of a deadly epidemic: What Ebola has taught us about holistic healing*. Santa Barbara, CA: ABC-CLIO/Praeger.

Kuriansky, J. (2017). *Transforming trauma to healing, hope and happiness.* [video file]. Access at https://www.youtube.com/watch?v=bfdJ9jAomMw.

Kuriansky, J. (2018). Psychotherapy, psychology, psychiatry and international policy: Professional contributions and personal experiences about mental health and wellbeing, psychosocial resilience and peace. *The International Journal of Psychotherapy, Counseling & Psychiatry: Theory, Research & Clinical Practice. Special Volume of the World Council for Psychotherapy 8th World Congress of Psychotherapy Conference 2017,* April 2018. Accessible at: http://ijpcp.com/journal03/j03a06.htm.

Kuriansky, J. (2019a). Climate change and disaster recovery: Activities, achievements and contributions of the IAAP New York Team. *Applied Psychology Around the World., 1*(3), 50–56.

Kuriansky, J. (2019b). Expressive arts for helping children heal in crisis and disaster. In I. A. Serlin, S. Krippner, & K. Rockefeller (Eds.), *Integrated health care for the traumatized: A whole person approach.* Lathan, NY: Rowman & Littlefield.

Kuriansky, J. (2019c). Migration, refugees and mental health at the United Nations: Activities, achievements and contributions of the IAAP New York Team. *Applied Psychology Around the World, 1*(3), 79–87. Retrieved from https://iaapsy.org/site/assets/files/1082/apaw_vol1i3.pdf.

Kuriansky, J. (2019d). Reinforcement des Capacities dans les Pays à Faibles Resources après une Catastrophe: Empathie, Empowerment et Estime de Sol, dans la Formation des "Consolateurs": Le Cas de Haiti (Building Capacity in Low Resource Settings Post-Disaster: Empowerment, Empathy and Esteem in Training Comforters in the Case of Haiti). In W. Jean-Charles (Ed.), *Empathie en Haïti de la Recherche à la Practique à la Croieée de Différents Regards* (pp. 207–230). Coconut Creek, FL: EducaVision.

Kuriansky, J. & Jean-Charles, W. (2012). Haiti Rebati: Update on activities rebuilding Haiti through the global kids connect project. *IAAP Bulletin of the International Association of Applied Psychology, 24*(2–3), part 21, 116–124. Retrieved from https://iaapsy.org/site/assets/documents/July2012.pdf.

Kuriansky, J., LeMay, M. & Kumar, A. (2015). Paradigm shifts in nature and well-being: principles, programs, and policies about the environment and climate change with actions by the United Nations for a sustainable future. In Nemeth, D.G., and Kuriansky, J.. *Ecopsychology: Advances in the Intersection of Psychology and Environmental Protection (11: Interventions and Policy),* Santa Barbara, CA: ABC-CLIO/Praeger.

Kuriansky, J., Margevich, A., Jean-Charles, W., & Daisey, R. (2018). Resilience and recovery in natural disasters and epidemics: Comparisons, challenges, and lessons learned from train-the-trainer projects. In G. Rich & S. Sirikantraporn (Eds.), *Human strengths and resilience: Developmental, cross-cultural, and international perspectives*. Lanham, MD: Lexington Books. https://www.amazon.com/Human-Strengths-Resilience-Developmental-Cross-Cultural/dp/1498554830/ref=mt_hardcover?_encoding=UTF8&me=.

Kuriansky, J., & Nemeth, D. G. (2013). A model for post-environmental disaster wellness workshops: preparing individuals and communities for hurricane anniversary reactions. *Ecopsychology, 5*(S1), S-38–S-45. https://doi.org/10.1089/eco.2013.0006

Kuriansky, J., Polizer, Y., & Zinsou, J. (2016). Children and Ebola: A model resilience and empowerment training and workshop. In J. Kuriansky (Ed.), *The psychosocial aspects of a deadly epidemic: What Ebola has taught us about holistic healing* (pp. 175–214). Santa Barbara, CA: ABC-CLIO/Praeger.

Kuriansky, J., Wu, L.-Y., Bao, C., Chand, D., Kong, S., Spooner, N., et al. (2015). Interventions by international and national organizations for psychosocial support after the Sichuan earthquake in China: A review and implications for sustainable development. In D. G. Nemeth, J. Kuriansky, & R. Hamilton (Eds.), *Ecopsychology: Advances in the intersection of psychology and environmental protection, volume 11: Interventions and policy* (pp. 171–231). Santa Barbara, CA: ABC-CLIO/Praeger.

Kuriansky, J., & Zinsou, J. (2019). Mental health and Well-being at the United Nations: Activities, achievements and contributions of the IAAP New York team. *Applied Psychology Around the World, 1*(3), 16–28. Retrieved from https://iaapsy.org/site/assets/files/1082/apaw_vol1i3.pdf.

Kuriansky, J., Zinsou, J., Arunagiri, V., Douyon, C., Chiu, A., Jean-Charles, W., et al. (2015). Effects of helping in a train-the-trainers program for youth in the global kids connect project after the 2010 Haiti earthquake: A paradigm shift to sustainable development. In D. G. Nemeth, J. Kuriansky, & R. B. Hamilton (Eds.), *Ecopsychology: Advances in the intersection of psychology and environmental protection* (Vol. 11, pp. 135–169). Santa Barbara, CA: ABC-CLIO/Praeger.

Luce, J. (2010). Dr. Judy on the Trauma of Disasters - like Haitian Earthquake. *Huffington Post.* Access at https://www.huffpost.com/entry/dr-judy-on-the-trauma-of_b_461779.

Maeda, M., & Oe, M. (2015). The great East Japan earthquake: Tsunami and nuclear disaster. In K. E. Cherry (Ed.), *Traumatic stress and long-term recovery: Coping with disasters and other negative life events* (pp. 71–90). New York: Springer.

Masangkay, M. (2015). Psychologist connects disaster-affected children around the world. Japan Times. Retrieved March 31, 2019 from https://www.japantimes.co.jp/news/2015/03/23/national/psychologist-connects-disaster-affected-children-around-the-world/#.WlKcemEbMyk.

Merriam-Webster. (n.d.). Definition of policy. Access at https://www.merriam-webster.com/dictionary/policy.

Nemeth, D. G., Hamilton, R. B., & Kuriansky, J. (2015). Reflections and recommendations: The need for leadership, holistic thinking, and community involvement. Pp.419-424: The role of Community in the Intersection of psychology and environmental protection, contribution by Kuriansky. In D. G. Nemeth, J. Kuriansky, & R. B. Hamilton (Eds.), *Ecopsychology: Advances from the intersection of psychology and environmental protection* (Vol. 2, pp. 419–424). Santa Barbara, CA: Praeger, ABC-CLIO.

Nemeth, D. G., & Kuriansky, J. (Eds.). (2015). *Ecopsychology: Advances from the Intersection of Psychology and Environmental Protection* (Vol. 2). Santa Barbara, CA: Praeger, ACB-CLIO.

Nemeth, D. G., Kuriansky, J., Reeder, K. P., Lewis, A., Marceaux, K., Whittington, T., et al. (2012). Addressing anniversary reactions of trauma through group process: The hurricane Katrina anniversary wellness workshops. *International Journal of Group Psychotherapy, 62*(1), 129–142.

Nemeth, D. G., & Olivier, T. W. (2017). *Innovative approaches to individual and community resilience: From theory to practice*. London: Elsevier.

Nemeth, D., Kuriansky, J., Olivier, T., Whittington, L. T., May, N., Hamilton, J., & Steger, A. (2011). Group Interventions for Disaster/Trauma Anniversary Reactions. Global Horizons: *The Center for Policy adn Resilience, 4*(1), 51–64.

Norris, F. H., Friedman, M. J., & Watson, P. J. (2002). 60,000 disaster victims speak: Part II. Summary and implications of the disaster mental health research. *Psychiatry: Interpersonal and Biological Processes, 65*(3), 240–260. https://doi.org/10.1521/psyc.65.3.240.20169

Norris, F. H., Friedman, M. J., Watson, P. J., Byrne, C. M., Diaz, E., & Kaniasty, K. (2002). 60,000 disaster victims speak: Part I. An empirical review of the empirical literature, 1981–2001. *Psychiatry, 65*(3), 207–239.

Phillips, J. R. (2015). Natural disasters: On wildfires and long-term recovery of community-residing adults. In K. E. Cherry (Ed.), *Traumatic stress and long-term recovery: Coping with disasters and other negative life events* (pp. 25–36). New York: Springer.

Rich, G., & Sirikantraporn, S. (Eds.). (2018). *Human strengths and resilience: Developmental, cross-cultural, and international perspectives*. Lanham, MD: Lexington Books.

Silva Brown, J. (2015). On tornadoes: Storm exposure, coping styles and resilience. In K. E. Cherry (Ed.), *Traumatic stress and long-term recovery: Coping with disasters and other negative life events* (pp. 37–55). New York: Springer.

Silva, J. L., Marks, L. D., & Cherry, K. E. (2009). The psychology behind helping and prosocial behaviors: An examination from intention to action. In K. E. Cherry (Ed.), *Lifespan perspectives on natural disasters: Coping with Katrina, Rita and other storms* (pp. 219–240). New York: Springer.

UNISDR. (2015). Sendai Framework for Disaster Risk Reduction (2015–203). Retrieved May 3, 2019, from: https://www.unisdr.org/files/43291_sendaiframeworkfordrren.pdf.

United Nations. (2018). Global Compact for Safe, Orderly and Regular Migration - Intergovernmentally Negotiated and Agreed Outcome, 13 July 2018. Retrieved July 30, 2019 from https://refugeesmigrants.un.org/sites/default/files/180713_agreed_outcome_global_compact_for_migration.pdf.

United Nations General Assembly. (2019). A/RES/74/2. Political declaration of the high-level meeting on universal health coverage. Accessed November 1, 2019 from: https://undocs.org/en/A/RES/74/2.

United Nations General Assembly. (n.d.). Session 56 *Resolution 195*. A/RES/56/195. 21 Jan 2002. Retrieved July 21, 2019 from http://www.gdrc.org/doyourbit/disasters-A-RES-56-195.pdf.

United Nations Office for Disaster Risk Reduction. (2015). *WCDRR: The Global Kids Connect Project: A model programme to promote resilience, Judy Kuriansky* [Video File]. Retrieved from https://bit.ly/2MALyw2.

Weems, C. F., & Banks, D. M. (2015). Severe stress and anxiety disorders in adolescence: The long-term effects of disasters. In K. E. Cherry (Ed.), *Traumatic stress and long-term recovery: Coping with disasters and other negative life events* (pp. 177–194). New York: Springer.

Weems, C. F., & Overstreet, S. (2009). An ecological-needs-based perspective of adolescent and youth emotional development in the context of disaster: Lessons learned from hurricane Katrina. In K. E. Cherry (Ed.), *Lifespan perspectives on natural disasters: Coping with Katrina, Rita and other storms* (pp. 27–44). New York: Springer.

Index

© Springer Nature Switzerland AG 2021

K. E. Cherry, A. Gibson (eds.), *The Intersection of Trauma and Disaster Behavioral Health*, https://doi.org/10.1007/978-3-030-51525-6

F

G

H

Y

The manufacturer's authorised representative in the EU is Springer Nature Customer Service Centre GmbH, Europaplatz 3, 69115 Heidelberg, Germany. If you have any concerns regarding our products, please contact ProductSafety@springernature.com

Printed and bound by CPI Group (UK) Ltd, Croydon, CR0 4YY
30/09/2026
02225695-0012